Measuring Benefits

of Government Investments

Studies of Government Finance

TITLES PUBLISHED

Measuring Benefits

of Government

Investments

ROBERT DORFMAN, *Editor*

Papers presented at a conference of experts
held November 7-9, 1963

GREENWOOD PRESS, PUBLISHERS
WESTPORT, CONNECTICUT

Library of Congress Cataloging in Publication Data

Dorfman, Robert, ed.
    Measuring benefits of government investments.

    (Studies of government finance)
    Reprint of the ed. published by Brookings
Institution, Washington.
    Includes bibliographical references and index.
    1. United States--Social policy--Case studies.
2. Evaluation research (Social action programs)--
United States--Case studies. 3. Cost effectiveness
--Case studies. I. Title. II. Series.
[HN58.D6 1980]   361.2'5'0973   79-28577
ISBN 0-313-22307-6 lib. bdg.

Reprinted with the permission of The Brookings Institution.

Reprinted in 1980 by Greenwood Press,
a division of Congressional Information Service, Inc.
51 Riverside Avenue, Westport, Connecticut 06880

Printed in the United States of America

10 9 8 7 6 5 4 3 2 1

THE BROOKINGS INSTITUTION is an independent organization devoted to nonpartisan research, education and publication in
economics, government, foreign policy, and the social sciences generally. Its principal purposes are to aid in the development of sound public
policies and to promote public understanding of issues of national importance.

The Institution was founded December 8, 1927, to merge the activities of
the Institute for Government Research, founded in 1916, the Institute of
Economics, founded in 1922, and the Robert Brookings Graduate School of
Economics and Government, founded in 1924.

The general administration of the Institution is the responsibility of a self-
perpetuating Board of Trustees. The Trustees are likewise charged with maintaining the independence of the staff and fostering the most favorable conditions for creative research and education. The immediate direction of the
policies, program, and staff of the Institution is vested in the President, assisted by the division directors and an advisory council, chosen from the
profesional staff of the Institution.

In publishing a study, the Institution presents it as a competent treatment
of a subject worthy of public consideration. The interpretations and conclusions in such publications are those of the author or authors and do not purport to represent the views of the other staff members, officers, or trustees of
the Brookings Institution.

BOARD OF TRUSTEES

Foreword

A LARGE PROPORTION of government expenditure is devoted to the construction of roads, hydroelectric projects, and other physical facilities that promote the efficiency of private economic and social activities. Lately, expenditures on research, health, education, welfare, and other services have also come to be regarded as significant public investments. Since funds are limited, the problem of allocating them on the basis of sensible and consistent criteria of project worth has become increasingly important.

Few such criteria have been developed except in the field of water resource development and defense planning. Evaluation of project worth—particularly expenditures for services—is exceedingly difficult, partly because it is difficult to foresee the consequences of such projects and partly because it is hard to appraise their social values.

To assist in the development of new techniques for measuring the benefits and costs of public investments, a conference was held at the Brookings Institution on November 7-9, 1963. The participants, who were selected from government agencies and academic and research institutions, discussed papers on the benefits of seven programs chosen to illustrate the problems of evaluating government investments. This volume incorporates the seven papers and comments on them by selected discussants, as well as an introduction by Robert Dorfman of Harvard University, who acted as conference chairman and editor of the volume.

The manuscript was prepared under the direction of Joseph A. Pechman, Director of Economic Studies at Brookings, and reviewed by an advisory committee consisting of William M. Capron and Samuel M. Cohn of the Bureau of the Budget; Otto Eckstein of Harvard University and the Council of Economic Advisers; and Charles L. Schultze of the University of Maryland.

The papers have also had the benefit of criticisms and suggestions from a number of professional specialists, including E. J. Kane, Princeton University, who provided mathematical assistance to Frederic M. Scherer; Lyle E. Craine, University of Michigan, and Roland N. McKean, University of California at Los Angeles, who made helpful comments on the paper by Ruth P. Mack and Sumner Myers; Rashi Fein, the Brookings Institution, and W. Lee Hansen, University of Wisconsin, who commented on the paper by Burton A. Weisbrod; and Richard Hahn, John Hume, and Edwin Mills, the Johns Hopkins University, Oscar Jones, U.S. Public Health Service, and Mr. Fein, who advised on or reviewed the paper by Herbert E. Klarman. The manuscript was edited for publication by Kathleen Sproul, and the index was prepared by Helen B. Eisenhart.

The conference was part of a special program of research and education on taxation and public expenditures, supervised by the National Committee on Government Finance and financed by a special grant from the Ford Foundation.

The views expressed in this book are those of the authors and are not presented as the views of the government or private organizations with which the authors are affiliated, or the views of the National Committee on Government Finance or its Advisory Committee, or the staff members, officers, or trustees of the Brookings Institution, or the Ford Foundation.

ROBERT D. CALKINS
President

February 1965
The Brookings Institution
1775 Massachusetts Avenue N.W.
Washington D.C.

Studies of Government Finance

Studies of Government Finance is a special program of research and education in taxation and government expenditures at the federal, state, and local levels. These studies are under the supervision of the National Committee on Government Finance appointed by the Trustees of the Brookings Institution, and are supported by a special grant from the Ford Foundation.

MEMBERS OF THE ADVISORY COMMITTEE

Contents

Authors' Tables and Figures

ROBERT DORFMAN*

Introduction

THE PAPERS AND DISCUSSIONS in this volume were presented
in a conference held at the Brookings Institution in November
1963 to explore the problems of appraising the benefits that are
likely to accrue from proposed public investment projects. To throw
light on this question from several directions, papers were com-
missioned to deal with it in seven contexts: research and development expenditures, outdoor recreation projects, education programs, federal aviation expenditures, highway programs, urban
renewal projects, and public health programs. The writers of the
papers were primarily academicians, most of them, however, also
with other experience in their fields of competence. To insure
pointed discussion, most of the discussants were practitioners,
chiefly from federal government agencies.

The conference was planned with the hope of inducing all
participants to talk about the same subject and to expose their
divergences of view freely. And, whatever else may be said about
the meeting, it accomplished those ends. We did talk about the
same topic, and we also disagreed with each other—at times with an
intensity that is still detectable in these published proceedings, even
after all the filtering that editorial tact imposes.

* Department of Economics, Harvard University.

1

The battle lines were clearly drawn. The authors of the papers and most of the other academicians among the conference participants were hopeful that the benefits of government investment projects could be appraised objectively and even quantitatively. They placed their primary reliance on the skillful application of benefit-cost analysis, which has been developed in the past thirty years mostly for application to water resource development projects.

The practitioners were very skeptical and inclined to doubt whether the most important social effects of government investments could ever be appraised quantitatively by cost-benefit analysis or any other formalized method. One of them likened the problem to appraising the quality of a horse-and-rabbit stew, the rabbit being cast as the consequences that can be measured and evaluated numerically, and the horse as the amalgam of external effects, social, emotional, and psychological impacts, and historical and aesthetic considerations that can be adjudged only roughly and subjectively. Since the horse was bound to dominate the flavor of the stew, meticulous evaluation of the rabbit would hardly seem worthwhile.

The advocates of benefit-cost analysis did not allow this to pass unchallenged. It was, they said, a counsel of despair. They agreed that many important aspects of a public investment project are not amenable to quantitative appraisal in the light of objective standards. In fact, at present this is painfully true, since benefit-cost analysis is a largely undeveloped art except in the context of water resource development, but even if one takes the most hopeful view of the future, some significant consequences of public undertakings must, it seems, always elude the craft of the quantifier. Nevertheless, the process of political decision can be sharpened significantly by removing as many aspects as possible from the realm of unsupported opinion and emotive rhetoric. In the field of water resources, public decisions have been made noticeably more rational and consistent through submitting all project proposals to the discipline of comparing measurable costs with measurable benefits. At the very least, such a process enables attention to be focused on the question of whether the unmeasurable benefits are deemed impressive enough to justify sustaining the measurable costs that they entail.

The debate at the conference thus revolved around conjectures about the relative sizes of the horse and the rabbit, not only now but in the future. We debated again and again whether the economist, the statistician, and the political scientist working in cooperation could devise means for quantifying a usefully large proportion of the consequences of public investment undertakings.

The debate was not conducted in a vacuum. We had before us seven skillfully prepared papers devoted to advancing the technique (or art) of benefit-cost analysis. Frederic Scherer proposed an ingenious technique for quantifying one aspect of the benefits of expenditures on technological development; Ruth Mack and Sumner Myers formalized the procedures needed to estimate the benefits of government investments in outdoor recreational facilities; Burton Weisbrod evaluated numerically a program for reducing the high school dropout rate; and so on through all the papers. Not one of the problems undertaken was amenable to cut-and-dried procedures. Each required fresh and inventive thought to adapt techniques previously used or to develop new ones. The result of these efforts is a case-study compendium which valiantly suggests how the cost-benefit approach can be applied in widely differing contexts, what difficulties it encounters, what expedients can be used to overcome the difficulties, and, perhaps, the ultimate limitations of this approach.

Both the designated discussants and many of the other conference participants constantly subjected the papers as well as the whole benefit-cost approach to searching criticism focused on the issue of whether this elaborate and laborious enterprise is worthwhile in the first place, in view of the inevitably huge uncertainties, errors, and omissions. We did not resolve this issue. But we did express it clearly, and the test cases that were presented, around which our discussions centered, may help to make subsequent debate more fruitful.

The Benefit-Cost Framework

Since the benefit-cost framework of project analysis forms the common background of all the papers and discussions in the volume, a sketch of it is appropriate here.

Incentives for Government Enterprise

Governments rush in where businessmen decline to tread. As a general rule, if a good or a service is desirable, it will also be profitable and thus will be provided by private enterprise. But there are important exceptions to this rule. Government initiative is, for example, called for in cases where investments that businessmen would deem unprofitable are socially worthwhile. These exceptions can arise from a number of circumstances: some relate to the conditions under which a product is distributed and consumed; some to the conditions of its production; some have other justifications.

CONDITIONS OF CONSUMPTION. The circumstances that favor government provision cluster around the concept of collectability or, rather, uncollectability. In the usual economic transaction, the user is charged for the good or service he consumes, the amount he is willing to pay measures the value of the commodity to him, and, since his use of the commodity precludes anyone else from benefiting from it, the value of the commodity to the user is also its value to the entire society. This standard analysis of social value is not strictly valid for any transaction (there are always side effects), and for a few types of transactions it is too wide of the mark to be acceptable. The most important of these latter cases are (1) collective goods and (2) goods that are characterized by external economies of consumption: in neither case is the provider of the good able to collect from beneficiaries a charge commensurate with the benefits conferred.

In general, a *collective good* is a facility or service that is made freely available to all comers without user charge, either because to assess a charge on each occasion of use would be excessively cumbersome or because use is not voluntary or even clearly definable. It is, for example, not feasible or desirable to levy a charge on every shipmaster who sees a lighthouse, on every householder whose door the patrolman passes, or on every housewife when a health officer inspects a food market.

With rare exceptions, collective goods cannot be provided by private firms, because they do not induce a flow of income to the provider. Therefore the responsibility for providing them falls frequently—but not invariably—to the government. In addition to the examples cited above, some important collective goods are national defense, civil and criminal justice, streets and most highways, and outdoor recreational facilities. Other significant categories include the findings of scientific research, since even patents usually do not enable the scientist or his employer to collect from beneficiaries more than a portion of the benefits conferred, and aesthetic amenities, such as a fine building or a landscape that confers benefits on passersby for which they cannot be assessed. The important feature of collective goods for our purpose here is that, since they are not sold, there are no market prices to assist in appraising their value.

Collective goods are allied to *external economies of consumption* but the latter come into being in a different way. When a man is treated for a communicable disease the relief afforded him is only part of the social value; every resident of his community benefits from the reduced danger of infection. When a high school student is dissuaded from leaving school, his neighbors receive benefits (among them a reduction in potential delinquency) over and above the direct benefit to the boy and his family. In other words the consumer of a good or service is not the sole beneficiary, and the amount he is willing to pay does not measure the entire value of the good to society. The act of consumption, in effect, creates a collective good. In such instances it may be socially desirable to provide the good, although the amounts collectable from direct users do not suffice to cover production costs. Market prices are not, in those cases, adequate measures of social value.

CONDITIONS OF PRODUCTION. The circumstances that conduce to government initiative relate to economies of scale. Some activities can be performed economically only on such a very large scale that for private enterprise to undertake them is infeasible or undesirable. For example, without invoking governmental power, it is not practicable to assemble the large areas of property required for highways, urban redevelopment, or hydroelectric projects. Some natural monopolies, such as water supply, are retained by the government as an alternative to regulating private development. It is probable, however, that whatever the condition of production the government would not undertake a project unless important collective goods or external benefits were involved.

OTHER INCENTIVES. The supposition that private investors may take an unduly short view of the consequences of their investments is an important justification for all investments having to do with the preservation of natural resources and their orderly exploitation. It is pertinent also to investments in urban renewal and improvements. In a way, the justification is related to the collective goods theme: there is little reason to believe that current prices for urban property or rural woodlands reflect adequately the importance of these depletable resources to future generations. Society as a whole may assign quite different values to such resources than do the participants in current markets.

And sometimes government undertakings are stimulated by an incentive of a very different sort: the desire to influence the distribution of income. The desire for a regional redistribution, for example, is one of the explicit motivations for the Appalachia Program, and the desire for socioeconomic redistribution plays a large role in urban renewal programs. Appraising the social value of such redistributions presents peculiarly difficult problems.

Formulas for Comparing Benefits and Costs

As the above discussion of incentives suggests, the government tends to intervene in precisely those markets in which prices are either lacking or are seriously divergent from social values. It is inherent in government enterprises, therefore, that market prices cannot be used in appraising their social contributions. Still, some economic basis is needed for judging which potential government undertakings are worthwhile and which are not. Benefit-cost analysis provides this basis.

Benefit-cost analysis is closely analogous to the methods of investment project appraisal used by businessmen. The only difference is that estimates of social value are used in place of estimates of sales value when appropriate. There are in use a number of slightly different formulas for comparing benefits and costs of government undertakings; two of the most popular ones are sketched below.

The starting point of any of the formulas is a projection of the physical output of the undertaking, either in each year of its life or in some typical year of operation. If the undertaking is a highway, for example, there would have to be estimates of the number of passenger-car miles, truck miles, and bus miles to be traveled on it in each year or in a typical year. Next, there would have to be estimates of the unit social value of each of these physical outputs, be they passenger-car miles, kilowatt hours, or whatever. These two estimates induce at once an estimate of the gross social contribution of the enterprise in a single year.

At this point the different formulas begin to diverge. One approach is to perform the gross benefit calculation for a typical year and to make a parallel computation for social costs in a typical year. The costs consist of two major components: current costs—the typical year expenditures for operating and maintaining the facilities; and capital costs—a charge levied against a year's operations to amortize the initial expenses of construction and installation. The ratio of gross annual benefits to total annual costs is the benefit-cost ratio. This formula amounts to a businessman's calculation of the ratio of sales to cost of goods sold, or of a profit-sales ratio, except, of course, that the value of output used in a benefit-cost computation is the social, rather than the market, value.

An alternative formula subtracts current costs in each year, or in a typical year, from gross benefits to obtain an estimate of current net benefits. The current net benefits for each year are discounted back to the date of inception of the project and added up to obtain an estimate of the present value of discounted net benefits. The ratio of this figure to the estimated capital cost of the project is then the benefit-cost ratio. This formula is therefore analogous to a businessman's calculation of the rate of profit that can be earned by capital invested in the undertaking. Indeed, in sophisticated variants of this approach, an internal rate of return, which is a strict analog of a rate of profit, may be computed.

These brief sketches have excluded many technical questions that may have significant, and even decisive, influence on the outcome. For example, what rate of interest should be used in amortizing the capital expenses or in discounting back the net benefits? What account, if any, should be taken of off-site and secondary benefits? What allowance should be made for consumer surplus? How should any effects on the distribution of income be taken into account? Such questions are important and arise in all formulations of benefit-cost comparisons, and some of them are discussed in the papers in this volume.

Quite clearly, the precise formula used for consolidating and expressing the results of the analysis is a relatively superficial matter: properly interpreted all the formulas lead to the same conclusions. The heart of the matter lies in deciding what benefits should be included and how they should be valued. The debate about benefit-cost analysis centers on the question of whether the social value of benefits can be estimated reliably enough to justify the trouble and effort involved in a benefit-cost computation.

This issue cannot be resolved categorically. It is no accident that benefit-cost analysis had its origin and highest development in the field of water resources. That is the field in which government operations are most analogous to private business and in which the highest proportion of outputs—water and power—are saleable commodities bearing relevant market prices. And nonpriceable, almost intangible consequences, though present, are less obtrusive than in other spheres of government activity.

A Chairman's Reactions to the Conference[1]

In planning the conference we deliberately avoided the "easy" case of water resources. Rather, our intent was to explore how

[1] Since this concluding section is a reaction to what was read and said at the time of the conference, it is phrased in the past tense. The papers and comments to be found in this volume are, of course, substantially what we heard at the conference, but they have profited by certain emendations resulting from the free-swinging discussions that occurred on the scene.

far the concepts and methods developed in the context of water resources could be extended to other areas of government investment.

The papers we heard were courageous and ingenious attempts to extend those methods to other, more difficult fields. Although the subject matter of the papers ranged widely—from air traffic control to the mechanism of urban blight—a strong strain of similarity in treatment suggested that benefit-cost analysis may be the same wherever it is applied. Typically, a paper started out by listing the various social and economic groups likely to be affected by the particular problem or program under investigation. In Herbert Klarman's paper this was short and informal; the population groups affected by a syphilis prevention program are too obvious to warrant extended discussion. In Jerome Rothenberg's paper the listing of the affected socioeconomic groups and the discussion of the distinctions among them were very elaborate, almost the heart of the paper. But, long or short, a categorization of affected persons served as a point of departure.

Next came a listing of the various impacts the program being evaluated made on the groups that had previously been enumerated. This listing, too, varied substantially in explicitness and elaborateness from paper to paper, depending on the nature of the problem and of the difficulties encountered.

Then came an attempt to place a unit value on each of the impacts on each of the groups. In the more numerical papers this was the most detailed, elaborate, and laborious (in the old-fashioned sense of the word) section of the analysis. This was also the critical stage. When it came to assigning social values, certain of the impacts had to be put aside because current methodology did not provide any way for making estimates of them that could be compared with the values of other impacts or with costs. Worse than that, there were often consequences whose values could not be compared quantitatively with each other or with costs, even in principle. Some day we may be able to measure the economic contribution of slum clearance to the reduction of fire hazards, but how can we ever hope to appraise its contribution to improving the quality of urban life?

Following this critical stage of his endeavor, each author had relatively clear sailing. He had only to add up the measurable costs and benefits and compare the results, with due allowance for the unmeasurable components.

The reactions to the papers were also remarkably similar, when account is taken of the diversity of subject matter. Most of the debate revolved around the relative importance of the types of benefits which had been included in the value-totals as against the types that had perforce been excluded. That something had been included and something had been left out was agreed in every instance; and just as uniformly there was disagreement as to whether enough had been included to make the enterprise worth the effort.

The discussion, then, stayed very close to conceptual and fundamental issues and only occasionally turned to questions of detailed method and technique. The contrast between this course of discussion and what would probably transpire at a similar symposium on benefit-cost appraisal in water resources was striking. When water resource specialists get together they no longer debate the relative importance of measurable and nonmeasurable impacts. They turn, rather, to such questions of technique as how to select an appropriate rate of discount, and how to estimate the benefits of a project by calculating the costs of alternative methods for providing an equivalent service.

In particular, it is hard to conceive of a paper on the benefits to be expected from a water resource development proposal that would not specify the proposed program quite carefully, so that it could be compared with alternatives. This specificity was (and is) lacking in most of the conference papers, and this, I think, compounded the difficulties encountered—since one of the best clues to the benefits of any project is the cost of attaining the same ends by other means. Scherer exploited that approach in his appraisal of research and development projects, but none of the other authors did.

This preoccupation with conceptual problems and comparative neglect of technical expedients is probably a symptom of the youngness of the field and of the opportunities still remaining for its development. The work of extending the methods of benefit-cost analysis and of criticizing and appraising these extensions has only just begun. It is the hope of all the members of the conference that this record of our proceedings will contribute to this endeavor and to the art of appraising public investment projects—to the end that we may increase the wisdom with which our public resources are spent.

FREDERIC M. SCHERER*

Government Research and Development Programs

INVESTMENT BY THE FEDERAL GOVERNMENT in research and development has certain features which distinguish it in degree and perhaps in kind from most other investment activities. One is the pervasive influence of uncertainty. Uncertainty is of course a companion to all human endeavor. But research and development programs attempting to create what has never been contrived before for adaptation to an ever-changing environment face uncertainties of exceptional magnitude. To illustrate: historical studies of completed U. S. weapon systems research and development efforts reveal errors in project technical output parameter predictions as high as 100 percent, while cost prediction errors of 200 percent and more were not uncommon.[1]

A second distinguishing variable is the degree to which gov-

* Department of Economics, Princeton University.
[1] See M. J. Peck and F. M. Scherer, *The Weapons Acquisition Process: An Economic Analysis* (Harvard Business School Division of Research, 1962), pp. 17-54 and 428-438, for a quantitative and qualitative analysis of the "unique environment of uncertainty" in military research and development programs. See also A. W. Marshall and W. H. Meckling, "Predictability of the Costs, Time, and Success of Development" in Universities-National Bureau Committee for Economic Research, *The Rate and Direction of Inventive Activity* (Princeton University Press, 1962), pp. 461-476.

ernment research and development programs have as a primary objective political and other intangible benefits difficult to translate into the convenient terms of a monetary common denominator. Military research and development programs aim ultimately at preserving or winning peace and freedom according to the American model; the Apollo program sought (at least up to early 1964) the prestige of being first (or in any event not a poor second) to reach the moon and the satisfaction of man's curiosity about the universe in which he dwells; and so on. Still, the intangible nature of these ultimate goals is not in itself a source of difficulty. A great proportion of all productive activity, including much investment by both private firms and government, is ultimately directed toward gratifying consumer wants which as a rule have roots no more tangible than the desire for peace and freedom and knowledge of the universe. The real difficulty is that in most areas of government research and development there exists no actual or potential market to impute through the price mechanism economic values reflecting the independent preferences of millions of private citizens. Some kind of nonmarket system of benefit imputation or estimation is required.

Third, research and development investment decisions often encompass important "tradeoffs" that have few close parallels in nonresearch activities. In particular, I have in mind decisions to trade resources for time—e.g., incurring higher development costs in order to insure earlier development project completion.

This paper is concerned mainly with the second and third of these distinctive aspects of government research and development. It examines the problem of measuring research and development program benefits in a nonmarket environment, paying special attention to the role played by benefit estimates in time-cost tradeoff decisions. In my focus on these topics the uncertainty element will be slighted, entering into the analysis only as an incidental (and usually disturbing) variable. This is clearly a significant limitation, for I believe that uncertainty is in many ways the most important single feature of research and development programs. Yet, to have the courage to attempt any analysis of the problem at all, one must assume that cost and benefit estimates can be and are made, despite all the uncertainties, and that decisions are taken, with or without risk aversion adjustments, on the basis of those estimates. That assumption is accepted here, at least provisionally, subject to empirical verification in a manner to be outlined.

The Objectives of Government R&D Programs

To limit the analysis further, it is useful to consider briefly the actual pattern of government research and development expenditures in recent years. One means is to distinguish between research (including the subcategory, basic research) and development. In fiscal year 1963, estimated government obligations of $12.7 billion for research and development were divided as follows: roughly 11.9 percent for basic research, 23.5 percent for applied research, and 64.6 percent for development.[2] Although in practice distinctions become quite blurred, *research* is officially defined as "systematic, intensive study directed toward fuller scientific knowledge of the subject studied," the subcategory *basic research* receiving the additional qualification that its aim be knowledge per se rather than practical application thereof. *Development* is defined as "the systematic use of scientific knowledge directed toward the production of useful materials, devices, systems, or methods, including design and development of prototypes and processes."[3]

These definitions suggest that the further one moves toward the development end of the R&D spectrum, the closer is the relationship between technological inputs and the expectation of concrete practical benefits as outputs. In part, this progression merely reflects the value systems and objectives of the scientific workers and sponsors involved. But it is also related to the degree of uncertainty associated with specific technical problems. When a problem can be clearly defined and when likely solutions can be identified, a development project often becomes appropriate. But when the problem itself is not well understood and/or when insight into promising solutions has not yet been gained, the

[2] National Science Foundation, *Federal Funds for Science: XI, Fiscal Years 1961, 1962, and 1963* (1963), p. 93.
[3] *Ibid.*, p. 77.

more unstructured, knowledge-seeking approach of research is generally adopted.

Indeed, one often hears that business firms presumably motivated by economic gain decide to support basic research projects largely on the basis of faith that a particular researcher or group of researchers working in an interesting field will sooner or later come up with something useful, while the literature on development project selection places much more stress on evaluating methods and market potential than on evaluating the man.[4] It is perhaps for similar reasons that academic research on research and particularly on "invention" has tended to be the domain of sociologists, while the attention of economists has generally gravitated toward the development end of the spectrum. And, at least for the present paper, I accept this division of labor. It may well be possible to find an aggregative relationship between total inputs into basic research and the overall benefits therefrom, but at the level of individual projects any such relationship is likely to be drowned out by uncertainties. Therefore attention will be directed in this paper toward development and the most applied of the applied research efforts, where the prospect of relating specific project benefits to costs appears much more solid.

A further reason for this emphasis deserves brief mention. "Knowledge"—supposedly the main output of research projects—comes tolerably close to being a free good once it is created and made public. As a result the problem of tracing the external economies from research projects becomes especially severe.[5] The problem is a good deal less serious, although not entirely absent, with development projects whose primary outputs are the engineering drawings and organizational know-how required to adopt or produce a specific new process or product.

Applied research and development projects are sponsored by the government to secure a variety of benefits. Some, such as the

---

[4] Unfortunately, semantic difficulties make this statement somewhat less than accurate. When people writing about "research project selection" specify what kinds of technical activity they are considering (and this happens all too infrequently), it is usually evident that they are really referring to development work, as defined by the National Science Foundation.

[5] See Richard R. Nelson, "The Simple Economics of Basic Scientific Research," *Journal of Political Economy,* Vol. 67 (June 1959), pp. 297-306.

TABLE 1. Departmental Allocation of Federal Expenditures for Research and Development, Fiscal Years 1961–1963[a]

(dollar amounts in millions)

Expenditures and Allocations to Departments	FY 1961	FY 1962	FY 1963
Total Expenditures	$8,714	$9,522	$11,282
Percentage Allocations			
Department of Defense	74.6%	67.4%	62.3%
National Aeronautics and Space Administration	7.1	11.3	17.2
Atomic Energy Commission	9.8	11.0	10.1
Department of Health, Education, and Welfare	3.9	5.5	5.6
Department of Agriculture	1.6	1.6	1.4
All Others	3.0	3.2	3.4
Total	100.0	100.0	100.0

Source: National Science Foundation, *Federal Funds for Science: XI, Fiscal Years 1961, 1962, and 1963* (1963) pp. 98–99.
[a] The data for fiscal years 1962 and 1963 are estimates. The accounts exclude outlays on research and development plant.

bulk of the applied research supported by the Department of Agriculture, are directed toward increasing productivity. Here the benefit measurement problem is generally straightforward, although it would be foolish to pretend that it is easy.[6] Other projects, such as the work sponsored by the National Institutes of Health, seek to enhance the domestic welfare. Still other programs are concerned with reaping benefits in the sphere of international power politics and national prestige. The most obvious examples are the Defense Department's vast R&D effort, the military bulk of the Atomic Energy Commission's R&D activities, and a substantial fraction of the National Aeronautics and Space Administration's rapidly growing program.[7] It is in this last sec-

[6] It is not even certain that increased productivity in agriculture is a net social benefit, given the immobility of agricultural resources and the income redistribution effects which, unless checked by controls or the accumulation of surpluses, result from innovations increasing output in the face of severely inelastic demand.

[7] Of course, products and processes which increase productivity or demand in other sectors of the economy may arise from military- and prestige-oriented programs. But these benefits—commonly referred to as "spillover"—are strictly incidental to the main purpose; there is also some reason to believe that the spillover from military and space R&D programs has been quite modest in recent years. See Robert A. Solo, "Gearing Military R&D to Economic Growth," *Harvard Business Review*, Vol. 40 (November-December 1962), pp. 49-60; Lawrence Galton, "Will

tor that the theoretical and practical problems of benefit evaluation appear most difficult and challenging. Partly for this reason and partly because, as Table 1 shows, defense, atomic energy, and space programs account for roughly 90 percent of all government R&D outlays, this paper emphasizes the benefit measurement problem in those programs whose ultimate objectives are essentially political.

The Economizing Problem in R&D Programs

Even though government research and development programs serve such esoteric goals as national security and national prestige, there is a definite need for economizing and for economic analysis in program decisons. To economists this is so obvious that it hardly merits saying, but it has not always been fully appreciated by the pure scientist or the career military leader.[8] And certainly there are instances in which a healthy skepticism is warranted.

The argument for economizing rests on two propositions: (1) resources, and especially highly trained and skilled technical talent, are scarce; and (2) the possibility of substitution exists, so that resources can be shifted to areas in which relatively urgent needs and wants are left unsatisfied from areas in which the unsatisfied needs are less urgent.

A casual glance at the situation in 1963 seems to validate both propositions. Colleges and universities have experienced increas-

Space Research Pay Off on Earth?" *New York Times Magazine*, May 26, 1963, pp. 29 and 93-95; and Richard Witkin, "Space Technology Is Producing Some Commercial Applications," *New York Times*, September 15, 1963, p. 75.

In the period since this paper was presented, NASA began to stress building a "capability" for future military and civilian space efforts as a major benefit of the current program. There is undoubtedly some merit to this line of reasoning, although the future benefits from possessing the purely physical outputs of existing programs (e.g., specific boosters, guidance systems, etc.) are likely to be small as a result of technological obsolescence. The building of skills and knowledge useful in future programs is perhaps more important. Ultimately, however, the problem of measuring the benefits of those future programs must be faced.

[8] At least, until Assistant Secretary Hitch's book became unofficially required reading in military circles. See Charles J. Hitch and R. N. McKean, *The Economics of Defense in the Nuclear Age* (Harvard University Press, 1960).

ing difficulty, as a direct consequence of government research and development program demands, in retaining for teaching purposes the faculty man-hours required adequately to educate a rapidly expanding student population. Valuable contributions to the economic development of emerging nations could be made by the talent now allocated to military and space R&D programs. It is possible, although no convincing evidence is available, that government programs have absorbed resources which otherwise might be employed in privately sponsored research to increase industrial productivity and create new products. And of course, within any given government agency, the use of scarce talent on one project necessarily means the unavailability of that talent for the many other projects which never seem to receive the support agency chiefs would like to give them.

Yet one hazard in this line of argument must be faced. In principle, the possibilities for using scarce technical resources to serve alternative ends are virtually infinite; but would the substitution process break down in practice? If the government employed fewer resources in space and military programs, would they in fact be absorbed by the colleges and universities, technical aid programs, and private R&D projects? Or would political taboos, factor price rigidities, and risk aversion or liquidity preference in the private investment sector lead to their unemployment, or to a displacement process with the same ultimate effect? The alternative to massive government R&D spending could conceivably be a massive recession. If so, the pump-priming benefits of such spending cannot be ignored. But my own belief is that there are many alternative uses crying out for the resources presently allocated to defense and space R&D, and that these resources could be transferred if appropriate fiscal and monetary measures were pursued. If this is not true, then the analysis which follows is not nearly as important as I consider it to be.

Assuming then that economizing is desirable and necessary in government research and development programs, several kinds of economizing decisions can be identified:

1.   Most obvious, decisions must be made about which projects to support and which to reject.

2.   In programs whose purpose is to develop operationally

useful equipment or "hardware," end-product quality objectives must be set. By and large, the higher the technical performance and reliability sought, given the existing set of technical
possibilities, the more development resources a program requires.[9] It also appears generally true that sooner or later diminishing marginal returns set in as more and more resources
are devoted to increasing any particular quality variable.[10]

3. When the decision-making problem is viewed from a
broader perspective which includes the eventual procurement
of operational equipment, the quantity of items to be procured
must be determined. The larger the quantity of items procured, the greater will procurement and operating costs be.

4. A decision must be made as to when to start research
and/or development. As time passes knowledge accumulates, better techniques become available, the unknowns in a given problem area decline, and the cost of accomplishing a specific development objective falls.

5. The government must decide how rapidly to execute the
development effort. As I shall argue later, within a certain
range development time can be reduced by increasing the total
amount of resources devoted to development.

6. Finally, a decision related to all of the foregoing—how
large the total program, agency, or government-wide research
and development budget will be—must be taken.

For decision-making of the type outlined here, the benefits of
alternative actions must be weighed against their costs. A business firm investing in research and development presumably de-
fines the benefits in dollar terms: revenues, quasi-rents, profits,
and so on. But when the federal government seeks to satisfy such
intangible goals as national security and prestige, it is denied
in many programs the convenience of a dollar common denominator. Its position is more like that of a consumer seeking to
maximize satisfaction or utility, subject to a fixed budget constraint or a fixed set of preferences regarding leisure as opposed
to additional income.

[9] It is assumed that unambiguously inferior alternatives are excluded from consideration.
[10] See Peck and Scherer, *op. cit.*, pp. 467-472.

To explore this analogy, let us assume that the government has an expected utility function, U, which is to be maximized subject to certain constraints. The nature of the utility represented by this function will be considered subsequently. As a first approximation, let us suppose that in the ith research and development program expected utility, U_i, is a function of $j = 1,...,m$ end product quality variables, q_{ij}, the quantity Q_i of units eventually to be produced, and the time, t_i, at which the R&D effort will be completed so that useful end products can be produced. Thus:

(1) $$U_i = U^i(q_{i1}, \cdots, q_{im}, Q_i, t_i).$$

As suggested previously, the expected total cost, C_i, of the program is a function of the same variables:

(2) $$C_i = C^i(q_{i1}, \cdots, q_{im}, Q_i, t_i).$$

Total cost, C_i, may not exceed and is assumed just to equal a provisionally imposed budgetary constraint, R. Then utility is maximized by maximizing the Lagrangian function:

(3) $$L = U^i(q_{i1}, \cdots, q_{im}, Q_i, t_i) + \lambda[R - C^i(q_{i1}, \cdots, q_{im}, Q_i, t_i)].$$

In addition to the budgetary constraint, there will be $m + 2$ first-order conditions of the form:[11]

(4) $$\frac{\partial U^i}{\partial q_{ij}} = \lambda \frac{\partial C^i}{\partial q_{ij}}.$$

Rearranging this, we find that:

(5) $$\lambda = \frac{\dfrac{\partial U^i}{\partial q_{ij}}}{\dfrac{\partial C^i}{\partial q_{ij}}} = \frac{\partial U^i}{\partial C^i}.$$

Intuitively, λ equals the increase in utility afforded by the last dollar spent on increasing quality variable, q_{ij}, or on any other quality or quantity or time variable. In other words, λ reflects the marginal utility of money and ultimately of resources allo-

[11] No attention will be paid at this point to second-order conditions. In reality the various relationships all appear eventually to exhibit the sort of diminishing marginal returns necessary to meet second-order conditions for a maximum.

cated to program *i*. As the amount of resources allocated to the program increases, the value of λ should as a rule be expected to decrease. By comparing the value of λ in program *i* with the analogous values for other programs (and perhaps also for non-government R&D possibilities) government decision-makers might decide rationally how large the budget for program *i*—and indeed for all programs—should be. For an optimal allocation of resources, the λ's for each program must be equal, assuming one common utility function to be maximized and one common scarce resource to be economized.

Direct Measurement of Program Benefits

It is all very well to write down the equations for utility maximization by the government and to speak of the marginal utility of money. But what in fact does utility mean in this context? Can it be measured in any conventional sense? And can it be made commensurable with more customary economic value indicators, such as the dollar?

Certainly it seems next to impossible to translate directly into dollar terms the value of national sovereignty or the prestige value of being first to land a man on the moon and to bring him home safely. Or to put the matter in a form more useful in marginal analysis, it would be most difficult to evaluate in dollar terms a 1 percent decrease in the prior subjective probability of achieving some political goal, such as deterring nuclear war or winning a conflict which has already broken out.[12]

There are, however, at least two ways out of this dilemma. Neither is completely satisfactory, but both go a considerable distance toward making rational resource allocation feasible. To contemplate the first approach we revert to the analogy of consumer choice. The theory of consumer choice does not require

---

[12] A social welfare function with variables for "the probability of victory," the number of lives lost in combat, and the sacrifices of present and future consumption as a result of victory-probability-increasing armaments expenditures is proposed by Jacob Marschak and M. R. Mickey in "Optimal Weapon Systems," *Naval Research Logistics Quarterly*, June 1954, pp. 116-140.

that utility be measured in any cardinal sense for a maximum to be achieved. It is only necessary that the decision-maker be able to measure utility ordinally—that is, to determine for all pairs of alternatives which of any two is preferred. Then, given a budget constraint, utility can be maximized by making small changes in the variables upon which it depends until no change can be found which is preferred.

This principle can be extended to decision-making in government R&D programs, and there is reason to believe that the process in fact operates approximately as the theory suggests. Three broad types of decision-making activities are observed when federal R&D budgets are hammered out. They occur simultaneously, each being linked to the others through feedback and readjustments as the final project structure is approached in as many as thirty successive iterations. Tentative budgets are assigned to lower echelons, which trade off the specific variables under their command until a most-preferred combination is struck. The budgets of the individual decentralized units are at the same time adjusted by higher authority, which seeks to establish a situation in which no small change in budget allocations can be found which is preferred (and, implicitly, has higher utility) to the status quo. Ideally, these mutual adjustments should lead to a condition in which the λ's confronting each decentralized unit are all equal. At a still more global level, the highest authority (presumably, the President and/or Congress) decides how large the overall government research and development program shall be, weighing at the margin the benefits which could be gained by allocating more resources to it against the benefits attainable if incremental resources were allocated to aid for emerging nations or to education—or left for research and development by private industry. Changes in the highest authority's choice of an optimal λ may in turn lead to further changes in the detailed choices of subordinate echelons.

Nevertheless, circumstances are bound to arise in which this essentially ordinal maximization process proves ineffective or breaks down in a program as large and complex as the government's R&D effort. Subordinate preferences inconsistent with the overall national interest and external effects deserve mention.[13] But here a third complication is of greater interest. The theory of consumer choice has tended to assume a simple decision-making unit with a well-defined set of preferences, paying little attention to how the diverse members of a spending unit manage to reconcile their frequently disparate individual preferences. This job of mutual accommodation is apparently accomplished with tolerable success in the typical spending unit (a family), partly at least because effective means of communication have been developed. However, in elaborately structured organizations, and especially in the research and development field, the necessary communication is much more difficult. Interviews with government personnel have suggested that decision-makers at the operating levels of technical programs can make tradeoffs fairly well among the variables directly within their field of expertise. But a concomitant of technical complexity and specialization is the need to pass the numerous choices that involve more than one organizational entity on to higher authority for decision.

The communication problem becomes acute at this point. Higher level decision-makers commonly have neither the time nor the knowledge required to understand thoroughly the technical details of a choice which implies, say, the sacrifice of a quality increment to reduce the R&D budget. With communication at a purely technical level thwarted, subordinates are forced to predict the implications of proposed changes in terms of some value system which defies articulation. In large business organizations the profit motive serves an important role merely by facilitating communication. By analogy, the government decision-makers' task of equating the λ's in various sub-budgets would be much easier if subordinates could measure and communicate in some meaningful sense the benefits related to particular budgetary changes.

Also, strictly ordinal measurement is likely to be inadequate for making decisions as to which large, discrete programs should be supported and which should not. In this case a criterion may

[13] For a further discussion, see Hitch and McKean, *op. cit.*, pp. 128-131, 161-165, and 396-402; and Peck and Scherer, *op. cit.*, pp. 476-478.

be needed to determine whether a potential program affords benefits sufficient to justify its costs. The notion of "benefits sufficient to justify costs" implies, not only cardinal measurement of utility, but also measurement commensurate with resource costs—e.g., in dollar terms.

There is another way of approaching the benefit measurement problem which avoids some of these difficulties. It is based upon the so-called cost-effectiveness technique used fairly extensively in military weapon systems program analysis. The cost-effectiveness approach itself is straightforward. By means of some decision-making process which may defy rational calculation, it is stipulated that certain definite, tangible objectives are to be attained. In this context an objective may be the maintenance of an offensive capability to destroy 140 strategic targets under a particular set of assumptions concerning defensive measures employed by the targets and prior damage to the offensive force; it may be the operation of a communication system linking certain areas with a specified minimum information capacity and maximum probable outage time; and so on. Given the objective, the costs of various alternative programs for achieving it are compared. And, at least as a first approximation, program choices are then based upon the criterion of minimizing the cost of achieving the specified objective.

Now there are many kinds of decisions which cannot be resolved adequately through the cost-effectiveness approach. It is useless for determining whether an objective is worth achieving. For instance, a minimum-cost program for detecting all submarines within 200 miles of the United States coast may be identified by cost-effectiveness analysis, but the analysis says nothing about whether the effort is in fact worthwhile. Similarly, the approach is of little assistance for analyzing changes in objectives. The cost of increasing the probable proportion of attacking bombers or ballistic missiles intercepted under some set of assumptions from 70 to 80 percent can be estimated, but much more fundamental value judgments are needed to resolve whether the expected saving of lives in the event of an attack is worthwhile. Again, cost-effectiveness techniques afford little guidance when military officials must choose between concentrating their resources on one seemingly most efficient weapon system and authorizing the development and production of two or more technologically different systems capable of accomplishing the same mission as a hedge against possible miscalculations or changes in the strategic environment.

Nevertheless, there are undoubtedly decisions in which limitations of this sort do not seriously impair the direct applicability of cost-effectiveness comparisons. Especially in the military field, which consumes the lion's share of government research and development resources, one can often find a wide variety of alternative weapon systems which are reasonably close substitutes for performing some broad mission generally considered essential. To cite perhaps the clearest example, the deterrent mission of maintaining a strategic bombing capability can be performed more or less effectively with B-29's, B-47's, B-52's, B-58's, B-70's, carrier-based attack aircraft, Polaris missiles of diverse ranges, mobile medium-range missiles emplaced in a frontier area, hardened Minuteman missiles, and so forth practically ad infinitum. It is also possible at a lower level of analysis to differentiate programs through differences only in some key variable. For instance, a substantial difference in reliability will affect the number of units required to perform a given mission, and hence the cost of performing that mission. Similar relationships may be found for changes in payload, dash speed, range, etc. What varies from alternative program to program in such cases is primarily (although never exclusively) the cost of accomplishing the mission. When this point of view is adopted, we see a close analogy between research and development programs serving such intangible goals as strategic deterrence and programs whose purpose is to create new cost-saving industrial processes. As long as we assume that capabilities for a mission will be acquired in any event, it is possible to quantify more or less explicitly the benefits expected from a new R&D program without attempting to judge the utility of the mission to be performed. Our estimator of program benefits in such instances is the cost savings the program permits over existing inferior alternatives.

That this estimator is a very imperfect one cannot be denied. There will always be differences between programs which cannot be resolved into monetary terms. For example, in the strategic bombing illustration, the loss of military lives is apt to be greater both in training and action with old-style bombers than with unmanned guided missiles. One can only hope that the errors attributable to ignoring such differences are relatively small and tend to impart no consistent biases. And even when all other variables are held constant, the inherent and institutionally generated uncertainties connected with program outcome predictions will necessarily lead to errors in estimating cost savings. In the past, military program cost estimates have been badly biased on the optimistic side due to the efforts of contractors and military advocates to win support for their pet projects.[14] For any meaningful analysis of cost savings to be made, the estimates would have to be prepared by a group which combines considerable competence with objectivity. Recently the Office of the Secretary of Defense has begun to build up a body of independent expertise for unbiased program cost estimation, and therefore might meet this requirement.

If only a few (e.g., roughly ten) programs of divergent size can be isolated in which the barriers to making reasonably good cost-effectiveness comparisons are not insuperable, and if we are willing to be content with the stochastic character of the benefit estimates, we may be able to secure a foothold from which considerable leverage can be exerted on the more general benefit estimation problem. But one further complication must first be examined.

The complication can be demonstrated through an example which will also more rigorously illustrate the approach suggested thus far. Suppose we have two alternative processes or systems X and Y for performing some function or mission considered essential over the next s years. By careful analysis we estimate two streams of costs over time—one stream, $x_0, \cdots, x_t, \cdots, x_s$, associated with performing the mission with X and another, $y_0, \cdots, y_t, \cdots, y_s$, with performing exactly the same mission with Y. Now let us arbitrarily take X to be the standard for comparison, perhaps because it is already available, not incorporating the advanced technological features which must be developed for Y. We define the benefit,

[14] See Peck and Scherer, *op. cit.*, pp. 411-424.

b_t, from employing Y rather than X in year t to be the difference $x_t - y_t$. If a research and development program is required before Y becomes available, it is likely that for the first few years the b_t will be negative, turning positive only when Y displaces X and begins to show annual cost savings.[15] We define

$$B_Y = \sum_{t=0}^{s} b_t$$

to be the total of the benefits associated with choosing Y rather than X. If our alternatives are in fact directly comparable, we would not choose Y unless B_Y is positive. And usually a more demanding criterion of choice will be imposed: Y will not be selected unless

$$\sum_{t=0}^{s} \frac{b_t}{(1+r)^t}$$

is positive, where r is an appropriate discount rate.[16]

So far this is all very elementary. But let us now assume that a third alternative Z is considered, with a cost stream $z_0,..., z_t,...,z_s$ such that

(6)
$$\sum_{t=0}^{s} \frac{z_t}{(1+r)^t} < \sum_{t=0}^{s} \frac{y_t}{(1+r)^t} < \sum_{t=0}^{s} \frac{x_t}{(1+r)^t} \, .$$

[15] If X is the use of a system already available while Y is the development of a new and more effective system, it may be necessary to include in the cost stream y_t some costs of using X as a stopgap until Y becomes available. The main thing is for the two streams to reflect exactly the same capabilities.

[16] As always, the determination of r poses problems. In the present case we are attempting to get around some of the difficulties connected with approaching program choices as a constrained utility maximization exercise. One of these difficulties is deciding upon the appropriate value of λ. But the discount rate r used here must reflect the opportunity cost of employing resources, which in turn is closely associated with λ. Presumably, a proper discount rate would be one which tends to set the present value of the expected net benefits of marginally desirable programs equal to zero. In other words, it would be the internal rate of return on a marginally desirable program. This begs the question of which program is marginally desirable. A subjective value judgment seems inescapable on this point, at least for the types of programs under consideration here. For a more general exploration into this largely unexplored problem area, see W. J. Baumol and R. E. Quandt, "Investment and Discount Rates Under Capital Rationing: A Programming Approach," *Economic Journal* (forthcoming).

That is, Z is more efficient than Y, which is more efficient than X. From the standpoint of making a program choice, this complication poses no new difficulties. If the relationships are as shown in inequality (6), we will choose Z. But how do we measure the benefits associated with Z in this case? If we compare Z's cost with the cost of the next-best alternative Y, as suggested in some texts,[17] we will obtain one value, while we will get a different value if X is selected as the basis for comparison. The benefits measure is not absolute—it has no definite and unambiguous zero point. Therefore, measurements of benefits from one cost-effectiveness analysis are not likely to be directly comparable with measurements from others unless some further calibration criterion is imposed.

If we are interested only in choosing most efficient or effective programs for any given mission, this lack of comparability need not concern us. However, for the step to be taken in a moment, comparability of benefits between analyses is essential. Therefore, let us adopt the following rule: when evaluating the benefits of a proposed R&D program, the most efficient system or method already being utilized should be taken as the standard alternative. While this is a plausible enough convention, we should emphasize that it is completely arbitrary and perhaps not even correct in a strict theoretical sense. We must be wary of the difficulties into which it may lead us.

A More Generalized Benefit Measurement System

Let us assume tentatively that with a diligent effort, reasonably meaningful, accurate, and comparable benefit estimates can be made through the cost-effectiveness method for ten or so R&D programs in a specific sector. The defense sector seems most likely to fill this bill of specifications, since its basic missions are better established than, say, those in the space field and since it has the largest stock of programs from which to draw a sample suitable for cost-effectiveness analysis. Then it may be possible to exploit this

[17] See, for example, Erich Schneider, *Wirtschaftlichkeitsrechnung* (Bern: Francke, 1951), p. 109.

information to estimate the benefits of a much larger sample of programs.

The proposed procedure is as follows. A panel of experts (i.e., senior military officers and high-level civilian operations analysts, if weapons R&D programs comprise the sample) would be selected. Panel members would be asked to rank through a paired comparisons technique some thirty or more programs currently in the R&D stage, including the ten or so programs on which monetary benefit estimates made through the cost-effectiveness method are available. For each possible program pair,[18] a panelist would be asked to make a choice of the following sort: "Which program offers the greater long-term net benefits or value, taking into account the program's expected operational functions, the probable costs of executing those functions, and the alternative means of accomplishing the functions?"[19] To guard against misinterpretation, factual material on the status and plans for each program would be disseminated and studied before rankings were attempted. In certain cases special assumptions not entirely consistent with current plans might have to be spelled out to make the procedure work.[20]

[18] The number of different pairs of n objects is $n!/(n-2)!2!$ If 30 programs are to be ranked, panelists would be required to make 435 paired comparisons. Some duplicate pairs might be added, as in psychological scaling experiments, to test the stability of panelists' choices.

[19] Some other, and possibly more elaborate, statement of the question might be found necessary to ensure appropriate communication of what is wanted.

[20] Assume, for instance, that Program A, which is being evaluated, offers a substantial improvement over existing systems, but that Program B, also in the research and development phase, offers a still greater improvement. Both systems are scheduled for eventual use. The benefits imputed to Program A are apt to be small owing to the competition from B. But, relative to currently available systems, A offers important benefits—and this is the standard chosen for the cost-effectiveness benefit estimates. In such a case we might want to stipulate that only A will be put into operation, so that panelists will clearly evaluate it against the status quo. But on the other hand, it might be still better to include both B and A in the rankings and to let B capture most of the imputed benefits from the $A + B$ package. I must admit that I am troubled about this kind of interdependence problem, which is fundamentally related to the problem of finding an unambiguous zero point from which to calibrate the benefits measure. The whole approach may break down as a result of this limitation.

By techniques which are well known,[21] the paired comparisons made by individual panelists can be aggregated to obtain an overall ranking of programs from "most valuable" to "least valuable" for each panelist. The choices of individual panelists can also be aggregated by means of scaling factors into a group consensus ranking. For any given program A the scaling factor is defined as:

$$\frac{\text{Number of times } A \text{ was chosen over other programs}}{\text{Total number of choices involving } A}.$$

It can vary from 0 (no choices in which A was preferred) to 1.0 (A preferred in every choice). The program with the highest scaling factor is assigned the highest rank (1), and so on. It may also be possible to go somewhat further than a group consensus ranking with the scaling factors. Suppose that three programs A, B, and C, have the scaling factors .667, .659, and .608. There are no other programs with scaling factors within this range. Then we would rank A first, B second, and C third within the subset. But further consideration suggests that we have something more than an ordinal measure. The difference between the scaling factors for A and B is .008, while the difference between B and C is .051. It would appear that in the group consensus the preference for A over B was much weaker than the preference for B over C (as well as for A over C). By aggregation of individual choices into the group consensus an interval measure of the differences in value between programs is approximated. One may of course raise philosophical objections to this transition, but even if the operation should be declared invalid the procedure which follows loses only precision and not validity.

Now let F_i be the scaling factor for the ith program belonging to the subset for which benefit estimates made through cost-effectiveness analysis are also available. Let B_i be the monetary benefit estimate for that program. Then performing a conventional least-squares regression (or possibly an orthogonal regression, since the F_i are also subject to errors) of the general form

(7) $$B_i = f(F_i) + e_i,$$

[21] See M. G. Kendall, *Rank Correlation Methods* (Griffin, 1948), pp. 121-138. For an extensive application, see Peck and Scherer, *op. cit.*, pp. 543-580 and 668-711.

we obtain an estimate of the relationship between the scaling factors and the benefit estimates. The most suitable specific form of the relationship would have to be established experimentally; *a priori* reasoning indicates only that B should be monotonically increasing with F. Equation (7) can then be employed to estimate the benefits of programs for which we have only scaling factors and not cost-effectiveness estimates. So unless something has gone wrong along the way, we secure estimates in dollar terms of the benefits anticipated from programs in which the benefits could not be directly quantified.

I would be the first to admit that many things could go wrong along the way. A few can be mentioned here; the comments following this paper provide an ample addendum. The method of setting a zero point for the cost-effectiveness benefit estimates is arbitrary and could in some cases be inconsistent with reality, as perceived by the panel of experts. The reality perceived by the panelists might in turn differ from some other group's reality. The whole concept of net benefits and value from programs with essentially political goals is fraught with hazards, and so panelists may find themselves unable to make meaningful and consistent judgments. Special difficulty might be experienced in comparing cost-saving programs (those suitable for the crucial cost-effectiveness benefit estimates) and programs with more intangible goals. Uncertainty is an inescapable water-muddier. And finally, the procedure of pooling individual judgments to obtain a consensus measure of programs' relative value may not be exempt from the contradictions suggested by Arrow's work on democratically derived social welfare functions.[22]

[22] See K. J. Arrow, *Social Choice and Individual Values* (Cowles Commission Monograph No. 12; Wiley, 1951). This aggregation problem was raised in an ingenious study by W. Giles Mellon, *An Approach to a General Theory of Priorities: An Outline of Problems and Methods* (Princeton University Econometric Research Program, Research Memorandum No. 42, July 1962). Mellon is concerned with determining how far purely ordinal measurements of value obtained by the group consensus method can be employed to make optimal military choices, especially in the logistics area. On p. 10 he quotes a letter from Professor Arrow on the applicability of Arrow's impossibility theorem to the pooling of expert rankings: "I feel that the question of pooling experts' opinions is different from that of a consensus of welfare judgments because presumably they are all judging the same underlying reality. I must admit that this gets into complicated epistemological questions if pushed far enough. However, if we postulate the existence of an objectively valid

To the best of my knowledge there is no body of theory which permits one to predict confidently whether an approach to benefit estimation of the sort proposed here will or will not break down. The question appears to be essentially an empirical one. The methodology suggested here will provide at least three explicit checks on the validity of the whole procedure. Trouble will be signaled if any of the following phenomena emerge:

1.   If the ranking of the subset of programs obtained through cost-effectiveness analysis disagrees significantly with the ranking of the same programs obtained by paired comparisons.

2.   If there is significant evidence of intransitivity in the individual panelists' paired comparisons; that is, if panelists frequently indicate that A is more valuable than B, B is more valuable than C, but C is more valuable than A.

3.   If there is little agreement among the various panelists on their rankings of the different programs.

Regression equation (7) supplies checks on the first problem. Obtaining either a negative slope or a low coefficient of determination with (7) would warrant rejection of the procedure. The paired comparisons technique yields built-in checks for the second and third problems by means of coefficients of "consistence" and "agreement" and their related significance tests.[23]

Results of the meager empirical work already done on the ordinal measurement of military utility or value are cause for at least mild optimism about the proposed method's prospects. Two recent studies found that different military personnel agreed rather well in judging the priority of various electronic and submarine spare parts items.[24] Still the evaluation problem in these studies was a good deal simpler than the one implied here. Per-

---

scale of some kind in the universe and assume that each expert's opinion is a random function with the objective scale or a parameter and with a probability distribution which has some degree of specification *a priori*, then we have a manageable statistical problem—that of estimating the underlying scale."

[23] See Kendall, *op. cit.*

[24] See R. J. Aumann and J. B. Kruskal, "Assigning Quantitative Values to Qualitative Factors in the Naval Electronics Program," *Naval Research Logistics Quarterly*, March 1959, p. 15; and Henry Solomon *et al.*, "Summary of a Method for Determining the Military Worth of Spare Parts," *ibid.*, September 1960, pp. 221-234.

haps more relevant is the following experiment reported by M. J. Peck and myself.[25]

Some forty-four persons from government, industry, and academia active at high levels in military R&D programs made paired comparisons of eight weapon system programs on four attributes: state of the art advance, value to the national arsenal, government performance, and contractor performance. Intransitive choices were infrequent on all four attributes and least frequent on the value attribute, despite ambiguities in the way "value" was defined. Agreement among panelists was highest on the value attribute—the Kendall coefficient of agreement was .348.[26] This does not suggest extremely strong agreement, but it also does not signify strong disagreement. To give the result more intuitive content, if a panel of thirty persons had divided by a twenty-four to six margin on each paired comparison, a coefficient of .338 would have been attained. Reasons for the disagreements among panelists included military service biases and lack of knowledge, and there was evidence that these could be limited greatly through selection of better-informed and more objective panelists.

In sum, there seems to be no compelling reason for rejecting out of hand the benefit estimation system proposed here. It might not work at all and it certainly cannot generate estimates of any great precision, but there is a reasonable chance that it could yield considerable insight into the benefits of government research and development programs. And through extensions the approach might help fill a major analytical void hampering present government R&D program decision-making. As Oskar Morgenstern has observed of the military choice problem:

> To develop a method (for determining military worth) would be more important than to introduce some further, purely technological, advances in some weapon system. It would improve the use of all these.[27]

The method suggested here at least seems worth an experimental attempt, especially if cost-effectiveness benefit estimates can be

[25] See Peck and Scherer, *op. cit.*, pp. 543-580 and 668-711.

[26] This is not the coefficient reported in Peck and Scherer, p. 556. A different coefficient was developed there to permit greater comparability of results from sub-panels.

[27] Morgenstern, *The Question of National Defense* (Random House, 1959), p. 205.

obtained as a largely free by-product of existing analytical work by the Defense Department.

One additional limitation should nevertheless be noted. If workable at all, the technique outlined here is likely to be useful only in estimating the benefits of programs ranked lower than the highest-ranked program for which a cost-effectiveness benefit estimate is available and higher than the lowest-ranked cost-effectiveness program. Extrapolation with equation (7) would be most hazardous. This automatically excludes massive efforts such as the Apollo program, where the need for meaningful benefit estimates is especially vital. Some other approach must be taken in such cases. I have toyed with the idea of polling a random sample of citizens, asking which they would prefer—the incremental goods which their tax contribution to the Apollo program would buy, or seeing the United States first to land a man on the moon.[28] But even if a majority of persons were in favor of sacrificing personal consumption (or savings) to pursue the program, the welfare implications of not compensating the opponents would have to be faced. And perhaps more troubling, I am not certain that complex and farsighted decisions of this nature can or should be made through democratic processes. Statesmanship—or in other environments dictatorship—may be called for when Apollo-like decisions have to be made. Therefore, I can only commend the problem to others.

Time-Cost Tradeoffs:
A Special Problem in Benefit Assessment

Let us change our perspective now. In order to examine a class of decisions in which benefit estimates appear to play an especially crucial role, we shall assume that the benefits of government R&D programs can in fact be estimated, however crudely.

[28] A casual effort of this sort was reported in a letter to the *New York Times*, Feb. 18, 1964, p. 34. A North Carolina resident asked 100 fellow citizens whether the Apollo project (at an estimated cost of $20 to $40 billion) should be continued or abandoned; 97 respondents favored abandoning the project. It is possible that posing the question in terms of individual tax incidence instead of overall costs would have elicited different results.

FIGURE 1. Development Possibility Curve

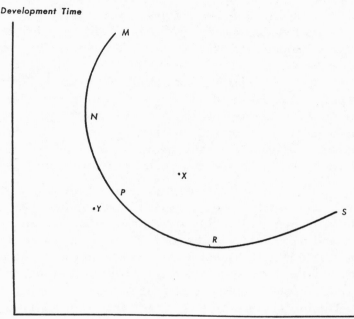

Specifically, we shall explore the problem of time-cost trade-off decisions—decisions to sacrifice development resources in order to reduce development time, or vice versa. Although the concept of such decisions was by no means original to M. J. Peck and myself, I shall here at least initially follow the approach suggested by us in 1962.[29]

Assume that at some moment in time, with some given stock of technical knowledge, a project is initiated to develop and design an end product which will have certain specified technical characteristics—that is, the end product quality is assumed to be constant.[30] To avoid the complication of production costs, it is as-

---

[29] Peck and Scherer, *op. cit.*, pp. 254-287.

[30] The assumption of constant quality, like all *ceteris paribus* assumptions of partial analysis, does violence to reality. (See Peck and Scherer, pp. 492-500.) It is made only to focus attention on a specific aspect of the problem and can easily be relaxed (with an attendant loss of insight) through the more general approach suggested in equations (1) through (5) above.

sumed that the output of the development project is the set of designs and know-how necessary to reproduce the contemplated end product in the desired quantities. Then it is postulated that there exists a function such as *MNPRS* in Figure 1 which relates expected development time to the expected total quantity of resources employed in a development effort.[31] Resources are measured in terms of a dollar cost common denominator.[32] The heart of the proposition advanced here is that within the range between *N* and *R*, the time-cost function has a negative slope; time can be decreased only by increasing total cost.

A number of reasons for this negative slope assumption were presented by Peck and me; here a brief summary must suffice:

1. As the development schedule is compressed, more and more tasks must be conducted concurrently rather than in series. Since each task yields knowledge useful in other tasks, more and more tasks must be conducted without all the knowledge that might be available in a series approach. The result is more false starts, tests which investigate the wrong variable or an unimportant one, wasted designs, etc.

2. Diminishing marginal returns eventually set in as the number of technical workers assigned at any moment to a given task is increased in order to complete the task more quickly.

[31] To guard against misinterpretation, it must be stressed that the development possibility function is a subjectively estimated *ex ante* relationship, and therefore entails uncertainty of measurement. I assume that the uniqueness of the values shown is due to the use of expected values. A modal approach is equally plausible. (See Peck and Scherer, *op. cit.*, pp. 300-309.)

It should also be noted that the function is shown as two-valued over much of its range. The segments *NM* and *RS* are inefficient in the sense that (for instance) at *S* the same development time could be achieved at less cost by moving westward onto the *PR* segment. The function turns upward through *RS* because of managerial breakdowns when a project is conducted on too massive a scale. It turns backward through *NM* because economies of specialization are lost when a project is carried out on too small a scale. (See Peck and Scherer, *op. cit.*, pp. 263-266.) Obviously, point *X* in Figure 1 is also inefficient in this strictly technical sense. Point *Y*, on the other hand, represents an infeasible quality-time-cost combination.

[32] This aggregation assumption of course raises difficulties which no "production function" analysis can really escape. See Joan Robinson, "The Production Function," *Economic Journal*, Vol. 65 (March 1955), pp. 67-71. But I agree with Mrs. Robinson that the technique is nevertheless useful analytically.

3. When uncertainties preclude identification of the best or an acceptable technical approach or design, several different approaches can be supported to ensure that the solution will be available when desired. Similarly, test program delays can be minimized by producing "backup" test vehicles to use in the event of an accident or failure. Hedging activities of this sort reduce the expected value of development time through an increase in expected development cost.

To the extent that the development possibility function does in fact possess a negative slope for these or any other reasons, government decision-makers have a definite choice problem: they must decide how far to carry the sacrifice of scarce development resources in order to reduce development time.

Does the Time-Cost Tradeoff Problem Really Exist?

I advance the problem in this tentative, insecure fashion because the more confident statement by Peck and me of the tradeoff relationship between time and cost has met with considerable reproof. In private conversations, military R&D officials have vigorously denied that any relationship of the form shown in Figure 1 exists. Their perception of the problem seems to be summarized in this excerpt from a trade journal editorial: "A technical program that does not run at its fastest feasible pace is the most wasteful of all."[33] A reviewer of the book by Peck and myself protested, without offering any positive alternative explanation, that "the authors' assertion that the evidence shows the time-money curve (money on the horizontal axis) to be not only downward sloping but also convex to the origin ('diminishing returns') is dubious indeed."[34]

Let us see at the outset whether some light can be shed on the apparent points of contention. Resolution hinges on understanding the relationship between the *rate of spending* at any moment in time during a development effort and the *total amount of*

[33] Robert Hotz, "Apollo and Its Critics," *Aviation Week & Space Technology,* April 29, 1963, p. 17.
[34] Review by Thomas Marschak, *Journal of Political Economy,* Vol. 71 (June 1963), p. 303.

spending over the course of the development effort. Most military officers with whom I have discussed the matter agree that within limits imposed by the specific technical problems and management's ability to control the technical effort, development time decreases as the rate of spending is increased and increases as the rate of spending is decreased. What they will not concede is that the same relationship holds true with respect to total development cost. They argue that by spending money at a lower rate and increasing development time, there is more time during which money is spent, and so more money is spent in total. Or by spending money at a greater rate and decreasing development time, there is less time for spending, and so less is spent in total.

We must observe at once that these suggested relationships are not necessarily congruent or reversible. Consider the case in which a development effort is started at one level of annual budgetary support and then, for some reason, the budget is "cut back" and the schedule must be "stretched out." It is possible—although not necessary—that the only thing which will be eliminated is direct technical effort, overhead expenditures remaining at their usual merry level. If fixed overhead comprises a high proportion of the development organization's costs, a modest budget cut may lead to a condition in which virtually no technical work gets done. Only the paper shuffling and brass polishing continue. This is rather extreme, but it must be admitted that, at least in some instances, decreases in the rate of spending can lead to an increase in a program's money cost.

There are still further weaknesses in this very popular argument.[35] If overhead costs really are fixed, they are irrelevant from an economic decision-making point of view. Whether incurred by an in-house government laboratory or by a captive contractor, they will be borne by the government in any event.[36]

---

[35] One is that such a situation is obviously inefficient, which violates one of the premises of the development possibility function concept. But I do not wish to stress this point, since I have maintained elsewhere that inefficiency is very common in contracted-out government R&D programs. See Scherer, *The Weapons Acquisition Process: Economic Incentives* (Harvard Business School Division of Research, 1964).

[36] The case of contractors with both government and commercial work in the

It is the resources with alternative uses which must be economized. In principle truly fixed costs should not be counted at all in estimating the development possibility function. And even if we ignore this vital point, the overhead absorption argument does not necessarily hold good for all types of decisions. It seems most appropriate for decisions to cut back the rate of spending (or time pattern of spending rates); in other words, to change the planned spending rate in a particular way.

But how about decisions to *increase* the planned rate of spending? One might argue that increasing the rate of spending permits broader absorption of overhead costs, and hence a reduction in total cost over the life of the program. But if this is true, overhead must have been inefficiently high in the first place. Why should the government choose to conduct its R&D programs in organizations with inefficiently high overheads? There may be good reasons—for example, a policy of maintaining the mobilization base. A rational accounting system would allocate excessive overhead costs to that policy account rather than to specific program accounts. However, as a rule we should expect the government to choose R&D contractors whose resource structure is well matched with the expected time pattern of spending rates in the specific program to be conducted.

If we can at least try to keep an open mind on whether overhead does or does not dominate all time-cost tradeoff decisions, then the relationships among time, rate of spending, and total cost can be analyzed more rigorously to good advantage. We begin with a geometric illustration. Assume that spending in a development project starts at a zero rate per unit of time and increases linearly to a peak at the time production designs are delivered.[37] Then if the length of the development period is OT_1, as in 2a of Figure 2, total spending will be the area OT_1A. Now assume that the rate of spending is increased in order to compress

---

same overhead center is somewhat different from a pecuniary, but not a real resource, point of view.

[37] This is not a bad approximation to reality, which appears to follow a buildup curve similar to the Gaussian normal curve in a great many cases. See Peck and Scherer, *op. cit.*, pp. 309-314.

FIGURE 2

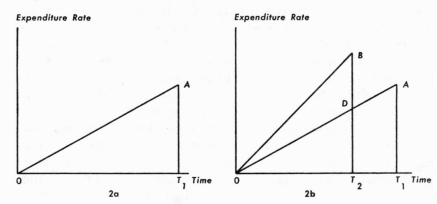

the development schedule to OT_2, as shown in Figure 2b. The new total spending is the area OT_2B. Is total spending more or less with the OT_2 schedule than with the OT_1 schedule? The answer depends upon whether the area T_2T_1AD, representing the savings due to spending over a shorter time span, is less or greater than the area OBD, representing the increase in outlays during the development span common to both schedules.

Now let C be the *total* cost of development, c the rate of spending per unit of time, T the amount of time spent on development, and t a running time variable. The rate of spending is evidently a function of both the total amount of time and the particular moment in time. Thus

(8) $$c = c(T, t).$$

Total cost is the integral of this function over time:

(9) $$C = \int_0^T c(T, t)\, dt.$$

For the Peck-Scherer hypothesis to hold, there must be some range in this function where $dC/dT < 0$; total cost decreases as time increases. Differentiating (9) once, we obtain:

(10) $$\frac{dC}{dT} = \int_0^T \frac{\partial c(T, t)}{\partial T}\, dt + c(T,T).$$

The first term corresponds to OBD in Figure 2b; the second term to T_2T_1AD. The second term is always positive for increases in T. If the Peck-Scherer hypothesis is to hold, the first term must be negative with an absolute value larger than that of the second term. Let us call this Condition I for supporting the hypothesis.

To specify completely the proposition advanced by Peck and myself, higher-order conditions must also be stated. Figure 1 clearly has the assumptions of smoothness and continuously diminishing returns built in, and so one may wish to insist that $d^2C/dT^2 > 0$ throughout the range $MNPR$. This condition, which reflects the letter of Figure 1, can be called Condition IIA. But here I must own up to past oversimplifications. In my mind Figure 1 has never been more than an idealization. What is crucial for an analysis of the time-cost tradeoff phenomenon is simply the assumption that the development possibility function have no point at which time and cost simultaneously achieve their minimum values. That is, there must exist two or more time-cost possibilities such that either time or cost must be sacrificed in moving from one combination to the other. Call this weaker case, which expresses the spirit of the analysis, Condition IIB.

Whether these conditions are satisfied in reality is largely an empirical question. Let us attempt to identify the principal cases in which some or all of the conditions do not hold and to see how consistent they are with what we can definitely observe about development program behavior. First, in Figure 3, 3a, 3b,

FIGURE 3

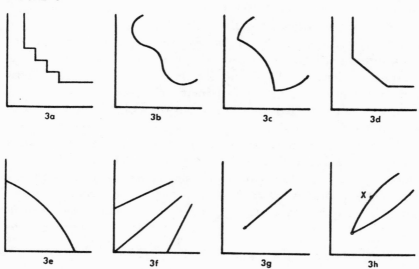

3a   3b   3c   3d

3e   3f   3g   3h

3c, and 3d all satisfy Condition IIB, but none satisfies Condition IIA and 3a does not meet Condition I. Therefore, all violate the letter of the hypothesis, even if not the spirit. All four appear more or less compatible with what one sees in actual R&D programs. Second, 3e and 3f illustrate cases which violate all three conditions. Both appear incompatible with reality, since they imply that R&D projects can be completed in zero time or at zero cost or both. In fact, one never observes R&D projects so effortlessly executed, unless we define "project" to include trivial cases such as those in which all the work has been done by someone else willing to give it away.

But 3g and 3h are more interesting. They plainly violate Conditions I and IIA. They also violate Condition IIB, since both figures indicate that there is some unique minimum cost/minimum time point. Thus they breach the spirit of the hypothesis. Now it is such cases that military officers seem to have in mind when they assert that the minimum time program is also the minimum cost program. How consistent are they with reality? One feature immediately arouses suspicion. Both 3g and 3h require that the first term in equation (10) become more and more negative up to the minimum development time and then abruptly increase. With Alfred Marshall, I am inclined to believe that activities involving large numbers of individuals exhibit more continuity. But this is no proof. More important, as the problem is formulated thus far, there is only one efficient combination in either 3g or 3h. Any possibility other than the minimum time/minimum cost combination involves the unnecessary and therefore inefficient sacrifice of one variable. Therefore, we would as a first approximation expect most or all programs to be conducted on a minimum time basis. But this surely does not correspond to reality, as anyone familiar with research and development in either government or industry will testify. There was a world of difference between the Polaris IRBM development on the one hand and the Talos antiaircraft missile development on the other. Similarly, the former technical director of the Apollo program asserted that the program's schedule in 1963 was "not the fastest and not the slowest," since the development organization was "dollar limited on how fast (it could) proceed."[38]

We can nonetheless find three possible explanations for the existence of less-than-minimum time development programs compatible with the assumptions of 3g or 3h. First, budgetary decision-makers may be irrational. But I doubt that psychopathic behavior is widespread enough in R&D decision-making to account for the large number of programs conducted at less than the most rapid pace. Second, a longer development period may have been preferred for some reason to a shorter one. This could be true in isolated instances, but I for one have seen very few development programs in which decision-makers did not want to decrease development time, other things being equal.

The third possibility is by far the most significant. The definition of efficiency assumed thus far—minimizing the total cost of a given time-quality outcome—leaves out one important element: the opportunity cost of using scarce resources in a program when attractive alternatives are left unsupported. In other words, it ignores the λ factor found in equations (3) through (5). The λ there is an essentially static concept. It can be represented dynamically by a time discount rate, r, so that the corrected total development cost function becomes

$$(11) \qquad C^* = \int_0^T c(T, t)e^{-rt}\, dt.$$

The first derivative is

$$(12) \qquad \frac{dC^*}{dT} = \int_0^T \frac{\partial c(T, t)e^{-rt}}{\partial T}\, dt + c(T,T)e^{-rT}.$$

Since the second (positive) term is discounted more heavily than the first term, addition of the discount rate has made it easier to achieve a negative slope for the (discounted) total cost func-

[38] "Moon Shot Costs Put at 20 Billion," *New York Times*, March 28, 1963, p. 1, quoting testimony of D. Brainerd Holmes before a House committee. Holmes stated further that in contrast to some military programs in which "money is a secondary factor, . . . money is dictating the time performance" of Apollo.

tion. This strengthens the case Peck and I have argued. At the same time, it also affords some support to the military officers' case. One *can* conceive of situations in which rational decision-makers, anxious to reduce development time, would nevertheless choose to operate at an *undiscounted* total cost point like X in Figure 3h—when resources are so scarce relative to needs that decision-makers would prefer higher total costs, falling due predominately in the more distant future, to lower (undiscounted) total costs accruing in the nearer future.[39]

The only verdict which can be drawn from *a priori* theorizing combined with minimum, virtually indisputable empirical observations is a Scotch one. The military officers' unique time-cost minimum cannot be ruled out. The Figure 1 configuration, along with Figure 3's several variants—3a through 3d—also cannot be ruled out. My own conviction is that at least Condition IIB usually holds and that Conditions I and IIA often hold; i.e., that there exist substantial possibilities for making efficient time-cost tradeoffs in most development programs. This conviction rests on three foundations: the reasoning given on pages 36-37 above concerning "concurrency vs. series scheduling" and diminishing returns; the meager empirical evidence cited by Peck and me on pages 259-263 of our book; and the explanatory powe: of the assumption, to be demonstrated in a moment. But I also believe that we badly need some inspired empirical research which will

[39] In a numerical model packed with insights, S. C. Daubin shows among other things that in certain cases it may be optimal to protract a project so that development time exceeds the combination at which cost is at a minimum—e.g., operating in the *MN* segment of Figure 1 or around point X in Figure 3h. (See "The Allocation of Development Funds: An Analytic Approach," *Naval Research Logistics Quarterly*, September 1958, pp. 263-276.) Assuming (without being very explicit about it, since he is mainly concerned with spending rates rather than totals) a time/total cost relationship similar to *MNPR*, Daubin varies an overall budgetary constraint to see what happens to the optimal level of budgetary support in four different R&D projects. He finds that when the budget is very tight, the rate of spending may be held below that level which would lead to a minimum cost program. But as the budget constraint is relaxed, the optimum very quickly shifts to the negatively sloped segment of the time/total cost tradeoff function. From then on, the larger the budget is, the lower is the optimal development time and the greater total development outlays. Daubin also demonstrates that the greater the benefits a project offers relative to development cost, the lower is the optimal development time and the greater the total development outlays.

show more conclusively what the nature of the time-cost relationship is in research and development projects and (assuming the relationship to be convex) how wide the range of efficient substitution of time for cost tends to be. For if, as some have asserted, chipping two or three years off a more leisurely Apollo schedule is going to require an incremental $10 billion or so of our national treasure, we surely ought to know what we are sacrificing so we can decide whether the benefits justify the sacrifice.

Experimental Estimation
of the Development Possibility Function

How can the development possibility function be estimated on a practical basis? I have no really imaginative new suggestions. The basic problem is that R&D projects are unique and difficult to replicate. We want to vary the rate of spending per unit of time and observe what happens to total development time and hence total development cost, holding all other variables constant. But except in the most stringently controlled experiment, it would be very difficult to hold constant such variables as the quality goals, the technical approach pursued, the state of knowledge, and the quality of the resources employed; the variance in the total cost variable due to any one of these may well be greater than the variance due to differences in rate of spending. Nor can we allow unwanted variables to vary and observe through regression analysis a net relationship between the rate of spending and total cost, since there are undoubtedly more relevant quality variables than there are R&D programs, and so we would have too few degrees of freedom. Two experimental approaches offer at least some promise. Both require breaking a development effort down into its many component activities and steps, analyzing the time-cost relationships in those steps, and then aggregating the observed relationships through a network analysis of the PERT type.[40]

---

[40] PERT is translated "Program Evaluation Research Task," the name given it when it was first applied in the Polaris program in 1959. A point I had hoped to discuss in detail must unfortunately, for brevity's sake, be condensed into this footnote, which will be intelligible only to those already familiar with PERT. I believe

One approach involves nothing more than performing Gedan-kenexperimente—asking the engineer in charge of an individual activity what the most likely time span would be if he had diverse levels of resources at his command. Something like this is apparently being done or at least being contemplated in connection with the application of the PERT/Cost system to defense and space R&D programs. But the estimates obtained in this manner are certain to be replete with strong biases of unknown direction. If company "customer relations" (sales) staffs lay a heavy hand on the estimation procedure, as they often do to convince the government that a program is worth getting into, the estimates will tend to be strongly optimistic. Predicting the direction of bias is much more difficult when the responsible project engineers are given a free hand in the estimation process. They seem to gravitate toward extremes—either the optimism of underestimating a problem's difficulty or the pessimism of hedging against all possible contingencies to avoid blame for future failures. To obtain useful time-cost tradeoff estimates through Gedankenexperimente, it seems essential to divorce the estimators from the consequences of their estimates. This condition is seldom met with in the information systems presently applied by the government, but it could be met in an experimental study whose direct data were kept out of the hands of company and government program monitors.

The other approach would be to select a variety of representative technical problems and tasks of modest scale and to assign each subproject or task to several isolated task groups differing only (insofar as experimental limitations permit) in size. This

the range of potential efficient time-cost substitution in a development project is significantly related to the structure of the PERT-type network of tasks. Ideally, resources should be provisionally allocated in a program so that all tasks are at their minimum discounted total cost level. Then resources are added only to critical path tasks in order to reduce overall expected development time to the optimal level. Now if the critical path shifts widely among tasks in this process, the elasticity of time-cost substitution is apt to be high. If the critical path involves pretty much the same tasks throughout the process, the elasticity of time-cost substitution will be relatively low, but will be higher, the higher the proportion that critical path task expenditures are to total development expenditures.

approach would clearly be expensive, but the information it would generate could be most valuable in estimating more generally the possibilities of substituting resources for development time.

The Dependence of Time-Cost Tradeoff Decisions
on Benefit Estimates

The discussion of time-cost substitution possibilities thus far has been a necessary preliminary to attacking our main concern: how time-cost tradeoff decisions are influenced by the estimates of an R&D program's prospective benefits. It has not been proved conclusively that there *is* a time-cost tradeoff problem, but let us assume that there is. And having gone this far, let us go a step farther and provisionally postulate smoothness and generally diminishing returns in the development possibility function. Thus, $dC/dT < 0$ and $d^2C/dT^2 > 0$. Finally, we assume with no more justification than the experiment outlined in this paper that benefit estimates can be translated into dollar terms, by fair means or foul, subject of course to more or less uncertainty.

In connection with the experiment it was proposed that benefits be measured net of all costs, including R&D costs and the outlays on equipment required to exploit project results. Here a different definition will be adopted to keep the several components of the picture distinct. Benefits, b_t, in year t will be defined in a gross sense—that is, before any allowance for research and development and investment costs is made. For the simple case of cost-saving innovations, the benefit in year t is the gross operating cost savings in year t, with no allowance for amortization or depreciation except for clear-cut user costs. The level of benefits realizable may well vary with the moment in time when they are realized, and so we write

(13) $$b_t = b(t).$$

This is illustrated in Figure 4, which shows realizable benefits declining as time passes—the usual case of technological obsolescence. Since in the case illustrated here the benefits may be reaped throughout many time periods after the development is completed at time, T, we define total benefits realized, B, to be

(14)
$$B = \int_{T}^{H} b(t)\, dt.$$

The upper limit of integration, H, can have several possible interpretations. It may connote the time when the benefits from a project decline to the zero point, as shown in Figure 4; or the time when the product of an R&D effort is expected to be retired from service; or (when benefits are expected to be non-zero for an indefinitely long time span) the decision-maker's horizon. None of these concepts is without difficulties, but some limit must be assumed when dealing with practical problems.

FIGURE 4

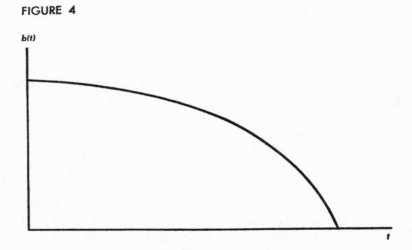

As before, total development cost is written as C, and we assume that its first derivative with respect to development time, T, is negative through a significant range. In addition, we must consider any investment, I, in equipment required to exploit the project results. For simplicity, it is assumed that any such investment is made at time $t = T$, just when the development effort yields its results.

Then we might impose a decision-making criterion which requires that the net surplus, S, of benefits over development and investment costs be maximized with respect to the development time variable:

(15)   Maximize $S = B - C - I$ with respect to T.

However, this formulation ignores the scarcity of resources implied by λ in equations (3) through (5). Since λ is a static concept, it is preferable to employ the time discount factor r which is related in a dynamic way to λ, and then to maximize the discounted net surplus, S^*:

(16)   Maximize $S^* = \int_{T}^{H} b(t)e^{-rt} - C^*(T, t) - Ie^{-rT}$.

The first-order conditions for a local maximum are

(17) $\dfrac{dS^*}{dT} = - b(T)e^{-rT} - (-)\dfrac{dC^*}{dT} + rIe^{-rT} = 0$.

Rearranged, we have

(18) $b(T)e^{-rT} = \dfrac{dC^*}{dT} + rIe^{-rT}$.

This means roughly that the discounted increase in benefits due to completing the development effort one period earlier must at the optimum equal the discounted increase in development cost due to compressing the development schedule plus the discounted opportunity cost of having to invest resources in exploiting the results one year earlier. For a local maximum, the second derivative must also be negative. Differentiating (17) again, we obtain

(19) $\dfrac{d^2S^*}{dT^2} = rb(T)e^{-rT} - \dfrac{db(T)}{dT}e^{-rT} - \dfrac{d^2C^*}{dT^2} - r^2Ie^{-rT} \overset{?}{<} 0$.

The last term is definitely negative. The third term is negative if diminishing marginal returns prevail in the development possibility function. The first term is definitely positive, assuming that benefits in year T are not negative. And the second term will also be positive, if benefits decline as time passes, e.g., because of technological obsolescence. Therefore we cannot be sure that a local maximum will be found. If the sum of the absolute values of the first two terms in (19) exceeds the sum of the absolute values of the last two terms, a corner solution may arise—the development will be conducted optimally in the minimum time.[41]

However, there are at least two reasons for believing that (19) is in fact negative. First, it is generally accepted that beyond some point pouring on more and more resources will not decrease development time at all. Therefore, d^2C^*/dT^2 must eventually approach infinity as T is reduced. But the first two terms on the right-hand side of (19) are not likely to behave in this way. Assume the strongest possible case—that benefits, $b(T)$, increase at an increasing rate as T is compressed. If T could be compressed to zero the first term of (19) would take on a maximum value of $rb(T)$. Since r is hardly likely to be infinite, this term could not approach infinity unless the benefits realizable in period 0 approached infinity. This is clearly implausible in the nonsupernatural world. And if the first term cannot climb to infinity, the second and related term also cannot if the $b(t)$ function is smooth. Therefore, as time is decreased, sooner or later (19) must become negative. Secondly, we observe in practice that many programs are conducted at less than the fastest pace. If decision-makers are rational in the sense of striving to maximize something like (16), they must assume that they are faced with a (19) which is negative.

Now if we grant the negativity of (19), a very interesting result emerges. Reconsideration of (18) discloses that for any given development possibility function, set of investment requirements, and discount rate, the deeper the stream $b(t)$ of expected

[41] Corner solutions will also tend to emerge if the development possibility function lacks the continuous diminishing returns attribute, as in Figures 3a through 3d. However, the corner may be somewhere other than the minimum time position.

realizable benefits is, and hence the larger $b(T)$ is for any T, the less the optimal development time, T, will be. This result is useful in explaining a phenomenon which is very evident in both military and commercial R&D programs: programs offering the greatest benefits tend to receive the most generous budgetary support and to be conducted in the shortest time, while programs offering meager benefits are supported only at austere levels, if at all.

Thus far the analysis has presumed a continuous stream of benefits realizable once the development project is successfully completed—the "continuous output" case of capital investment theory. A "point output" case might also be imagined: e.g., in some military situations, where the product of a development effort is for some reason assumed to yield benefits only in the event of actual combat. If we suppose the benefits to have the fixed value K and the time when they can be realized to be known with certainty, marginal analysis breaks down. The development effort should be conducted so that the end product becomes available just when it is needed, assuming that the discounted benefits exceed discounted total cost so that the effort is worth while at all. Consider, however, a somewhat different case. It is known that benefits K will be realized if the end product is available during the period when it is needed, but the time of need is not known with certainty.[42] Instead, there is a continuous prior subjective probability distribution $\varphi(t)$ describing the decision-maker's expectations as to when the product will be needed. In this case the discounted expected value of gross benefits in any period t_j is defined by $\varphi(t_j)Ke^{-rt_j}$. The conditions for maximization of net expected benefits (the expected value of gross benefits less cost) in this instance are virtually identical to those of equations (18) and (19). For an expected net benefits maximizer, the greater the benefit K, given the probability distribution and a smooth convex development possibility function, the shorter the optimal development schedule will tend to be.

[42] This is basically the case stressed by Peck and me, except that the point output benefit was assumed to decline as a function of time. That is, the later the benefit happened to be reaped, the less it would be due to gradual obsolescence. In such a case K would simply be replaced by the function $k(t)$.

It is nevertheless questionable whether military decision-makers do or should maximize the expected value of net benefits in important R&D programs. Rather, if point output benefits depend upon the time of need and the time of need is uncertain, the decision-makers may tend to adopt a more conservative minimax-like strategy, accepting the increased cost connected with minimizing development time in order to be prepared for the worst. Again marginal analysis breaks down, unless some kind of risk aversion function can be introduced into the model.

The possibility of a significant conflict element in such situations raises the problem of rivals' reactions to one's R&D decisions. In arms races, space races, and oligopoly, decisions to carry out a "crash program" whose output will encroach upon rivals' benefit streams are bound to incite competitive reactions from the rivals. One way of analyzing the problem is to redefine the benefits function originally formulated in equation (13). Recognizing that the pace at which Nation A's development programs are conducted affects the decisions of rivals, we define the level of benefits realizable by Nation A at any moment to be

$$(20) \qquad\qquad b = b(t, T).$$

This concept is illustrated by Figure 5, which is simply a modification of Figure 4. For any given set of actions by Nation A—

FIGURE 5

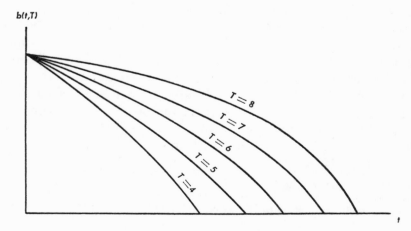

say, a development pace $T = 8$—a definite time flow of benefits is expected. The pattern shows declining benefits over time to reflect the gradual improvement in rival system effectiveness. But as Nation A speeds up its development pace from $T = 8$ to $T = 4$, rivals find the effectiveness of their existing systems threatened and speed up the development of improved systems, causing deterioration of the relative effectiveness and hence the benefits of the end product from A's development effort. The smaller A's T is, the more rivals are spurred to speed up their own developments, and so the more rapidly the benefits from A's program erode. In year $t = 0$ increases in A's development pace are not likely to affect rivals' posture due to time lags, but as time passes the adaptation of rivals becomes more and more complete. Therefore we may write:

$$
(21) \qquad \frac{\partial b(t, T)}{\partial T} = 0 \qquad \text{for } t = 0.
$$

$$
(22) \qquad \frac{\partial b(t, T)}{\partial T} > 0 \qquad \text{for } t > 0.
$$

$$
(23) \qquad \frac{\partial^2 b(t, T)}{\partial T \partial t} > 0.
$$

As in equation (14), total gross benefits realized from A's program are defined by

$$
(24) \qquad B(t, T) = \int_T^H b(t, T) e^{-rt} dt.
$$

When we differentiate this with respect to T, we obtain an additional term which did not appear in equations (17) and (18):

$$
(25) \qquad \frac{\partial B(t, T)}{\partial T} = \int_T^H \frac{\partial b(t, T) e^{-rt}}{\partial T} dt - b(T, T) e^{-rT}.
$$

The second term in this expression is identical to the first term on the right-hand side of (17)—that is, to the first derivative of

(14). It can be called the duration effect, for it shows that when development time is decreased, benefits are realized over a longer period. The first term in (25) can be called the shift effect, reflecting the overall deterioration of benefits due to rivals' reactions to a reduction in A's development time. Now when rivals do react to changes in A's T, equation (25) must be substituted for the left-hand side of the maximization equation (18). And if the absolute value of the first term of equation (25) exceeds the absolute value of the second term because rivals' reactions to A's development efforts are strong, the pattern of optimal behavior for A will change markedly. For now decreases in development time will bring decreases in overall benefits realized rather than increases, and A will have a definite incentive to stretch out its development program until costs are minimized, or even not to conduct the development at all.

This result is obviously at odds with what one observes in the real world of arms and space races. Something is wrong, either with the world or the model. Both criticisms are probably appropriate. Military decision-makers have, in my opinion, been prone to ignore the benefit-eroding effects of rival reactions in planning new weapons research and development programs; in other words, to ignore the shift effect in (25). But the model also has a weakness. It assumes that rivals behave more or less passively; that they will not increase the tempo of their development efforts unless provoked by A's tempo increase.[43] Yet in actuality the officials responsible for our military and space research and development programs have tended to fear the worst: an uncountered rival innovation which would cause severe embarrassment or worse to the United States.[44]

[43] It does not assume complete passiveness. The declining slope of the $b(t)$ function presupposes some activity in the enemy camp.

[44] Whether this fear is rational or not should be a crucial question in any policy debate. My own feeling is that for possibly sound conservative reasons we are prone to be overly suspicious. For instance, our fear of a German atomic bomb in World War II was the main reason for conducting the Manhattan Project on a "crash" basis. There was of course evidence which supported this fear, but there was also a tendency to downgrade any evidence of what in fact proved to be true. The hydrogen bomb case was more complex, as I have tried to show in an unpublished historical study, "The Nuclear Arms Race: A Case Study of Behavior Under Conditions Analogous to Oligopoly" (1962). Recently it appears we have been persistently

TABLE 2. Gross Benefits for A and B in Year t_k

		B's Choice	
A		Minimize Time	Minimize Cost
A's Choice — Minimize Time		0 0	−100 +100
A's Choice — Minimize Cost		+100 −100	0 0

For a flagrant oversimplification, consider the game matrix (Table 2) of incremental gross benefits before amortization of development cost in year t_k (where $t_k > T_{min}$) for rival nations A (below the diagonals) and B. If A minimizes development time, so as to have a new and more effective weapon system available in year t_k, and B does the same, the new weapons will neutralize each other and no benefit will be realized. If A minimizes time and B doesn't, being caught short in year t_k, A will gain 100 and B will lose the same amount. Conversely, if B minimizes time and A does not, A will suffer losses of 100 and B will gain 100; while if both rivals stick to a leisurely development pace, the status quo in forces will be preserved and incremental benefits will be zero. This is a constant sum symmetric game, and a pure strategy minimax-maximin saddle point exists. If a minimax psychology prevails, both A and B will choose minimum time developments.

The picture changes when we consider, as undoubtedly we should, net rather than gross benefits by the simple expedient of

---

unwilling to believe reports that the Soviets do not want, or at least have not yet decided, to engage in a race to the moon. See "U.S. Aide Rebuffs Soviet's Moon Bid," *New York Times*, Sept. 18, 1963, p. 11.

TABLE 3. Net Benefits for A and B in Year t_k

		B's Choice	
A		Minimize Time	Minimize Cost
A's Choice — Minimize Time		−30 \ −30	−110 \ +70
A's Choice — Minimize Cost		+70 \ −110	−10 \ −10

deducting an allowance for amortization of development costs. If the development possibility function is in fact convex to the origin, the minimum-time program will tend to cost a good deal more than the minimum-cost program. Assume that the appropriate amortization allowance is 30 for the minimum-time program and 10 for the minimum-cost program. Then the game matrix is as shown in Table 3. Now a constant sum game no longer exists. Both A and B could make the best of their rivalry by conducting minimum cost developments (or none at all) if they could overcome their mutual suspicions. But if they continue to fear the worst and behave like minimaxers, both will end up conducting costly crash programs. And to inject one further note of reality, they may suffer further mutual benefit losses if the weapons whose development has been accelerated fray, as many modern weapons have, the already slender hair holding Damocles' sword. As Secretary of State Dean Rusk remarked in 1962, while nations are "pouring more and more resources and skill into improving armaments, they are, on balance, enjoying less and less security."[45] Such is the benefit pattern of arms races.

[45] "Rusk Sees Growing Peril of War Set Off by Mistake," *New York Times*, June 17, 1962, p. 1.

Conclusion

Perhaps I have strayed in the last few pages above rather far from the direct problem of measuring government R&D program benefits, but I believe the excursion has been pertinent. If the assumption of a convex development possibility function is correct, the costs of reducing development time cannot be ignored. In certain cases—notably in programs which are not focal points of arms or space races or the like—the choice of an optimal time-cost combination depends directly upon the assessment of benefits flowing from the programs. Generally, the deeper the stream of benefits, the more rapidly the program should be conducted. In other cases (e.g., when rivals may be expected to react in an unfavorable manner to one's technological leadership) program benefit estimates must take into account shifts in the benefit function due to rival reactions. Inattention to this problem is likely to result in decisions which are not only wrong in an economic sense, but also accelerate the arms race. Finally, decisions in programs which directly affect the vital interests of rival nations may tend to be dominated by institutional and psychological variables, but even then benefit estimates are required to assess the consequences of alternative actions.

It is important therefore to develop effective methods for measuring the benefits of government R&D programs and the costs of alternative development time schedules. To date, little work has been done in these directions. In this paper I have proposed some modest experiments designed to determine whether in fact benefits can be measured in monetary terms and to shed light on the possibilities for time-cost substitution. I cannot be certain that the experiments will be successful, for the problem is extremely difficult. But better information on the benefits and costs of research and development programs would most assuredly lead to improved resource allocation decisions, and so I believe that an experimental effort is well worthwhile.

Comments

EDWIN MANSFIELD, *Wharton School of Finance and Commerce, University of Pennsylvania*

My first reaction to Scherer's very interesting (and controversial) paper is admiration for his courage. Although my obligation is to point out some of the shortcomings of the paper, it is important to recognize at the outset that the problem which Scherer tackles—that of measuring the benefits from military and space R&D projects—is about as difficult a one as can be imagined. If he has not gotten very far with it, this is not surprising.

Defects of the Model

Scherer begins by using the conventional static theory of consumer behavior under certainty to describe the problem facing the government in deciding how much to spend on various research projects. Although at first this model seems to be introduced only for illustrative purposes, we soon find that Scherer views it as a reasonable approximation to the problem actually facing the government. And in this latter context the model has at least three important defects.

First, it omits from consideration the crucial variable of uncertainty. As Burton Klein has emphasized, the problem of carrying out research and development is much more akin to a problem of sequential decision under uncertainty than to the problem of the consumer under static certainty.[46] Second, as Scherer formulates the problem, the implicit assumption seems to be that there are no important effects of the characteristics of other projects on the utility of the R&D project under consideration. This is unfortunate, since the prospective benefits from a project often depend heavily on whether certain competitive or complementary projects are carried out. Third, the formula-

[46] Burton Klein, "The Decision-Making Problem in Development," in Universities-National Bureau Committee for Economic Research, *The Rate and Direction of Inventive Activity* (Princeton University Press, 1962).

tion does not take into account the choice between beginning the project now and postponing it to a later date. This may be important, because there are often significant benefits derived from waiting: better information may be available and mistakes can be avoided.[47]

Next, Scherer claims that "there is reason to believe that the process in fact operates approximately as the theory suggests." He is, of course, saying that improvements are sought only up to the point where the extra cost equals the extra utility. But if we do not know whose utility function is the relevant one and where the decision-maker obtains the relevant information regarding costs, it is difficult to know what this means. Moreover, in view of the omission from the theory of such vital elements as uncertainty, interdependence, and postponability, it is difficult for us to believe that this theory can be a faithful description of the existing process.

Turning to the task of deriving operational measures of the benefits from government R&D programs, Scherer begins by describing the cost-effectiveness technique, which uses as an estimate of the value of a project the difference between (1) the discounted costs of doing a particular job with the methods derived from the project, and (2) the discounted costs of doing the same job with the best alternative method currently available. As Scherer himself points out, this technique has a disadvantage: it can be used only when an R&D program is designed to do essentially the same job as that which was done before. In addition, there are obvious, and important, problems in estimating the discounted costs of doing the job if the project is carried out, since the outcome of the project is uncertain.

Use of Panels of Experts

Assuming that these difficulties can somehow be resolved for a sample of R&D projects, Scherer goes on to suggest a procedure for using the results to estimate the benefits of a much larger

[47] For some interesting studies of investment decisions which take into account the possibility that projects may be postponed, see Stephen Marglin, *Approaches to Dynamic Investment Planning* (North-Holland, 1963), and George Terborgh, *Dynamic Equipment Policy* (McGraw-Hill, 1949).

class of programs. First, he would have a panel of experts rank all projects, including those for which the cost-effectiveness studies had been carried out, by their prospective value. Then, using the subset of projects for which the results of the cost-effectiveness studies are available, he would regress the scaling factors resulting from the rankings on the estimates of project value obtained from the cost-effectiveness technique. Finally, he would use the regression to estimate the project value for the projects for which there are no estimates from the cost-effectiveness technique.

I fear that he may be handing over much too much of the problem in an unstructured form to his proposed board of experts. It is unfortunate that he did not make more of an attempt to break the problem down and to construct a model which would suggest (1) more specific questions to be put to the experts and (2) ways in which the answers might be put together to produce a better overall result. Consider the question Scherer would ask: "Which program offers the greater long-term net benefits or value," considering the program's contemplated operational functions and the probable costs of executing those functions?

What does he mean here by benefits or value? Given a definition of "benefits," does he mean their expected value? If, as is often the case, the benefits from one project depend upon whether another project is carried out, what assumptions should be made in this regard? Isn't it necessary to have some estimate of the variance of the outcome? Should no account be taken of the fact that the net benefits may be greater if the beginning of a project is postponed? In view of the ambiguity of the question, the difficulties in the cost-effectiveness technique, and the obvious —and important—problems in finding competent and "objective" experts, I am much less hopeful than Scherer is regarding the usefulness of his proposal.

Time-Cost Tradeoffs

On the section of the paper that is concerned with time-cost tradeoffs I have only two comments, because here Scherer is dealing less with the measurement of benefits than with the effect of benefit estimates on development time. First, I agree with him that development time is likely to be a decreasing function of total cost and in most cases will be generally, although not continually, convex to the origin. Second, I share his misgivings regarding the likelihood of success of the experiments he suggests to measure this function. The Gedankenexperimente, if properly carried out, would reveal something about the opinions of the engineers, but these opinions may not be worth very much. The second approach he recommends would be extremely expensive.

WILLIS H. SHAPLEY, *U.S. Bureau of the Budget*

Everyone stresses the importance of taking account of uncertainty in the analysis of research and development. I think, however, that it is important to go farther and recognize that disagreements—fundamental disagreements on objectives, values, and so on—are an even bigger factor than technical uncertainties in limiting what can be done to develop a rational analytic approach to the measurement of research and development benefits. The assumption that a single objective measure of research and development benefits can be developed tends to sweep these disagreements under the rug. A useful analysis, it seems to me, must recognize explicitly the plural nature of objectives. Everyone has many different objectives, and ours is a very pluralistic society. Since benefits have to be measured in relation to a particular objective, the existence of the differing and conflicting objectives of the parties involved in the appraisal of R&D projects is of overriding importance, compared to other theoretical refinements.

This leads me to a general comment (and here I may be fighting the problem of the conference somewhat) on the whole concept of looking for measures of benefits and cost-benefit ratios. If the purpose of this search is, as I take it to be, to contribute, not just to a worthwhile intellectual exercise, but mainly to the decision-making process, then I would say that the use of a measure of the sort envisaged in Scherer's paper would tend to complicate decision-making rather than help it. In theoretical discussions, it is nice to visualize a curve, with "decisions" being made by picking the maximum or something that is lower or higher than something else. But for utility in the real world of decision-making, agreement is needed, not only on the many inputs to the measure, but also on the validity of the measure itself, including the underlying theoretical concepts and method of calculation. The possibilities of getting a measure with enough objectivity so that people whose interests are going to be adversely affected will accept it as a judge or as a major tool in the decision-making seem to me very small indeed.

Limitations of Cost-Effectiveness Studies

But enough of fighting the problem, at least on that front. Let me now comment on a few of Scherer's assumptions. I think, for example, that he overstates the remarkable things that can be done in the analysis of military programs, and especially overestimates both what can and what has really come out of cost-effectiveness studies. I have followed such studies fairly closely over the years, and certainly feel that the method is something that should be encouraged. I have, however, not yet found any case in which acceptable cost-effectiveness comparisons have been made between programs for different missions.

Everybody talks about analyzing different strategic retaliatory forces—how many missiles vs. how many aircraft—and a great deal that is valuable has been and can be done on this within the framework of particular assumptions, because these are things that can be compared. But I am not aware that anyone has yet found a measure which will compare programs for strategic retaliatory forces with, say, programs for the Army divisions needed for a conventional war. Each problem can be analyzed separately with some degree of meaning, but to get the peg points for the type of comparisons involved in Scherer's approach would call for cost-effectiveness comparisons between different areas of a sort which has not yet been done and does not seem to be in the cards. For this reason I feel that Scherer's theory assumes too much, and that the Department of Defense is not in a position to undertake the experiment he suggests.

Also, to the extent (and I may not be entirely clear on the

mathematics) that the paper's intent was to compare cost-benefit ratios for different kinds of development programs, I think his approach might lead to wrong results. Even if you disregard for the moment the problem of just what objectives to measure the benefits against, you are faced with programs that range widely in size. At one extreme are things measured in billions—such as a voyage to the moon or a ballistic missile program. At the other extreme are programs—such as those of the Quartermaster Corps —where the cost of development programs is very small and the benefits, measured on some particular scale, may also be fairly small. On the other hand, these smaller programs are often clearly sensible to carry out. Within a defense budget of a certain overall size, it may make sense to go on making minor improvements or to proceed with development programs where the cost is relatively small, even if the cost-benefit ratio in the particular case comes out lower than the cost-benefit ratio for an expensive program which cannot be accommodated within the total budget.

Importance of Short-Term Benefits

I join Edwin Mansfield in his comments on the question to be asked of Scherer's "panel of experts," and add a few comments of my own. The question to the panel is directed at an appraisal of *long-term* net benefits. To assume that the only matter for concern is the long-term effect of a program is, I feel, an excessive prejudgment. In many cases (as highly publicized in today's defense posture), what we are after is to develop options of one sort or another—that is, we are taking actions now to create conditions for future decisions. The decision is made, for example, to establish the *capacity* for producing a very large number of Minuteman missiles, and in the military R&D programs there are many similar instances. In such cases, the value of the program should be measured, not in terms of the long-term benefits of actually completing the project and proceeding with production and deployment, but in terms of the short-term benefits of having the option of making a future decision to do so.

We are also, especially in military or other competitive programs, at times very much interested in and concerned with short-term effects in another way (which Scherer alludes to in the latter part of his paper). If we announce a decision to go to the moon or to build a missile defense system, this may influence our competitor in deciding what he will do. These shorter-term benefits are a valid criterion against which the program should be judged, even if three or five years later we abandon the program and thus never achieve the long-term benefits of completing it. So I question whether Scherer's panel should address itself exclusively to the question of maximizing long-term benefits.

Relevant Measures of Cost

Furthermore, I am not entirely sure that the "total cost" that enters these considerations is the correct measure (aside from the point that advance estimates of total cost tend to be so unreliable). Total cost is not really relevant, it seems to me, except in comparing things that are clearly alternatives for the same purpose. In working on the budget, we find that the relevant thing to watch is the annual outlay—the appropriations, the decisions on the annual commitment of funds—rather than the total cost. This is perhaps a peculiar feature of the American way of budgeting, but I believe it goes a good deal deeper than that.

In the development area, for example, and especially at the moment, a great deal is being made of the total cost of the manned lunar-landing program. Total cost is an interesting question, but one that is not especially relevant to the decision about going ahead with the program—unless you also make an assumption that the question is something all by itself, and that once you have bought the program, as you would buy a can of beans off the shelf, nothing else will happen and everything else will go back to where it was before.

I think it is clear that if we decide to go ahead with the manned lunar program, we are thereby making a decision to proceed not only with that program alone, but also with future successor programs not yet defined, or, at the very least, we are creating conditions in which we are practically obliged to proceed with future programs. So in making a decision to proceed with a major development program, what should be counted as the total cost is not merely the total cost of *that* program, but a larger and indefinite total which includes the cost of future decisions that are the consequence of the decision at hand.

In practice, this works as follows: in major developmental and other decisions, the cost aspects must be considered within the framework of alternative overall cost projections of, say, the total budget of the federal government, or the total defense budget, or some other fairly large aggregation. The primary concern is the effect of the various choices that are before us on the total rate of the expenditures—i.e., the total outlays required each year for the foreseeable future. When we are looking six or seven years into the future, there is little to go on and everything gets quite hazy. The effects of the total ultimate cost of a particular program on overall future budgets are generally overshadowed by the inherent uncertainties and the effects of other decisions to be made between now and then. But I should repeat that in the narrower context of selecting alternative approaches to the same problem, where the estimates can be on the same basis, total costs are very relevant to the decision process.

Research and Development Decisions

It is worth pointing out that the panel procedure Scherer describes does approximate how R&D decisions are usually made. There is, in one form or another, a panel of experts—the staff of or advisers to the decision-makers—who sit around a table and make judgments of exactly the sort that he describes, namely, that Program A is more important than Program B. However, it seems to me much more appropriate that the experts on the panel should address themselves directly to the actual decisions and choices at hand, rather than to the problem of establishing a measure which will then somehow be used to help make the decisions.

On the constructive side, I would suggest an important need for a somewhat different approach than that taken by Scherer. The direction I have in mind is actually pointed to in his paper and is also alluded to in Mansfield's remarks. What each decision-maker really needs is a relatively simple but fair model relating the principal variables and factors that enter into the types of decisions he faces. The aim of such a model would be, not to devise a method for measuring benefits, but to bring to the attention of the decision-makers and their advisers the factors and interrelationships which they should take account of in making their decisions on a judgment basis. This kind of model will be very different from a research model for academic analysis of the problem. Among other things, it will have to relate considerations which the decision-makers themselves recognize—and believe to be important regardless of whether they are really important—as well as bring into the discussion the things which the professional analysts think should be important to them.

Finally, I would like to skim over very briefly what seems to me the most appropriate way to approach the broad decisions that are necessary concerning the government's research and development programs. To begin with, we must distinguish between (1) research, regardless of how it is defined, (2) general technological development, (3) specific technological development without immediate applications (something which is now being called advanced development), and (4) engineering development directed to defined objectives.

In the case of research, the decisions for the most part should be decentralized to the scientific community, to the people who are familiar with what is going on and with the technical possibilities. The scientific community should function as the "free market," to decide what is to be done and, for the most part, how *much* is to be done in various fields. The principal decisions the government must make involve the amount of resources to put at the disposal of the community and the organization of the skeleton (at least) of a decision-making process so that new ideas will get a hearing.

The same type of approach can be followed in the area of general technological development. Here again, it seems to me not appropriate to expect cost-benefit analyses on a refined basis. But in regard to major development projects of the third and fourth types—which are the subject of most of the discussion here today— the appraisal of benefits in relation to costs becomes a central consideration. Even here, however, I believe that developing a method for measuring the benefits of this type of project *in general,* as Scherer has attempted, is not as important and productive as dealing with specific cases. For each project, one can make a refined analysis of the objectives that are pertinent to it, and the benefits it is likely to contribute toward these objectives—and then consider these, on a judgment basis, in relation to the project's cost in the framework of a broader budgetary context.

Concluding Statement

FREDERIC M. SCHERER

Many penetrating suggestions have been offered by Edwin Mansfield and Willis Shapley, as well as by participants in the discussion that is not published here. In this concluding statement I wish to react to certain points which I consider most crucial.

Several commentators were troubled by the unstructured character of my approach to benefit estimation. Mansfield, for example, would prefer breaking down the problem, constructing a model which confronts the experts with detailed questions on each potential type of benefit. To some extent I concur with his desire. If I were a budgetary decision-maker, I should definitely want a thorough analysis of all the benefit categories. But I fear that such an approach, while essential to intelligent decision-making, is not apt to yield benefit estimates commensurable with program costs when program objectives are highly intangible.

When faced with decisions concerning goods that satisfy subjective wants, consumers generally appear able to choose which of two alternatives they prefer (that is, which one affords greater utility) even though they are unable to attach any numbers to the amount of satisfaction anticipated. The unstructured, global approach proposed here was motivated by the desire to hurdle the barriers to commensurability through demanding that experts evaluate program benefits with no higher degree of measurement capability than the theory of consumer choice assumes. I see no feasible way of keeping the individual judgments ordinal without resorting to a global approach. Suppose, for instance, we were to ask panel members to make paired comparisons in terms of each conceivable dimension affecting ultimate benefit expectations—e.g., cost savings, safety, flexibility, security of control, likelihood of triggering further innovations, and so on. How would we combine the resultant rankings into an overall estimate of benefits? The only meaningful system of weights would be one reflecting cardinally the assessed value of each sub-benefit. But to insist that the panelists articulate these weights would be to insist upon more than ordinal measurement. Of course, the panelists would be forced to assign weights implicitly before making paired comparisons in terms of overall benefits. Observation of consumer behavior suggests, however, that individuals are much better able to do this implicit weighting job than to quantify a system of explicit weights. Therefore, although information on the subdimensions of program value would clearly be useful, the basic approach proposed here seems to stand or fall on the possibility of eliciting meaningful global benefit comparisons.

There is, nevertheless, one area in which considerable structuring is warranted. Given the inevitability of benefit interactions among programs, it is vital that assumptions about the level and scope of each program in the sample (and also of related programs not in the sample) be distinctly specified. If the scope of Program A is altered, benefits in Program B may well be affected. This interdependence does raise practical problems. If decisions to change the initially assumed program structure were motivated by benefit estimates generated under that structure, the emergence of significant new interactions could destroy the continued validity of the original estimates. For practical applications the estimation process would have to be iterative.

The possibility of consistent bias in the panelists' judgments disturbed nearly all who commented on my proposal. For instance, discussants were afraid that choices might be biased by the panelists' propensity to avoid (or alternatively, to embrace) risks and uncertainties, by their perception of the tightness of agency budgetary constraints and the opportunity to devote unused funds to alternative projects, etc. To combat this problem I have considered altering the original proposal, but concluded that the change I had in mind would not work. Specifically, I had thought of asking panelists to compare programs on the basis of *gross* benefits rather than *net* (that is, gross benefits less costs). This would remove the panelists to some extent from the decision-maker's role of determining which programs were in fact worth supporting. But since the benchmark cost-effectiveness estimates presuppose a net benefits concept, the use of a gross benefits criterion in panel rankings would be infeasible.

As an alternative, one might cope with the panelists' possible risk-aversion biases by coupling the comparisons of the expected net benefits for each program pair with the answers to some such question as this: "Of which program's future value are you more confident?" These choices would lead to a ranking of programs on the confidence (or inversely, uncertainty) dimension. If a significant rank correlation between the benefit ranking and the confidence ranking emerged, one would suspect risk-aversion biases. But the procedure would of course not permit estimation of the degree of bias.

Other kinds of bias might enter into the benefit rankings. A panelist associated with a particular agency or branch of technology might favor programs advancing his own special interests. There could also be a tendency for cost-saving programs to be favored over programs with less mundane goals, or vice versa. Ideally, the analysis of disagreement among panelists would call attention to such biases, assuming that the panel includes a balanced diversity of special skills and interests. But discussants have pointed out some reasons for the possible failure of this check. If communication among panelists were permitted, expressed judgments might converge upon a consensus belying original beliefs. This difficulty might be minimized through the use of multiple independent panels kept (at least conceptually) in "isolation booths," as suggested by Charles Schultze. Yet, even under ideally controlled conditions, one could never secure conclusive proof that the consensus rankings of the panel members bore any meaningful relationship to reality. Indeed, it is not certain that such a thing as objective reality exists when one is attempting to measure the benefits of programs with goals as intangible as the national security. I am inclined to believe that there exists at least a semblance of reality in the defense world, but contemplation on the paradoxes of deterrence induces gnawing doubts. Certainly I cannot prove that an objective reality does exist in defense, and so a modicum of faith is required for the benefit measurement approach proposed here even to be tried. I am, however, optimistic enough to hope that someone will attempt the measurement task, if only to see what happens.

Finally, I should like to express mild disagreement with Edwin Mansfield's assertion that my model omits the crucial variable of uncertainty. It is true that I have reluctantly subdued that variable, in the interest of concentrating on my main problem, but it is there, if only in repose. In particular, the development possibility analysis explicitly takes uncertainty into account, since a major reason for the inverse time/total cost relationship is the added cost of concurrently supporting more than one technical approach toward an objective, to decrease expected development time when uncertainties preclude *ex ante* identification of the "best," or a satisfactory, approach.[48]

As my analysis has suggested, these additional costs will be accepted only when the stream of benefits is relatively deep. Budgetary decisions on how liberally to support a program at any moment in time necessarily entail this type of commitment. Likewise, the model presented here indicates that holding development expenditures to a low level while uncertainties are eliminated (the sequential decision approach advocated by Burton Klein and his RAND Corporation colleagues) is optimal only when the stream of expected benefits is relatively shallow— that is, when saving time is not urgent.

[48] One might also interpret the "parallel paths" strategy in terms of a more complex objective function. Increasing expected development cost by running additional technical approaches not only decreases the expected value of development time, but also increases the probability that the development will be completed within some maximum acceptable time span and reduces the variance associated with development time estimates. These latter reductions in risk might have utility independent of reductions in the expected value of development time.

RUTH P. MACK & SUMNER MYERS*

Outdoor Recreation

THE OBJECTIVE OF THIS PAPER is to explore concepts and methods for providing a useful measure of the benefit derived from government expenditure on outdoor recreation. We find that benefits need to be measured in terms of a merit-weighted service unit: merit-weighted user-days of outdoor recreation.

The benefit-cost calculation for this field is particularly intractable.[1] The benefits are complex and abound in noncommensurable aspects that sharply limit the theoretical and actual possibilities of measuring them on the basis of either free market prices or any other system purporting to express their value to the user and the state. Nevertheless a decision to purchase or develop land for recreational purposes actually implies that value has been attributed to it—value in excess of that generated by some alternative purpose to which the funds might have been put. What can the economic calculus contribute to the improvement of these attributions?

* Institute of Public Administration and National Planning Association, respectively.
[1] The dollars actually spent on recreation are, for the most part, spent for capital plant rather than for the direct purchase of current services. Therefore the appropriate benefit-cost comparisons involve calculating the present value of discounted streams of costs and benefits which are expected to accrue over a span of years. We focus on only one aspect of this total problem—the present value of service currently generated.

Recreation and Public Expenditure Decisions

It is traditional for economic value (or merit) to be assigned in terms of utility to the user or to "third parties" and for this utility to be treated as a stable function of money. Accordingly, we turn first to money measures and inquire whether recreational benefit can be given a price that seems reasonably comprehensive and reliable. Our conclusion is negative: in the absence of market mechanisms that function with sufficient breadth and depth, market prices for outdoor recreation do not serve the purpose; nor can other satisfactory dollar measures of merit be contrived. There is, however, a broad field for empirical work which could improve information about what various sorts of people are willing to pay for various sorts of recreation. This information, though too incomplete for attribution of merit, can form a sound basis for user charges—a field of growing public interest.

Our second major conclusion is that it seems possible to calibrate recreational benefit, in effect, directly along a utility scale. The "merit" in providing the service of recreation rests in this utility and its contribution to social welfare. The utility of a user-day of recreation differs, of course, depending on what sort of recreation is used by whom. These differences are taken into account in the weighting system whereby various sorts of recreation for various sorts of people under varying conditions of supply are converted to merit-weighted user-days. The weights imply the formulation of criteria for proper performance of government's function with respect to the provision of the service. The political process as well as explicit judgments are viewed as contributing to the formulation of the criteria.

These notions can help, we believe, to improve recreational programs at each level of government. And this alone is valuable. But it seems possible that choices between recreation and other public purposes may also be steadied by the sorts of comparisons that the analysis facilitates.

The Commodity: Outdoor Recreation

Outdoor recreation may be defined as "leisure time activity undertaken in a relatively nonurban environment characterized by a natural setting, for the primary purposes of enjoyment and physical or mental well-being."[2] Dollars spent on outdoor recreation provide a good characterized by utilities of three sorts: immediate enjoyment, long-term benefits for the individual, benefits to the nation as a whole.

Immediate enjoyment consists of the sense of pleasure experienced immediately before, during, and after participation in outdoor recreation. Whether or not people are enjoying themselves at a given time can be determined simply by asking them. Concerning the degree and quality of their enjoyment they can speak less adequately.

The second utility, long-term benefits, presents a more serious block to successful measurement. Long-term benefits are both physical and psychic. The physical benefits are of course not uniquely associated with outdoor recreation; strong bodies can be built at an indoor gymnasium. But society places particular value on outdoor recreation, not only because it builds strong bodies, but also because it does this in an environment that is credited with building healthy minds. We know, or at least think we know, that outdoor recreation is "good for people." The general proposition that it *is* good for people has been advanced but not examined in depth in various Outdoor Recreation Resources Review Commission (ORRRC) studies.[3]

What is there about recreation in the out-of-doors which yields the special psychic benefit that we intuitively hold to be so important? An investigation bearing on these matters made for the National Institute of Mental Health found that experiences in

[2] *The Future of Outdoor Recreation in Metropolitan Regions of the United States (A Report to the Outdoor Recreation Resources Review Commission,* No. 21, 1962, Washington, D.C.), Vol. II, p. 1. (Further references to these studies will be cited as *ORRRC Study Reports.*)

[3] See, for example, Lawrence K. Frank, "Outdoor Recreation in Relation to Physical and Mental Health," in *Trends in American Living and Outdoor Recreation (ORRRC Study Report,* No. 22, 1962), p. 215.

the out-of-doors confer a unique and long-lasting psychic benefit on the participants because of taking place in a *natural environment*.[4]

Obviously, such benefits do not develop from one isolated experience with the natural environment. To attain them, a continuing relationship must be established. In this sense, the special psychic benefits of outdoor recreation are cumulative and compound with experience; furthermore, they develop their own momentum. According to a survey by the University of Michigan's Survey Research Center, "once a person has acquired experience with an activity he is more likely to continue as he grows older than people who did not engage in this activity in their youth."[5] Tabular material in the survey shows that the difference is substantial. Similar findings apply to campers in wilderness areas.[6]

The third utility—benefits to the nation as a whole—includes two chief sorts. There are, first, the third-party benefits which result from the advantage to *all* people, whether or not users of outdoor recreation, of living in a country where more rather than fewer people are educated in the ways of the out-of-doors. Here again, the benefit rests on the assumption that the enjoyment of outdoor recreation is meritorious, just as is the enjoyment of good books, art, music, or any sort of extension of one's capacity to appreciate the fullness of life. Second, there is the conservation aspect of outdoor recreation resources. Historically, interest in conservation has been the mainspring of government concern in the acquisition of recreational lands. The conservation movement expresses the belief that the growth of industry and cities should not be allowed to despoil major natural beauties, and the belief that the nation should maintain its natural wonders and offer them to its inhabitants for their pleasure.

Granted, then, that the commodity, outdoor recreation, possesses the varied utilities sketched above, what is the govern-

[4] Harold F. Searles, M.D., *The Nonhuman Environment* (International Universities Press, 1960).

[5] Eva Mueller and Gerald Gurin, *Participation in Outdoor Recreation: Factors Affecting Demand Among American Adults (ORRRC Study Report*, No. 20, 1962), p. 23.

[6] See *Wilderness and Recreation—A Report on Resources, Values, and Problems (ORRRC Study Report*, No. 3, 1962).

ment's responsibility with respect to it? Three general bases of public interest are relevant:

1. Government's paternal interest in the individual's general well-being. Government promotes, by private and public means, the consumption of a good to which superior merit is attributed.

2. Third-party interest in outdoor recreation and conservation. Since in this connection there is no private mechanism of response with respect to either demand or supply, governments must respond directly.

3. Government's interest in the efficient supply of resources. Outdoor recreation can sometimes be supplied substantially more economically by public than by private means.

In meeting these general responsibilities a government has a somewhat unusual relationship to the service—outdoor recreation. On the one hand, if outdoor recreation is recognized, as it now seems to be, as an integral part of the American way of life, government may have an obligation to guarantee a "minimum subsistence level" or "modest but adequate" level to all, as it has for many years guaranteed such minimums of other consumer goods and services. On the other hand, means tests are not applied to parks. And supply conditions have dictated that governments are the major suppliers of recreation in certain sorts of facilities—wilderness areas, wide stretches of shore front—at whatever level they are used by rich as well as by poor. Whether this supply should be without charge is a moot question.

Measurement and Classes of Public Decision

If outdoor recreation is acknowledged as an appropriate object of government expenditure, how much of the nation's resources should be invested in it? The theoretical answer (or perhaps another and indirect way of asking the question) is: funds should be invested to the point of maximizing total welfare. Other than that answer, which is a conceptual hornet's nest of many seasons, what can be said about at least increasing welfare via government action?

Welfare can be increased by making better rather than worse

decisions in the areas in which government must, whether by commission or omission, constantly make decisions. Major decisions bearing on outdoor recreation are:

1. How many dollars to spend (and to acquire in each of several ways) on all public purposes vis-à-vis those of the private economy. Measurements relating to this matter are not the concern of this paper.

2. How much to spend on outdoor recreation vis-à-vis other public purposes.

3. How to optimize the aggregate utility of all outdoor recreation resulting from each additional dollar of public expenditure on outdoor recreation.

Aid in deciding how much to spend on outdoor recreation vis-à-vis other public purposes (or on some particular park) requires a measurement of the benefit derived from spending a given number of dollars on recreation (or on a specific bundle of recreation) which may be compared with the benefit derived from spending the same number of dollars on other public purposes (or other bundles of recreation). Implied is some sort of calibration of benefits—a comparison of the relative value, the utility generated. No reasonably precise measure seems even theoretically possible.

Imperfect measures can nevertheless be useful, and we propose a test to determine whether they are or not. A measure is useful in proportion to: (1) the percent of total benefit that it adequately calibrates; (2) its capacity to uncover, focus, and steady the aspects of total benefit that it fails to calibrate; (3) its general acceptability; (4) its power to improve administrative or legislative action because of a capacity to focus debate on well-considered notions of the public weal.[7]

Note that a dollar measure of benefit that is perfect as far as it goes could conceivably get a negative mark on this test if it scored low on points 1 and 2, for then it might actually deteriorate legislative action by its seductive misinformation—the im-

[7] Point 3 has been added here in response to one of the many helpful criticisms of Lyle E. Craine (see "Comments" section below). We have also followed his suggestion of indicating at several points how the test is applied.

plication that the neat dollar figure measured the relative value of two alternatives. A measure must of course do better than simply stay in the black. It must pass the test at some acceptable level that takes into account, among other things, the cost of preparing it.

Outdoor recreation is, not a single homogeneous product, but a bundle of products. Consequently, we must compare utilities generated by wilderness and by Coney Island; recreation for different age-income groups (young and rich vs. old and poor); recreation for different interest groups (swimmers vs. bird-watchers); recreation provided free to the users vs. that at market or other prices. Each of these comparisons corresponds to actual decisions that must somehow be made.

The simplest and most desirable measure of benefit, other things being the same, is a dollar figure. It is more versatile than any alternative. A second possibility we explore in this paper is a physical service unit—"merit-weighted user-days" of outdoor recreation. The physical service unit is based on the notion that user-days constitute a common denominator for service rendered by government dollars spent on recreation. But a service unit based on this proposition alone would be far too crude to pass the "usefulness test." The unit can be made more productive by introducing a concept of *relative merit* intended to take account of the relative utility engendered by a wide variety of qualitative elements which make the difference between a "poorer" and a "better" recreation program.

If such a measure could be achieved, the decision-maker would be in a position to compare the social value of a given number of merit-weighted user days of current outdoor recreation that can be bought for the next million dollars of capital expenditure directly with, for example, the social value of a given number of merit-weighted vehicle-miles that a million dollars would provide. Comparisons could extend to other government programs for which the end-product measures can meet the "usefulness test" at an acceptable level. The functions of such a measure would be, not to replace, but to supplement, bring into focus, and provide a frame of reference for other considerations. The measure might also serve to improve the rigor and relevance of the descriptive evaluations and exhortations that inevitably are fed to legislators.

In the next section we will explore ways of applying the first principle—dollar values. Following that, we discuss how the second principle—merit weighted user-days—might be applied.

Dollar Measures

Can the relative merit of different sorts of recreation for different sorts of people be assigned a valid dollar figure? This is the central question that must be asked. And even if the answer to it is in the negative, we may find that dollar measures can aid in determining a fee that should be paid under specified circumstances for the use of publicly provided recreation.

Theoretical Limitations of Market Prices

In a free-enterprise economy it is assumed that the utility of a commodity is measured by the number of dollars that people are willing to spend on it at the margin. Herein lies the magic of the market place and the price system. But even if this system operated perfectly within its own boundaries (and needless to say it does not do so) it clearly would provide measures of benefit that are deficient for our present purposes in several serious respects:

1. Price in any event fails entirely to reflect the benefits of recreation and conservation to the nation as a whole. This is true of values in terms of the national image and the third-party benefits previously discussed.

2. For direct benefits to users, prices at best cannot reflect utilities that people fail to appreciate. Much of the "superior merit" is probably not recognized by the participant. People who are pleasure bent may ordinarily not think of the long-term implications of their enjoyment. Accordingly, the price they are willing to pay is likely to reflect at best only some portion of the total utility.

3. There is, of course, the standard problem of arriving at total or average benefit from data that theoretically measure benefit at the margin. On the other hand, in view of the im-

perfections of recreational pricing, it is not at all clear that the data actually do reflect a meaningful marginal value. All this involves, among other things, efforts to cope with consumer surplus; various methods have been suggested.[8]

4. Provision of recreation free of cost to the user generates consumer surpluses in connection with public recreation and depresses the price that can be won for privately provided recreation. Therefore actual amounts paid for such outdoor recreation as is bought privately distort the apparent relative utility of different sorts of outdoor recreation and generally distort (by underestimating) the utility of outdoor recreation relative to other consumer goods. Both distortions have further ramifications if the objective of maximizing social welfare is taken seriously. It is generally believed that something close to equal distribution of income is a necessary condition for maximizing social welfare. But in connection with outdoor recreation, the optimal distribution among the population of two other constraints must be questioned on the same grounds: leisure time (particularly the paid vacation) is one important determinant of participation in outdoor recreation; previous experience plus general educational level is another. In consequence, even a perfectly functioning price system, given the less-than-optimal distribution of income, leisure time, and education, would not adequately reflect the merit of outdoor recreation viewed in the proper social perspective.

These manifest theoretical deficiencies of the market price in metering the merit of outdoor recreation are compounded by practical difficulties in acquiring and interpreting such meter readings as exist. But despite all of these shortcomings, market prices would be useful partial indicators if they were available.

Indirect Approaches to Measurement

Most of the efforts to arrive at dollar values have used an indirect approach. They have evaluated recreation by attributing

[8] See J. Hirshleifer, J. De Haven, and J. Milliman, *Water Supply: Economics, Technology, and Policy* (University of Chicago Press, 1960), p. 97; and Roland N. McKean, *Efficiency in Government Through Systems Analysis* (Wiley, 1959), pp. 170-173.

the prices paid for some sort of related service or output. These efforts have not, however, met with wide acceptance to date. It is still true, as was stated in 1958, that "none of these attempts [to quantify recreational values] have completely satisfied their proponents or fully met the objections of affected interests and agencies."[9]

The attempts to measure the value of recreation have been grouped by Marion Clawson into four general catagories:[10]

1. Gross volume of business generated as a result of the availability of outdoor recreation opportunities;

2. Estimated value of the gross expenditures on outdoor recreation added to local business alone;

3. Demand for outdoor recreation, measured by the willingness of users to pay specified sums of money for travel to recreational areas;

4. Measurement of consumers' surplus resulting from the provision of recreation at no or low cost to persons who would otherwise be willing to pay relatively large sums for such opportunities.

Clawson assesses the merits and deficiencies inherent in these various indirect approaches, concluding that, while most of them lead up conceptual blind alleys, "it is both theoretically possible and practically manageable to put monetary values on outdoor recreation."

Perhaps, but certainly there is no royal road to this achievement. We have noted (usefulness test no. 3) that if dollar values are to be useful they must be widely accepted. Individually weak measures may earn acceptance if they yield results that tend to be consistent with each other. To further explore the possibility of consensus we submit (half-heartedly) three measures based on a different principle.

The fundamental notion underlying the procedure is that the average value of public outdoor recreation can be imputed from the average expenditure per unit of time spent on analogous recreational experiences. The calculation is based on the admittedly

[9] Albert M. Trice and Samuel E. Wood, "Measurement of Recreation Benefits," *Land Economics*, Vol. 34 (August 1958), p. 195.

[10] Clawson, "Methods of Measuring the Demand for and Value of Outdoor Recreation" (Reprint No. 10, Resources for the Future, February 1959).

shaky assumption that what people already spend for recreation for which they pay indicates the actual worth to them of recreation of comparable quality for which they do not pay.

Sufficient data, however crude, are available from three different sources to permit rough approximations embodying this general principle. The data provide information on:

1. *The average expenditure per .hour for all outdoor recreation, public and private.* According to the National Recreation Survey, expenditures away from home on vacations, outdoor recreation trips, and other outings amount to an average of $74.9 per person, and each person spends an average of 258 hours per year on outdoor recreation of all sorts.[11] According to these data, then, the average rate of expenditure on all outdoor recreation and associated services, public and private, is roughly 30 cents per hour.

2. *The average expenditure per hour for all away-from-home recreation.* People might be willing to spend for public away-from-home recreation what they are now spending on other recreational activities away from home, on the assumption that participation in public recreation activities provides them with at least as much pleasure as the other things they are doing. Surveys of income and expenditure for urban families show that average family expenditures on all recreation, other than for automobiles, was $225 a year.[12] An examination of detailed breakdowns of aggregate consumer expenditures suggests that about 40 percent of the total is for away-from-home recreation.[13] And applying this to the recreation total yields an estimate of $90 per family. To this figure must be added expenditure on pleasure driving, which the recent ORRRC studies

[11] *National Recreation Survey* (*ORRRC Study Report*, No. 19, 1962), p. 367, Table 5.41, and p. 361, Table 5.35.

[12] U.S. Bureau of Labor Statistics, *Survey of Consumer Expenditures, 1960-61* (Reports: Series No. 237, 1963).

[13] Aggregate consumer expenditure, as tabulated in U.S. Office of Business Economics, *Survey of Current Business* (July 1962), Table 14, was grouped in a parallel fashion to that of the recreation category in the *Survey of Consumer Expenditures* (see footnote 12 above). The categories for the aggregate data are sufficiently detailed to provide the basis of an informed guess as to how much of the total was spent on away-from-home recreation.

indicate is one of the most important forms of outdoor recreation.[14] The total, then, is $188 per year per family. A study made for the Mutual Broadcasting System indicates that 182 hours per year per person are spent on away-from-home recreation.[15] Using an average family size of 3.5 persons, we arrive at an estimate of about 30 cents per person hour.

3. *The average rate of expenditure for pleasure driving.* As previously mentioned, pleasure driving is the most popular form of outdoor recreational activity. It is, for the most part, privately paid for (except for highway subsidies). It seems reasonable to postulate that whatever people pay for pleasure driving has some exceedingly rough, perhaps minimal, correspondence to the value of other outdoor recreation experiences. Recreational auto trips usually involve an average of four persons per car. Using the previously mentioned figure of 5.74 cents per mile and assuming an average pleasure driving speed of 25 miles per hour (including stops for sightseeing, etc.), we estimate that 36 cents per hour per person is spent on pleasure driving.[16]

Actual and Artificial Market Prices for Recreation

Two approaches are possible. The first involves estimating prices for public facilities on the basis of the market prices for comparable private facilities. The second involves the use of experimental pricing and other fairly conventional market research

[14] According to data from the U.S. Bureau of Public Roads, the average passenger car travels 9,655 miles per year; 17.6 percent is for pleasure driving, including vacations. The average cost per car mile excluding depreciation, insurance, and other "sunk" costs is 5.74 cents per mile. Applying this to the mileage allocable to pleasure driving, we estimate that $98 per year of car expenses are allocable to outdoor recreation.

[15] From unpublished data in "A Nationwide Study of Living Habits" (a survey conducted for the Mutual Broadcasting System by J. A. Ward, Inc., in 1954), as shown in Sebastian de Grazia, *Of Time, Work, and Leisure* (Twentieth Century Fund, 1962), p. 444, Table III.

[16] The American Automobile Association plans trips for its members on the assumption that they will average 40 miles per hour over grade-separated highways, and 30 miles per hour over non-grade-separated highways. The estimate of 25 miles per hour allows for stops and assumes that some, if not most, of the pleasure driving is over winding, back-country roads.

techniques. Taken together the two approaches are likely to contribute to an acceptable system of prices that can be attributed
to public facilities. The information that is required is not now
available for either approach. However, in both cases it could be
acquired by the expenditure of money and energy. The following discussion, then, is simply an illustration of what might be done.

PRIVATE PRICES. Private prices might be used as a guide in setting the value of publicly provided outdoor recreation, even
though prices for some private facilities—notably camp sites—are
heavily depressed by underpriced public facilities of a comparable nature. However, this omnipresent influence is largely sup-
pressed with respect to private recreation facilities that are set
up for people willing to pay for exclusivity. A case in point is the
Idlewild Country Club—a private recreation facility operating as
a membership club in Bergen County, New Jersey, approximately one hour's drive from Manhattan. Idlewild encompasses
about twenty acres of rugged countryside. It features picnicking,
swimming, and other sports in an environment very similar to
that of a good quality state park. Members pay $2 per day to use
the park. The differentiating factor between Idlewild and comparable public facilities seems to be the "guarantee" to Idlewild's
members that the park is free of "undesirable elements."
  In any event, exclusivity may at least partially insulate the private price from the competition of the public supply. If so, the
price of two differentiated products may be roughly equated—
private parks, from which some people are excluded, and comparable public parks, to which all are admitted. Theoretically,
only the recreation portion of the total price should be attributed
to a public park from a comparable but "exclusive" private facility. The problem is to judge how large a portion of the joint
product—recreation plus exclusivity—adheres to the latter. It
seems reasonable to suppose that exclusivity is only a fraction of
the total. (Otherwise why not stay home and be really exclusive?) The matter could be studied, but for the moment a rough
judgment is all there is to go on.
  The gross comparison of public and private recreation must be
made on the basis of each major sort of facility. A distinction

would have to be made, for example, between ocean-front parks, parks with lakes, and parks without water.

Take the lake-front category. Parks would be ranked from best to worst on the basis of the extent and quality of the service they provide. It would then be necessary to find one or more private facilities to correspond with one or more of the public facilities. This would establish reference points with respect to price.

The rank positions, for example, of Idlewild Country Club on the private axis and Taconic State Park on the public axis are roughly comparable. Though the scale of the two parks is different, the services rendered to their respective visitors is approximately the same. Since Taconic State Park provides a better quality of physical facility than does Idlewild, it should theoretically be priced higher than Idlewild. Since Idlewild charges $2 per person, a price of $2.25 per person would not seem out of line for Taconic. However, the "value" of exclusivity has not yet been discounted. As a guess, we might allow 25 cents for exclusivity, bringing down the value of Taconic to $2 per person per day. But Lake Welch State Park is superior to Taconic State Park; thus it might be assigned a value of $2.25 per person per day. Similarly, Mohansic State Park, somewhat inferior to Taconic, might be assigned a value of $1.75 per person per day.

Beach-front parks need, for the most part, to be compared with beach clubs, of which there are, of course, a large number. Search in the metropolitan area yielded thirteen for which prices for a guest could be discovered; the price per weekend-day ranged from $6.60 to $3. There may also be other private beach-front facilities at a far lower price. Proper study of the matter could not be undertaken for the purpose of this paper; whether it would bear fruit is hard to say. Certainly, beach clubs offer not only exclusivity but "luxurious" environments to meet the status-seeking needs of their members. Therefore they are perhaps not too useful as a basis of judging services of public parks.

In general, we suggest a principle of comparison that needs work and judgment to apply. It has the particular advantage of tackling only one problem of pricing at a time, so that failures in connection with some sorts of recreation may not vitiate successes in others. This may be helpful if the policy of charging user fees for some sorts of publicly provided recreation were to be adopted.

OTHER APPROACHES. While outdoor recreation is undoubtedly a complex and subtle "product" to evaluate, the similarities in the general problems are strong enough to warrant adapting the market researcher's methodology. It is not at all uncommon for private industry to price its new products on the basis of such procedures in advance of the product's release to the market. Some market researchers have been remarkably successful in pre-setting market prices. In the case of outdoor recreation, of course, the market researcher's efforts would not be finally tested in the real world unless public attitudes about pricing outdoor recreation were to change drastically. Nevertheless, some experimentation might be possible. Research findings could be applied somewhat differently in different sorts of parks and in widely separated geographic areas.

One approach is to experiment with user-fees in public parks. The U.S. Forest Service has interested itself in this line of investigation. A beginning in the direction of other market research efforts was made on behalf of the National Park Service in 1958.[17] In a survey of the outdoor recreational activities and preferences of persons living in the Delaware River Basin area, respondents were questioned about their willingness to pay entrance fees. They expressed "a general willingness to pay an entrance fee for 'satisfactory publicly owned outdoor recreational areas' for day trips. . . . The median average amount they said they would consider a reasonable fee was 50¢ per person."[18] The interviewers noted, however, that this finding must be interpreted with some caution because people often express a willingness to pay before the fact, but are much less willing when the time comes. But those respondents who are puffing up their status by indicating a willingness to pay a fee are undoubtedly balanced by others who hesitate to admit such willingness lest officials might be moved to institute or raise fees. Such conflicting biases fatally weaken inter-

[17] Audience Research, Inc. *A Study of Outdoor Recreational Activities and Preferences of the Population Living in the Region of the Delaware River Basin* (prepared for the U.S. National Park Service, 1958).

[18] *Ibid.*, p. 10.

pretations that might be drawn from the data. But this does not necessarily signify that the approach itself is barren; rather, it may argue for a more searching methodology.

One standard technique is that of "projective" methods: a respondent may be asked what he believes *others* would spend for a service or facility. Asking a number of such questions in a highly specific way often eliminates the bias resulting from the desire of some people to appear more affluent than they are; it may even reduce the bias resulting from the fear that fees might be increased. In any event, it is better to have one bias, the direction of which is known.

Another technique entails the selection of three or four similar large sample groups who would be told that a proposed service or facility would cost a different price along an ascending scale. For example, one group might be told the admission charge was 50 cents, the next group that it was 75 cents, and so on. Each would be asked to indicate on a "reaction scale" the statement that best reflects their own attitude toward this price. The resulting data could indicate when a sizable "resistance point" is reached.[19]

There are, of course, all kinds of difficulties with these and analogous approaches, however expertly they are pursued. Perhaps the most serious difficulty is how to provide the interviewees with a controlled picture of what they are pricing and under what conditions. The knowledge that good quality public parks are available free will reduce what a respondent is willing to pay for outdoor recreation. Nevertheless, we believe that the approach is worth exploring more fully.

Evaluation

We have indicated that there are very serious theoretical difficulties in using market prices as an indication of the value of publicly provided recreation. But if, for some purposes at least, these difficulties may be ignored, several observations are in order.

Concerning the several indirect or omnibus approaches, there is

[19] Letter from Dr. Henry Ostberg, president of Blankenship & Ostberg, Inc., Market Researchers, August 19, 1963.

a magic figure of 30 to 35 cents per average person hour of average outdoor recreation that reappears in several contexts. Three different indirect calculations came up with figures in this range. The direct imputation of a price of $2 for a lake park facility, using a customary figure of an average six-hour stay, also comes out at about the same level. The convergence seems to encourage a more careful examination of the available and potentially available information.

This raises several questions. If more confidence in the average figure could be gained, to what use could the knowledge be put? What, for example, does an average of this sort mean? Who pays it and for what? The questions are important in considering the relevance of such a figure to a service provided by governments for all of a country's people.

It seems likely that the "representative family" paying this "average" figure is well above the "median family" of a statistical income array. Expenditure on all recreation is one of the few groups isolated in published data on family income-expenditure surveys for which the percent of income spent on it increases as income increases. This characteristic without doubt reflects increases both in the expenditure per recreation hour and in the number of hours spent in recreation (which have been shown to be strongly affected by leisure time and paid vacations). It seems reasonable to assume that this would also apply to outdoor recreation. If so, an arithmetic average tends to be pulled toward the upper portion of the income array. We have one piece of direct information on this, which shows in extreme form the heavy weight of high-income families in the average. The supplementary information applies to the average expenditure per hour of outdoor recreation based on National Recreation Survey questionnaires. It suggests that, though the average figure is 30 cents, the figure attributable to a median-income family is 26 cents.[20]

[20] *National Recreation Survey* (see footnote 11 above); calculations based on Tables 5.35 and 5.41. The figures for hourly expenditure show a gradual progression through most of the income range (ignoring the under $2,000 group). As family income rises there is a good deal of parallel increase in number of hours and expenditure per hour. At the $15,000 income level there is a heavy upward jump in the average hourly expenditure. The survey aimed at a difficult target and its findings are seriously in need of confirmation.

It seems not unlikely that a similar downward adjustment might apply to the other average figures if it was desirable to think in terms of market price that characterizes purchases of families in the middle of the income array. How this price would relate to the marginal users, other things the same, is indeed hard to say.

Assuming for the moment, however, that a perfect market price could be achieved, would this constitute an acceptable criterion of merit for publicly supplied outdoor recreation? The first category of our usefulness test, comprehensiveness, is met indifferently: a price that people are willing to pay excludes the benefit of which they are unaware or that they undervalue; it excludes third-party benefits and direct national benefits. The second category of our test is defaulted: a market price does nothing to uncover, focus, and steady the aspects of benefits that it fails to cover. Category three is passed with flying colors: dollars are the acceptable measure, par excellence. But category four is scored very indifferently: interprogram decisions may be aided by dollar comparisons of benefit, though some allowance would need to be made for omitted values; but for intra-program decisions the score is likely to be poor, since the omitted values are often those most critical to sound administrative and legislative consideration of alternative recreational projects.

Somewhat more helpful are the approaches that feature the pricing of specific sorts of recreational services either directly or by projective techniques. True, we do not now have market prices that can be applied to the yield of any considerable number of specific government recreational properties or sorts of recreation. However, this present negative conclusion could conceivably be modified if work which seems feasible were undertaken. We suggest an approach capable of developing market prices that bear on specific sorts of recreational facilities. This would be particularly useful in establishing user-fees for publicly owned parks or even for facilities within them. These measures pass the second and fourth usefulness tests somewhat better, and the others no worse, than the omnibus approaches. In any event, the prices could be helpful in setting user-fees.

Merit-Weighted User-Days

The foregoing section indicated that reliance on dollar measures imputed from market prices is precarious for interprogram decisions and virtually impossible for intra-program decisions. The present section explores the problems involved in providing a useful measure based on direct consideration of the merit of services rendered, rather than on their market-based dollar value. The unit of service is a "merit-weighted user-day."

A simple user-day of recreation is a crude unit of benefit. It is based on the thought that benefit is a function of the number of participants and of the duration of the recreational experience. However, the function is not always—or even usually—that of constant proportionality. The measure is, therefore, at best a gross one. Indeed, a moment's reflection indicates that it is too crude to be useful as a guide to decision.

"Merit" rests in the utility measured along a social welfare function. This may be roughly conceptualized as the sum of the utilities to individuals, after each utility is weighted in the context of the public weal. To this should be added such further public benefits as are external to the benefit to recreation users.

But the weight in terms of social merit that ought to be accorded to a user-day of different people under different circumstances varies widely. A day spent by a child at a day camp in the country has long-term value, as well as the ramifying effects discussed in our introductory section. An adult's day of picnicking in a crowded park has, on the average, few of these sorts of values. Thus, to assume that each day has about the same merit is incorrect, since the child's day will presumably produce benefits more nearly two or even three times those of the adult's day.

The merit-weight of a given volume and character of service also varies according to the vacuum produced by the extent of unmet need that already exists. The marginal utility of additional recreation declines as larger amounts are made available.

These statements imply a judgment about a number of things. Similar judgments are implicit (though there is an illusion of avoiding them) if by using unweighted user-days all sorts of user-days are, in fact, considered of equal merit. Use of a service unit implicitly and unavoidably involves a value scheme. So does a dollar unit which, if based on private prices, is woven by the shuttlecock of consumer choice. In the case of the physical service unit, the value scheme must be supplied, and a necessary safeguard is that it be explicitly formulated, exhibited, and discussed until some consensus is achieved.

The Value Scheme: Performance Criteria[21]

The value scheme rests on criteria of good performance for government action in the field of outdoor recreation. The criteria fall in several groups: public-private relationships; quality standards; balance among the kinds of recreation that are provided; distributive justice among present members of the community; justice with respect to present and future members.

Public-private relationships are complex. Outdoor recreation is not the type of service which only government can provide. Much of it is provided privately, and the relative amount can be increased or decreased. Public recreational resources serve two rather different functions: (1) they provide all or most of the recreation used by people unable to pay for it—a notion made familiar by our standards of minimum adequacy of family living; (2) they provide most of some sorts of recreation—wilderness or shorefront areas—for rich and poor alike. Obviously, many questions of policy are implicit. What that policy should be is a matter that needs to be determined by considered judgment and the political process. Needed are criteria that concern:

1.   Conservation as a public purpose.

2.   Government policy with respect to providing recreation where public supply price is markedly lower than private. (Linked with problems under point 5 below concerning appropriate user-prices.)

[21] See Leo Grebler, "Criteria for Appraising Governmental Housing Programs," *American Economic Review, Papers and Proceedings of the 72d Annual Meeting* December 1959, Vol. 50 (May 1960), pp. 321-339. Grebler outlined ten criteria for what he called "workable performance of the housing sector," and we have gratefully borrowed from his basic conceptions.

3.   Government provision of recreation as a method of setting standards for private provision.

4.   Public policy with respect to encouraging private recreation. (Frequently this can be done with far smaller expenditure than provision of comparable public recreation would involve.)

5.   The charge of a total or partial price for some sorts of recreation to, perhaps, some sorts of people. (The dollars received as user-payments would potentially provide a larger yield of benefit when applied to services of other kinds for other sorts of people.)

6.   Public provision of a "modest but adequate" standard for people unable to pay for it.

A second group of criteria concerns quality. Standards of adequacy have been developed in physical terms—use per acre, bacterial counts per liter, etc.—and these need to be considered in relation to the cost of achieving additional increments of adequacy. Moreover, the extent of use itself can convert one type of facility into another (too many people camping in a wilderness area destroy the essence of wilderness camping). The provision of services such as that of recreational counselors and leaders may greatly enrich the recreational experience. Quality standards, then, broadly defined, are essential; they would concern:

7.   The maintenance of a designated standard of care for grounds and equipment.

8.   The development of recreational facilities and program as a method of expanding and deepening use.

9.   Physical standards of use and adequacy including the control of overcrowding of what are, by definition, low density facilities.

A third criteria group involves the character of the recreation afforded. Outdoor recreation is different from most commodities in that its spectrum ranges widely—from hunting in the wilderness to playing shuffleboard in county parks. Some recreational experiences are deemed more worthwhile than others, and more educational in the broadest sense; provision for them should therefore receive more weight. Moreover, different people have different tastes, and any one person may require varied recreational fare. Criteria are needed for:

10. The maintenance of a well-conceived and well-balanced variety of recreation facilities which takes the differential merit and tastes into account.[22]

A fourth group of standards concerns distributive justice. If recreation is a "good thing," it is important that the state offer it to all inhabitants in equitable amounts. Equity requires that provisions bear a proper relationship to needs. Need in turn has some relationship to what people use, or would use were it available under appropriate constraints of time and other costs. Need also involves growing-space for what recreational education today will cause to be the use of tomorrow. Need reflects different intensities and qualities of the utility that different sorts of people are likely to derive from the out-of-doors. Also to be considered, then, are:

11. The geographic distribution of areas with respect to accessibility for after-work, weekend, and vacation use.
12. Distribution among people of different ages, education, and previous experience.
13. Distribution among people of different incomes and, consequently, different opportunities to use private facilities.

A fifth group concerns requirements of future generations. Since recreation involves long-term capital commitment, it could involve strong conflicts of interest between the people of today and tomorrow. Fortunately, conflicts in theory are, for three reasons, modified in practice. First, the urgency of dealing with the problem of the future vs. the present is usually argued on the proposition that we as a nation are running out of recreation land, or at least recreation land around metropolitan areas. With

[22] This criterion is heavily concerned with the utility considerations raised by Professor Craine and discussed in our note to his comments (see footnote 28 below). The criterion also encounters political obstacles, since it is likely to involve adjudication between interest groups that are heavily committed to some particular sector of the outdoor recreation spectrum.

some important exceptions—notably unused beachfront—this is simply not the case.[23] Second, assuring an adequate recreational environment for the future goes beyond the relatively simple act of spending money now for the purchase of land. Techniques may be developed to cope with the problems of the future in less expensive ways. Even at present there are all sorts of step-by-step procedures that are possible.[24] Finally, most parks have a great deal of underused capacity which can be brought into use in short order by building appropriate facilities; this also often provides time to meet the future as it unfolds. The performance standards implicit in these statements involve:

14. Meeting the requirements of the future by a variety of means appropriate to the uncertainty that surrounds prediction.

Devising Weights

Performance criteria are necessarily complex, intermeshed, and to some extent overlapping and even conflicting. Nevertheless, these must form the basis for evaluating which sorts of user-days have higher merit than others. A universal solution would involve a multivariate scheme of impossible complexity.

Fortunately, public choices (like private business choices) can often be made piecemeal. A total program of outdoor recreation can be built up by a series of specific choices between relatively few concrete propositions. Then only such criteria as are served differently by the alternative propositions need be considered and incorporated in the tally of weighted user-days. This incre-

[23] See *ORRRC Study Reports*, No. 8 and No. 21, Vol. II. Recreation land appears not to be lacking in the nation's most populous northeastern megalopolis stretching from Fairfield County in Connecticut through New Castle County in Delaware, according to a special study conducted by the Economic Research Service of the U.S. Department of Agriculture for the Institute of Public Administration.

[24] *ORRRC Study Reports*, No. 8 and No. 21, Vol. II, point to the following techniques: (1) to reserve land that is not self-reserving by nature of its topography or other physical features (there are a variety of ways in which this can be accomplished); (2) to conserve the esthetic qualities of land that is now and will continue to be in private ownership; (3) to rehabilitate and reuse now-spoiled recreation land, especially ocean-front land.

mental decision technique and suboptimization is, of course, familiar in business procedures.

Of two or more propositions, the one tentatively judged most desirable is the one estimated to provide the maximum number of units—merit-weighted user-days—for a given cost. If values not incorporated in the weighting scheme are relevant (values no doubt that seem too subtle to be even roughly weighed), they would be considered on a second go-round and a final judgment arrived at.

The tally of weighted user-days generated by each alternative is computed (1) by estimating the number of people that each park under consideration is likely to attract under given design specifications, and (2) by classifying the users on the basis of such performance criteria as are active in connection with the particular choice. The application of the value scheme rests on the belief discussed above, that some user-days are more meritorious than others. These relative judgments are incorporated in weights which are unity for the standard unit of use and more or less than unity for user-days of greater or less merit.

An Example

Assume that a region is considering the acquisition of park lands capable of accommodating 10,000 people on a weekend day. Their present facilities typically attract 90,000 people, so they now contemplate accommodating 100,000. Three sites are under consideration. To simplify the picture, assume that all are topographically and otherwise equally advantageous and that the cost of land and facilities necessary to accommodate and attract the 10,000 visitors is the same for each site.

The sites differ, however, with respect to operating criteria that have been generally accepted. They involve the questions of distributive justice mentioned above in 11, 12 and 13 in our listing of criteria. Most sharply different for the several parks is their power to promote a more equitable accessibility to recreational areas suitable for a day-trip of people who live in each of the five counties of the region (No. 11). However, other desiderata also have a different impact for the three alternatives. How are

TABLE 1. The Weighting Factor as a Function of Existing and Desired Levels of Service

County	Population (1)	Proportional Level of Service[a] (2)	Merit-Weighted Level of Service[b] (3)	Existing Level of Service[c] (4)	Implicit Merit Factor[d] (5)
A	400,000	20,000	20,000	30,000	0.7
B	160,000	8,000	10,000	5,000	2.0
C	1,040,000	52,000	50,000	40,000	1.25
D	200,000	10,000	15,000	10,000	1.5
E	200,000	10,000	5,000	5,000	1.0
	2,000,000	100,000	100,000	90,000	

[a] Number of persons from indicated counties expected to be visiting parks on an average weekend day if all counties were equally represented on a per capita basis at a 5 percent level.
[b] See text description of how the weighted level is determined.
[c] Number of persons from indicated counties visiting parks on an average weekend day. (Data obtainable through National Park surveys.)
[d] Column 3 divided by column 4.

these differences incorporated into the tally of weighted user-days?

Table 1 describes the derivation of the weights to be applied to the estimated user-days that each candidate park would probably attract from each of the five regions. The first approximation of gross equity is based on the prescription that the parks of the region accommodate people from each county in proportion to the number of people in the county. Column 2 of Table 1 shows how many of the estimated 100,000 visitors would come from each county if 5 percent of the county's population in each case visited some one of the region's parks on a typical summer weekend day.

But gross geographic equity is not the only factor to be considered. Other performance criteria point to deficiencies with respect to a distributive equity for Counties B and D and to relative surpluses for Counties C and E. Considerations such as the following might represent the judgments that were made and described in defending the final decision in terms of the accepted criteria:

1. On an average weekend day 30,000 persons from County A visit regional parks because many of these are very near at

hand. That is, visitors exceed the 20,000 that are expected on the basis of equal per capita attendance. Because County A is a normal county with no special recreation needs or private opportunities, the merit-weighted level of service for County A is set at 20,000.

2.   Only 5,000 residents of County B visit the region's parks, although 8,000 might be expected to do so. In addition, County B is relatively high density with poor alternative private opportunities. (Criteria 6 and 13 above.) The weighted level of service of 10,000 takes this into account.

3.   County C, like A above, is also considered to have no special recreation problem. The low use of parks can be attributed to difficult access problems. Its weighted level of service therefore is judged to be close to the proportional per capita basis.

4.   County D is adequately served on a straight per capita basis. But it is a low-income area with many young, large families. Its merit-weighted level of service is therefore raised to 15,000 persons per day. (Criteria 12, 13.)

5.   On a straight per capita basis, 10,000 residents of County E would be expected to visit parks. However, the weighted level of service for that county is judged to be 5,000, the present level, because County E residents are well-to-do and have excellent private alternatives in the form of numerous marinas and swimming clubs.

Note that the welfare scheme is embodied directly in the merit-weighted level of service shown in column 3. Considerations include primarily the kinds of values discussed in the performance criteria. But the notion of diminishing marginal utility is also present in that, other things the same, the weight should be somewhat higher when, for some particular class of service, the deficiency between what is required and provided is large rather than small. The last column displays the implicit weighting scheme which converts the existing to the desired level of service. Thus the final judgment of how welfare is to be furthered dictates that an additional visit from County B carries twice the merit as one from the wealthy County E and more than twice that of the already well-supplied County A. In some other instance, one might prefer to

TABLE 2. Values of Proposed Parks as Indicated by the Number of Merit-Weighted User-Days

County from Which Visitors Are Estimated To Come (1)	Merit Factor (from Table 1) (1)	Park I		Park II		Park III	
		Estimate of Future Daily Visitors from Each County (2)	Merit-Weighted User-Days[a] (3)	Estimate of Future Daily Visitors from Each County (4)	Merit-Weighted User-Days[b] (5)	Estimate of Future Daily Visitors from Each County (6)	Merit-Weighted User-Days[c] (7)
A	0.7	1,000	700	500	350	3,000	2,100
B	2.0	5,000	10,000	2,000	4,000	1,000	2,000
C	1.25	3,000	3,750	5,000	6,250	3,000	3,750
D	1.5	500	750	1,000	1,500	2,000	3,000
E	1.0	500	500	1,500	1,500	1,000	1,000
		10,000	15,700	10,000	13,600	10,000	11,850

[a] Column 1 times column 2.
[b] Column 1 times column 4.
[c] Column 1 times column 6.

arrive at the weights directly and from these derive the total merit-weighted product.

Table 2 applies these merit factors to the estimated drawing capacity from each county of each candidate park. Though our simplifying assumptions dictate that the total number of visitors and the cost will be the same for each candidate, the sums of weighted user-days will differ. Apparently the capacity of Park I to attract the poorer families of County B as well as its proximity to the large County C boost its tally of weighted user-days to the top figure. Unless noncommensurable values of some importance were to argue against this choice, Park I would be the winner.

Temporal and Spatial Cumulation of Decisions

The example illustrates one set of choices. A recreation program would, of course, involve a large number of such choices. Would the improved series of choices among competing candidates, each made by one group of decision-makers at a particular time, build toward a materially improved program of wider scope and longer duration?

With respect to duration, the method contains a self-adjusting attribute, since the weights change automatically as their purpose is served (providing the criteria themselves are not revised). They do so simply because the need declines and also because such need as remains has decreasing marginal utility. County B's underprovision would be damped by the choice of Park I and its weight in future decisions thereby reduced, other things being the same.

Successive consideration of performance criteria by administrators and legislative groups would, it seems likely, move toward some consensus with respect to the points at issue and the relative values to be served. Thus, over time, objectives as well as performance may be cumulatively improved and adapted to changing circumstances and values.

With respect to scope—maximizing between decisions about recreation made by different governmental jurisdictions—the process would not provide the same decisions that might emerge from the application to all decisions of a single optimal many-dimensioned value matrix. But there is a question whether such a thing can or should exist. Perhaps administrative and legislative decision in the small, provided it is guided and disciplined by the insistent demand that it be justified in explicit and hard-headed welfare terms, can itself build a judgment in the large which even theoretically is preferable to any preordained set of overall rules.[25]

In any event, in practice these subtleties may be of little importance. For one thing, often the greater desirability of one rather than another proposition is so evident that candidates win by a margin wide enough to insulate securely against minor distortion. Ignorance, in a sense, provides another form of insulation. Information and judgment are necessarily uncertain and as such cannot under any circumstances provide precise degrees of advantage, much less actual maximization.

It seems likely, then, that within the necessarily fuzzy contours

[25] The law provides an interesting precedent, since it has made an experimental laboratory of diverse jurisdictions. To do likewise in the field of expenditure decisions, we need to soften the irreversible characteristics of much government action and to devise ways to learn from experience.

by means of which a standard of excellence in such matters can be defined, the legislator or budget director would be in a position to assume that the recreational projects submitted as requiring an appropriation of funds would provide the highest recreational benefit which could presently be bought with that amount of money. The weighted user-days that it represented, along with further supplementary specifics, could then, as suggested earlier, be contrasted with the yield of the same amount of money spent on other government services.

The Usefulness Test

The first category, comprehensive calibration, is met reasonably well with respect to comprehensiveness and indifferently with respect to the quality of calibration—merit-weights provide a "shadow price" in somewhat flabby dollars. In the second category the method does well: the development of performance criteria is aimed very specifically at, among other things, uncovering noncommensurable values and placing consideration of them in proper relation to that of the relatively measurable benefits. Category three, acceptability, is a question mark: merit-weights are a coin that is stranger than a subway token. Learning to accept the coin, at least in government circles, will depend on the extent and quality of experiment with it and how satisfactory, relative to that of other ways of structuring rational choice, the experience proves to be.

The fourth category has a mixed score: with respect to inter-program decisions, merit-weighted service units can do no better than provide legislators with a choice between two or more marginal commodities—the benefit, for example, of the next $1 million, spent to maximum advantage, on this number of merit-weighted user-days of recreation, or on some specified number of merit-weighted alleviation of poverty days (description of both programs appended). With respect to intra-program decisions, it should channel debate and decision toward consideration of operating criteria which themselves require formulation and reformulation at each level of government. It seems reasonable to suppose that this would deteriorate the power of private or political advantage to influence choice and elevate that of qualitative and quantitative consideration of appropriate matters. In many cases, heroic judgments will still be required, and the best that the effort to provide merit-weights can achieve is a check list and some quantitative orders of magnitude of benefits and costs. These should aid in identifying and hefting salient characteristics of alternative proposals.

Conclusion

We have discussed two ways of attempting to arrive at an estimate of the current benefit to be derived from a given amount of current expenditure on outdoor recreation. The dollar-value approach seems seriously limited as a method for intra-program analysis. The basic difficulty is that market prices tend to be least suitable at the very point where, for the purpose of pricing an entire publicly supplied recreational program, we need them most. Also, the relationship between the price and total relevant benefit is drastically uneven for different sorts of recreation for different sorts of people, having different amounts and kinds of leisure time and education.

But the figure of about 30 cents per hour for an overall average value of all sorts of outdoor recreation seems to keep bobbing up in each of several calculations which use different methods and different data. Its persistence could be coincidental, but we ought to be able to compile information that would serve to dismiss it as an impostor or to gain the confidence and knowledge necessary to revise it and understand its meaning and limitations. At best, however, the figure would have serious theoretical shortcomings as a measure of the merit in publicly supplied outdoor recreation. This and especially more particularized market prices may afford a guide if user-prices are charged for some public recreational facilities.

The second approach, a measure in terms of a merit-weighted service unit, seems promising as a method of improving intra-program decisions. In addition to its power to utilize hard information it has the capacity to accommodate value judgments about noncommensurables, partly by submitting them to rough calibration and partly by simply leaving a space that must somehow be filled in by explicit judgments. Used in a decision matrix of the sort described, the weighting scheme is not static but takes changing supply and other conditions into account.

The measure may also aid in interprogram comparisons. It provides a summary, though supplementary, picture of the service benefit that proposed total expenditures on recreation would yield. This estimate could then be compared with analogous pictures of service benefits from spending the same amount of money for other end products of government. For the purpose of interprogram comparisons, one preferably should think in terms of the gross user-days that a given expenditure would buy rather than in terms of the sum of merit-weighted user-days (which could be a larger or smaller figure). The way programs were chosen (using the weighting system) would make it appropriate to state that "a million dollars spent on outdoor recreation would add most efficiently, say, 300,000 of the most beneficial user-days that can presently be designed." Further, a value estimate based on private prices for the average recreational experience, adjusted for its sins of omission and commission (and this is a large order), may at some time provide an estimate of total dollar value of the current output of dollars spent on recreation. Needless to say (see footnote 1 above), evaluation of a capital investment must then move to consideration of a stream of discounted outputs and inputs.

Comments

HOWARD E. BALL, *Bureau of Outdoor Recreation, U.S. Department of the Interior*

The professional economist continues to be impressed with the challenge of his research frontiers and the compelling necessity of developing new understanding and greater precision. I, on the other hand, as a practitioner, am impressed with our striking inability to apply effectively what we already actually know. My instincts and my experience both tell me that we can and should make considerable affirmative gains in the taking of rational resource allocation decisions, all within the present state of the art.

It might be most helpful to the purpose of this conference if I could state objectively, and with some authority, the present rating of the federal government's recreation resource planners as competent dollar allocators—either by a comparative rank ordering, or by a summary of the results of an impartial economic intelligence test. Being unable to do this, I must fall back on subjective judgment. Clearly, the energy economists, farm economists, water economists, and transportation economists are far ahead of us, in concepts and methodology and in application of the present state of the benefit evaluation art. I am not prepared to say who, if anybody, is running behind the recreation economist.

I hasten to add that there is no reason to be either embarrassed or ashamed of the current situation. There are sound historical reasons for it. First of all, we must understand the professional origins and make-up of the men who dominated the pioneer era during which public outdoor recreation evolved into an identifiable thing—a viable federal program. Second, we must appreciate the political climate which prevailed during this same era. Finally, we must consider the geopolitical happenstance that the North American continent was opened up from the Atlantic Coast westward, rather than from the Pacific Coast eastward.

The Conservation Era

I am talking, of course, about the Conservation Era, and the events which triggered its development. The men who led it were embattled missionaries. Their weapon was the massive public land withdrawal or reservation. Their motive was to protect and preserve the physical resource base, along with its flora and fauna and its outstanding esthetic attractions. Potential private or corporate profit was the threat. Time was running out. Their battle was fought and won long before sophisticated economic techniques had been thought up, much less applied, either by industry or government. These pioneers sought out and found a bold group of statesmen, in Congress and in the White House, who moved in blitzkrieg style to save much of the best of the remaining public domain. Cost-benefit analysis, in the sense it is being discussed here today, never crossed their conservationist, resource-preservation-oriented minds. Such cost accounting as took place, if there was any, was done at the ballot box, not at a cash register.

Thus, it happened that the vast bulk of the outdoor recreation potential of the United States came into being in our western states, a region which at that time had an extremely low average population density per square mile (and in some cases still does have it). Under conditions which amounted to a fantastic local and regional oversupply of recreation-carrying capacity, it is understandable how the tradition of a zero price became associated with recreation use of the federal public lands.

Only during my own generation of natural resource managers and planners have new factors converged on this inheritance from the Conservation Era, working both subtle and radical changes into the situation. One major factor has been the steeply mounting rate of federal investment in water resource development and highway building programs. Another has been the growing political and economic recognition of what I call the proximity factor (or the market-demand factor) in reviewing the strategy of federal outdoor recreation spending.

The Changing Pattern

We have yet to witness the submission of a federal outdoor recreation proposal which starts with an analysis of human behavior and winds up with an ideal site selection and facility installation plan; said plan having cranked in a national priority factor, and carrying with it a guarantee that it will deliver the most recreation-use units per taxpayer dollar spent. Most planning of federal outdoor recreation-carrying capacity today (I am referring to direct programs, not to programs of an incentive or assistance nature) operates under a set of built-in restraints which tend to dampen responsiveness to the pure recreation-demand component. Practically all of the major resource agencies are acutely aware of these inhibiting restraints, and within recent years nearly every one of the agencies has sought—and most of them have won—legislative or administrative liberalization with respect to its part of the public outdoor recreation supply picture.

With this increase in opportunity and the elevation of recreation to full respectability as a program or project purpose has come an increased obligation for the recreation planner to jump into the Darwinian struggle for money and a willingness to marshal economic justifications for his claimancy against resources. Headway is being made on many fronts. A joint interagency task force has been putting the finishing touches on an interim methodology for evaluating and calculating recreation benefits commonly associated with large federal impoundments. We are moving rapidly toward clarification of federal policy and practice on allocation of recreation's share of joint project costs in multipurpose projects. New ground is being broken in the matter of defining acceptable least-cost alternatives, and this has required significant changes in conventional thinking about such elusive matters as comparability, substitutability, and quality differentials.

As a final example, a study team, established by the Recreation Advisory Council to take up the matter of user charges for federal outdoor recreation areas and facilities, went out on thin ice and actually produced, not a set of principles and criteria for charges, but a concrete schedule of uniform federal recreation charges. Activation of the schedule depended on the outcome of congressional action upon the Land and Water Conservation Fund bill (HR 3846), in which a legislative history on the subject of user-fees for federally provided outdoor recreation was being written.[26] The subject is of course hotly controversial. Strangely enough, polls of consumer opinion show us that the citizen is of a fairly even temper on this matter. In contrast, legislators and experts within the federal bureaucracy tend to align themselves in rather strongly polarized fashion, either for or against "public recreation at a price."

Recreation Economics and a Few Sacred Myths

I have brought my background remarks around to the subject of user-fees, because so much of one's reaction to the Mack-Myers' thesis hinges upon whether or not you go along with the

[26] The bill was signed into law on September 3, 1964 (P.L. 88-578).

first two thirds of their paper. In those pages the authors systematically demolish hope for a market-price, dollar-measure evaluation approach, and thus pave the way for their "merit-weighted user-day" concept of measurement. I must postpone any critique of this concept as a yardstick for resource investment decisions until I have taken a closer look at what is alleged to be wrong with—or difficult about—the imposition of a price structure on federally provided recreation opportunities.

There is little point in reiterating the authors' rather convincing catalogue of problems presented by the market-price approach or the dollar-value concept. I could, without much effort, expand upon it—thereby, perhaps, reinforcing their own conclusions. But I would prefer to start with a bit of probing into some of the sacred folklore surrounding recreation economics which presently stands in the way of making fundamental headway on the benefits-evaluation problem. (I do not in any way imply that our two authors subscribe to any or all of this mythology.)

First, and foremost, I would like to point out that there never has been truly "free" or zero cost public outdoor recreation. What we really mean is that most of us (but not all) have helped pay for it, directly or indirectly, whether we consumed any of it or not. Some of it is paid for under flat-rate tax systems, some under excise-tax systems, some of it under general and progressive income-tax systems, and some of it comes out of the hide of stockholders. To be a beneficiary, however, you have to want public outdoor recreation and have the psychosomatic and financial wherewithal to go get it. It would not come to you, except vicariously. I won't mention more than fleetingly the hidden, but nevertheless real, opportunity costs wrapped up in our failure to keep a meaningful set of federal real estate books, and our unwillingness to impute costs to public land withdrawals and use dedications. I admit that there is a haunting imprecision or uncertainty about the magnitude of the subsidy involved, but that is mainly because we have been willing to put up with poor cost-accounting methods—and sometimes with none at all. This makes it difficult, for example, to pin down what is meant when people say they will go along with a system of "nominal" user charges, but then add that straight pay-as-you-go is out of the question.

Another deep-seated myth is that there are unique or special intangible qualities associated with the outdoor recreation experience which stand forever as insurmountable barriers to quantitative analysis. There are, of course, normative and value judgments associated with the esthetic and sociological facets of most outdoor recreation activities, but this is also the case with practically every other concern of human beings. Cosmetics, painting, music, drama, much of our food and drink, most of our clothing, certain features of our cars and our houses—all have wrapped up in them esthetic components of deep value and significance, or prestige and status and power components. Collectively, they represent a veritable spaghetti bowl of value-judgment factors— intangibles that no man in his right mind would attempt to submit to measurement by conventional quantitative techniques. Does this prevent their being dealt with intelligently and objectively from an economic point of view? Does this exempt them from the rationalities and irrationalities of market analysis, from pricing, from regulation, or from the more or less whimsical fluctuations in supply and demand? It does not. Even Ladies Aid Societies and Cub Scout dens have budgets.

By what mystique, then, can the serious recreation economist claim extraordinary treatment for his bailiwick? How can we presume to attach overriding importance to the psychic satisfactions of outdoor activities, as against the vast array of analogous satisfactions flowing from human behavior that takes place, for example, indoors? This seems to me a fallacy or a bias that we might well be rid of. It stems, I think, from the normal and unwitting tendency of the outdoor recreation advocate to project his own value-preference structure upon the total population. That other kinds of advocates all tend to indulge in similar behavior should not blind us to our own methodological errors.

Also in sore need of re-examination is the widely held concept of an absolute hierarchy or pecking order of quality attached to the spectrum of outdoor recreation activity. Advocates of this concept manage to suggest that there are "recreation Brahmins" who alone can appreciate certain forms of activity—such as communing with the wilderness at $30 a day—while down at the other end of the scale stand peons and untouchables who pant for the hot sand and hot dogs of public beaches. Between these extremes is a great army who want only physical release or possibly fatigue or adversity. But to try to cast the immense array of activities that make up outdoor recreation—and the vast variety of human tastes and choices—into a meaningful universal scalar system with cream on top and whey on the bottom seems a futile and probably nonsensical exercise. We should not confuse ability to pay, on the one hand, or the unit cost of providing units of supply, on the other, with the true common denominator of recreation satisfaction: that of individual fulfillment as judged by the consumer's own personal frame of reference.

Prices and Market Research

Speaking now more directly to the first half of the Mack-Myers' paper, I wish the rather promising concept of average expenditure per hour for "away-from-home" recreation had been explored more fully. Without the benefit of figures to back me up, I must still say that the "magic" 30 cents per hour is probably much too low, or certainly very conservative. To take only one component of their data, I cannot accept the $98 per year of auto expense allocated to outdoor recreation by the Bureau of Public Roads. This does not square with other information from the Bureau—that nearly 18 percent of all traffic on the interstate system today is generated by recreation passenger travel. Nor does it square with the "Red Book" figure cited by Herbert Mohring in his urban highway paper, where a value of $1.55 per hour for passenger car operation is discussed.

When Mack and Myers discussed actual and artificial market prices, I was glad to note that they recognized the price-depressing influence of zero, or underpriced, public recreation facilities and services of comparable nature. Nobody really knows the nature and extent of this factor at present, but it will be brought under careful study. The bold new program of the Department of Agriculture, designed to bring farmers and rural landowners prominently into the outdoor recreation supply picture, is an income- and profit-oriented program, as it must be. In other words outdoor recreation is being thought about as another agricultural or rural land product, with possibilities for commercial success.

And who knows? If this approach catches on, we could—since there are some 50 million acres of private land which are targets for diversion from surplus crop production—wind up stockpiling recreation!

I favor, of course, several of the recommendations made by the authors in their discussion of market research. But I see no reason for holding solely to interview techniques and testing of opinions about hypothetical situations. The federal government at present operates several thousand land and water recreation areas. These management units embrace the widest possible array of activities, environments, and locations, which offer limitless opportunities for the empirical or experimental approach. The Forest Service is in effect carrying on such experiments right now, within one or more of its regions. And the Park Service has amassed considerable practical experience on the point of consumer reaction to admission fees and charges, including those imposed both by concessionaires and by the government. A well-designed effort with scientific sampling and controls could lead to "before" and "after" results of tremendous utility to the benefit evaluator and public investment allocator.

"User-Days"

I turn now to the authors' concept of "user-days." I agree that we need clear and affirmative policy regarding the respective role of public and private parties in the provision of recreation supply. That ample capital and entrepreneurial skill is available and is moving responsively into portions of the demand spectrum is evidenced by the travel and vacation section in the Sunday *New York Times;* by suburbia's massive bowling alleys; and by the innumerable cooperative-type swimming pools, country clubs, and Little League diamonds. The policy problem is how far governments can and should go in furnishing extra-market incentives or subsidies to the private investor-manager, to induce him to move into that part of the activity spectrum which is more laden with risk and uncertainty, or which promises only an unexciting rate of return. Insurance schemes, low-cost credit, tax relief, and public-private consortiums may offer promise, but all of these cost money and have to compete with the force account alternative.

I also agree that some provision must be made for the welfare portion of the demand spectrum, although we know that this is a real Pandora's box, once we look under the lid. Further, some portion of our national historical, scientific, and esthetic shrines ought to be exempt from cost-benefit tests, if for no other reason than that our GNP is high enough to allow this nation to make such exceptions.

The discussion of distributive justice makes a good deal of sense, although I can vouch that the principle is much more easily formulated in words than achieved through bureaucratic deeds. In any case, the authors' five-county, three-park model makes a real contribution to the theory of comparative site selection. And I was especially interested to note that their weighting technique stemmed in the main from distributive justice and proximity factors, coupled with estimates as to supply excesses and deficiencies—rather than from some mystique of an arbitrarily ranked set of values attributed to types of activity.

The suggested matrix could, of course, be greatly refined and made more complex. For example, provision could be made for fractions of balanced user-days, with an underpinning of factual data showing the frequency distribution of actual attendance behavior by, say, tenths of a user-day. The supply surpluses and deficiencies could be shredded out along specific activity lines, and then pulled into a synthesis or summary expression.

The Bureau of Outdoor Recreation has recently constructed just such a matrix for the entire nation, broken down into the nine official Census regions. We have applied the matrix against the proposed fiscal year 1965 land acquisition expenditure pattern for all federal agencies. I can't divulge the results, but it is possible that they may influence budget decisions.

Mack and Myers have presented us with ideas that are both stimulating and useful. They have certainly demonstrated that recreation economics is not such an elusive beast after all. What is needed most is a will to attack the problems with the same rigorous methods which have proved so productive in other fields of evaluation. This is an exciting era of development in the field of outdoor recreation. Sound approaches to cost-benefit evaluation do not constitute a threat to the richness of the personal recreation experience. Indeed, I think we shall find that sound economic analysis can become the servant of the burgeoning outdoor recreation movement, and will permit the recreation planner and the resource allocator to take their places more effectively at the bargaining table. Getting funds for fun is a tough fight. We need all the weapons that can be mustered.

LYLE E. CRAINE, *School of Natural Resources, University of Michigan*

Ruth Mack and Sumner Myers were explicit that their objective was to explore concepts and methods of measuring benefits derived from government expenditure on outdoor recreation. But their contribution goes well beyond that, for they provide a perspective—a structure for critical thought—on the larger question of social responsibility for outdoor recreation.

The paper, in this perspective, presents two kinds of areas, each of which deserves critical review. One has to do with the methodological problems of measurement; the other with the conceptual orientation of outdoor recreation on which measurement approaches are based. I leave to others (with better credentials than mine) the job of commenting upon methodological questions, and devote my remarks here to some of the conceptual questions raised by the paper.

Conceptual Orientation

First, I want to summarize my understanding of what the authors are telling us about the conceptual content of the problem; then, within this framework, I will comment on several specific aspects of the paper. As I see it, the paper encompasses three approaches that are fundamental to the comments I want to make.

1. It describes outdoor recreation as a commodity possessing three kinds of utility: that derived from the immediate personal enjoyment; that derived from the long-term beneficial effect upon the individual; and that derived by the nation from hav-

ing more, rather than fewer, outdoor recreation opportunities as a part of the national scene. In terms of these utilities, the authors postulate three bases for the public interest in outdoor recreation: paternalistic provision of some level of recreation opportunity as a means of improving the individual's well-being; provision of the overall societal benefits which may stem from greater, or different, outdoor recreation opportunities; and promotion of efficient use of resources devoted to recreation.

2. The paper classifies the decisions that bear on outdoor recreation. Three decision factors are recognized: the distribution of capital investment funds between the public and private sectors of the economy; distribution of public investment funds between outdoor recreation and other purposes; and maximization of recreation utility from resources to be devoted to outdoor recreation. The authors consider the first type—public vs. private —as outside the scope of their paper. Their analysis, therefore, distinguishes between the other two as "interprogram" and "intra-program" decisions.

3. The paper analyzes two kinds of benefit measures. First, the authors test the usefulness of monetary measures. They believe that dollar valuations would be particularly useful in making interprogram comparisons, if values could be developed that are dependable. But they believe that monetary measures are useless in intra-program decisions, because the relation between price and benefit is drastically different for different sorts of recreation for different sorts of people.

We are told that we can't really depend upon price measures generally because (1) for immediate use we have no acceptable method of deriving price and (2), even if we did, price has serious theoretical shortcomings as a measure of recreation merit. Surprisingly, in spite of this bearish attitude about the usefulness of monetary measures, the authors proceed to devote over a third of their paper to different approaches for deriving a price. They show three different calculations by which one can arrive at a figure of 30 to 35 cents per person-hour as a value for outdoor recreation. They find the consistency of their findings under three different approaches a sufficiently hopeful "coincidence" to deserve further testing and development.

Nevertheless, having for the present found dollar values seriously deficient, they examine the usefulness of a user-day unit. They propose refining the crude user-day unit by weighting it with some twelve criteria of good performance for government action in outdoor recreation. By the application of these performance standards, the authors conceive a "merit-weighted user-day" which they propose as a measure of merit based directly on the judged value of services rendered, rather than on their market-based dollar figure. This is an intriguing idea, particularly for those who are becoming a little disenchanted with efforts to put a dollar-and-cent sign on every value of society. However, before I slip into further editorializing, let me turn to the three aspects of the paper about which I want to comment critically.

Test of Usefulness

In the statement of their objective, Mack and Myers emphasize that they are seeking a useful measure of recreation benefit. They make explicit their test of usefulness by postulating that a dollar measure is useful in proportion to (1) the percent of total benefit that it adequately covers and (2) its capacity to focus and steady the aspects of total benefit that it fails to cover.

I would applaud the emphasis upon criteria of usefulness of monetary measures of benefit, and the effort to make explicit the standards of usefulness. This seems to me a significant contribution. I am not aware of a "usefulness test" having been given explicit attention in other literature on benefit-cost analysis.

I am particularly pleased to see the emphasis given to the second criterion of usefulness—that is, the capacity to "focus and steady" those benefits that the measurement itself fails to cover. This gives overt attention to a contribution of economic analysis that both economists and noneconomists have been slow to acknowledge. (Too often there is an assumption that economic valuations are absolutes—either absolutely right or absolutely a fraud, depending upon one's predilections regarding the issue at hand.) This Mack-Myers criterion makes quite explicit what may be economics' greatest contribution in analyzing public investment: not its ability to show us by the most sophisticated calculations the right answer, but its ability to direct us to the right questions that will help isolate and order the noneconomic values for more effective subjective consideration and debate.

I am, however, disappointed on two scores in the treatment of the usefulness test. First, I miss a third criterion—acceptability. Yet the authors actually injected that criterion in discussing their indirect approaches to dollar measurements when they said, "But certainly if dollar values are to be useful, they must be widely accepted." I should like to see "acceptability" specifically incorporated as an additional criterion of usefulness. Second, I wish the authors had shown us more concretely how they applied their usefulness test. How, for example, does each alternative approach to measurement meet the test of usefulness? It is not always clear how the test affected the authors' judgment about each measurement approach they considered.[27]

The Physical Service Unit

My second comment has to do with the use of the physical service unit as a measure of benefit. With the best of economic analysis today—and probably for some time in the future—we still have many values entering into public investment that will not be reflected by economic measures. These values may play a decisive role in a public decision without actually being perceived. Their identity and their expression are often blurred in an undifferentiated alignment of special interests.

The concept of "merit-weighted user-days" as a measure of certain kinds of recreation service and similar kinds of physical service units may hold promise of adding another tool for sharpening the value questions involved in the political decisions. The very process of formulating these kinds of measures can contribute to public enlightenment about the values that are involved. The service-unit approach at least disciplines the expression of values and makes the value scheme explicit and observable, all of which helps to direct debate and negotiation to issues relevant

[27] Editor's note: As footnote 7 in the main text above indicates, the authors adopted the two suggestions made in this paragraph in the revised draft of their paper.

to value questions. This, I submit, would be a major contribution to more rational political decisions. It occurs to me that, over the immediate future, there may be a higher rate of return for time and talent spent in developing and applying service-unit measurements than for that spent in seeking refinements in monetary measures of recreation benefits.

The Utility Functions of Outdoor Recreation

My third comment concerns the authors' recognition that delineating the range of utilities which society receives from recreation is the starting point for any consideration of public values. The social utility derived from recreation is, after all, fundamental to the question of public interest and to the justification for public investment.

However, after the authors identified the utility functions of outdoor recreation they did not make clear how this process guided their thinking about measurements. If the utilities they cite are the end product for society, the measures of value should somehow be quite clearly related to the utility functions. I sense here a conceptual chasm between the idea of recreation utility and the subsequent approaches to value measurements. I wish we could find the bridge that would join together the end product in terms of social utility and investment criteria. I suspect that efforts to build such a bridge would give us a greater payoff than efforts to further refine economic measures of recreation benefits.

Let me illustrate with two further observations. Different kinds of recreation facilities clearly play different roles in providing each of the utilities from recreation. Take, for instance, the utility derived from the "long-term individual benefit." According to the characteristics ascribed by the authors to recreation producing this utility, it is likely to be greater from wilderness experience, let us say, than from shuffleboard. It would thus seem to follow (all other things being equal) that public investment seeking this particular utility would be more justified if applied to promoting wilderness use than to enlarging shuffleboard opportunities. I do not pretend to assess the methodological problems involved, but this line of reasoning suggests to me that a system of weighting physical service-units in terms of utility functions would be more responsive to the ultimate social question posed by recreation than the performance criteria used by Mack and Myers.[28]

Finally, I suggest that the utility approach would attach greater significance to the limitations of price and the deficiencies of consumer behavior as a basis for valuing recreation investment. The authors have recognized that the market does not "reflect utility that is not in some sense appreciated by the users." The same may be said for any measure derived from the recreationist's choice, including the various indirect ways that the authors have explored of imputing values. These valuations, even if reliable, measure only the utility that Mack and Myers call "personal enjoyment." Neither "long-term benefits" nor "third-party" values are going to be reflected by the usual recreation consumer. This logic, I submit, leaves us with little reason to look to prices, or any other measure based upon consumer choice, as a guide in seeking two of the three recreation utilities—yet these are the two that appear to be most important to public endeavor in the field.

To pursue this logic somewhat further, one may say that consumer valuations are applicable only to the achievement of "per-

[28] The authors agree entirely with Professor Craine's assertion that the differential utility of various sorts of recreation must be taken into account. They intended their criterion 10 to accomplish this task and have tried to make this clearer in the revised draft. Criteria 12 and 13, and occasionally others, also may have some relevance to the matter.

The authors believe that the best that can be done on this important question of taking "utility functions" into account is to say that, other things the same, one sort of recreation has roughly one, one and one-half, two, etc., times as much merit as some other. However, as usual, other things are seldom the same. For example—and this is an essential point—as in any schedule, value depends on supply as well as on hypothetical demand. That is why criterion 10 emphasizes "the maintenance of a well-conceived and *well-balanced* variety of recreation facilities" (italics added). Also, that is why our illustrative example has a built-in capacity to reduce the weight attached to user-days of a particular attribute of recreation as its supply is increased.

This joining of merit with supply considerations in a dynamic framework constitutes our effort, in Professor Craine's words, to "find the bridge that would join together the end product in terms of social utility and investment criteria." We certainly agree that these crude but necessary procedures seem to shoulder out "efforts to further refine economic measures of recreation benefits."

sonal enjoyment"; justification for public investment in pursuit of this utility is found only in gaining efficient resource utilization. This utility does not carry any obligation that it be supported from general tax revenues. Indeed, the logic suggests that, in the interests of equity, greater use be made of user-fees. The response to user-fees would, in turn, provide measures of consumers' valuations of the personal enjoyment utility and greatly reduce our problem of valuation. Perhaps one of the greatest uses of economic valuation is to determine a level of user-fees— a point Mack and Myers have made almost incidentally. My only purposes in pursuing this little exercise in logic are (1) to indicate that the concept of the utility functions of recreation can lead to significant policy questions about public investment, and (2) to suggest that if investment criteria are to be meaningful they need to relate to these questions.

BURTON A. WEISBROD*

Preventing High School Dropouts

EDUCATION IS A LARGE and rapidly growing component of the United States' gross national product. In 1961 formal education expenditures totaled $27.3 billion, or 5.4 percent of GNP compared to 2.9 percent in 1948 and 3.9 percent in 1954.[1] The growth reflects not only the rising school age population but also the increasing expenditure per child-year in elementary and high school ($179 in 1948; $265 in 1954, and $393 in 1961) and the increasing number of years of schooling attained per child.[2] Educational attainments of the population 25 years of age and over have risen sharply. In 1940, only 52 percent of the population in this age group had more than eight years of schooling; by 1950 the figure was 60 percent, and by 1960, 70 percent.[3]

The proportion of children beginning high school who graduate has also climbed steeply: of those enrolled in the ninth grade in 1924-25, only 33 percent graduated four years later, but of the ninth graders in 1958-59, 70 percent graduated four years later.[4]

* Department of Economics, University of Wisconsin.
[1] These expenditures include private schools and higher education; but the pattern for elementary and secondary schools alone is similar—1.9, 2.9, and 3.8 percent in 1948, 1954, and 1961, respectively. Computed from U.S. Department of Health, Education, and Welfare, *Health, Education, and Welfare Trends, 1962 Edition*, p. 60.

[2] Current expenditures only. See *ibid.*, p. 61.

[3] *Ibid.*, p. 53.

[4] From U.S. Office of Education, "Number and Percent of Public High School

In spite of this large achievement, an attitude of urgency is developing about those students who fail to complete high school. It is predicted that, during the remaining years of the 1960's, 7.5 million of the 26 million young labor force entrants will not have graduated from high school.[5]

Literature arguing the severity of the "dropout problem" and proposing solutions has reached mountainous proportions. One recent publication briefly describes fifty-five separate local programs to deal with the problem in various cities; a bibliography contains more than a hundred professional journal articles on the subject published within the last five years; the National Education Association has established a special research project for action, consultation, and clearinghouse operations regarding dropouts; and statements by the U.S. Commissioner of Education, the Secretary of Labor, and the President of the United States urging students to remain in school have become routine front-page news.

The objectives of this paper are (1) to examine the nature of the "dropout problem" and to see why it is apparently of national concern; (2) to analyze the kinds of costs and benefits of programs to diminish dropouts; and (3) to consider a specific program whereby the profitability of investment in dropout prevention might be estimated. Accordingly, this paper is divided into three main sections: a conceptual analysis of the social dropout problem; a consideration of means for quantifying returns from a dropout prevention program; and a presentation of a benefit-cost analysis for a specific prevention project.

The Nature of the Dropout Problem

What is "the dropout problem"? First, we had better know what a "dropout" is. According to a forthcoming publication under the auspices of the U.S. Office of Education:[6]

Pupils in Grade 9, in Selected Years, Who Did Not Graduate 4 Years Later" (table prepared by Elementary-Secondary Surveys Section, Division of Educational Statistics; April 1963, dittoed).

[5] And 2.5 million of these will not have completed eight grades of elementary school. See U.S. Bureau of Labor Statistics, *From School to Work* (March 1960), p. 1.

[6] "Pupil Accounting for State School Systems," reported in National Education Association, *School Dropouts* (Project: School Dropouts, April 1963), p. 2.

A dropout is a pupil who leaves a school, for any reason except death, before graduation or completion of a program of studies and without transferring to another school.

But is a dropout necessarily a "dropout problem"? The dropout rate continues to decline, now being less than half of what it was during the 1920's; even the absolute number is well below the level of the 1920's, in spite of the population boom. But the talk of a problem mounts. It seems that a dropout becomes a "problem" when society feels that he or she is unable to adjust to the society and the economy. If a dropout leaves school to work on the family farm, where he stays throughout his working life, then he remains part of the agricultural society and may be presumed to have no particular problems of integration with it. He may miss what some would call the "good life" which greater education might have brought, but this is after all a matter of taste. Thus, dropping out may represent the exercise of consumer sovereignty by a person whose marginal rates of time preference, income-leisure preference, or schoolwork preference, together with his market opportunities (including the value of additional schooling) and the opportunities provided in school (including the type of education and training available), dictate that he "consume" (or invest in) no more than the required schooling. Dropping out *may* be an "optimal" solution.

But what if the dropout has imperfect information or foresight regarding future employment and income opportunities, or has a set of preferences which will change as he grows older and which he will later regret having, as they led to his dropping out of school? In either case, the student will decide later that he erred. At that time, dropping out may constitute a problem for the dropout, but is it necessarily a problem for the society? If incomplete knowledge of available information was the root of the problem, then society may wish to improve the availability of information; if imperfect foresight was the trouble, then society may be unable to do anything, for it, too, lacks perfect foresight. If utility-function inconsistency existed, then the ethical problem arises—which of the individual's functions should prevail? And consumer sovereignty becomes an ambiguous guide to policy.

In any case, if the problems of the dropout are handled directly by the individual, or within the family unit, the "society" may be unconcerned; after all, freedom, or consumer sovereignty, seems to imply the right to make a decision which may be regretted later, assuming no significant external effects. But the matter becomes a *social* problem when the consequences begin to impinge seriously on others, or when the private, familial solution is seen to be unsatisfactory or impossible, so that solution at a more aggregative level is required.

This framework provides a basis for explaining why the "social dropout problem" is apparently at an all-time high, even though neither the relative nor the absolute magnitudes of dropouts are near the previous peaks. Recent and expected technological developments provide a powerful case for the view that future job opportunities and earnings prospects will be increasingly slight for the high school dropout, so that a dropout constitutes virtually *prima facie* evidence of either incomplete use of available knowledge, or of youthful disregard of the future for the pleasures of the present.

At the same time, dropping out is not always a matter of choice. It may, for instance, reflect the financial situation of the student, and, if that is coupled with capital markets in which borrowing for education is difficult, both the student and the economy may forgo a profitable social investment when the student leaves school. On top of this is the fact that, as the society has become increasingly interdependent, the individual in economic distress has become less and less able to be bailed out by his family, or to work on the family farm. As a result, the society as a whole has come to bear the burdens. Unemployment is no longer solely a family problem: through welfare assistance and unemployment compensation we all share the costs of maintaining the unemployed.

Dropouts and the Unemployment Problem

This line of discussion suggests that for a dropout to be a problem, and for the problem to be considered a social, as distinct from a private, problem there must be adverse consequences which the individual either cannot foresee, cannot control, or cannot adjust to, without involving people outside his family. This also suggests that the social dropout problem as it exists today is largely an unemployment problem—that is, an inadequacy of job opportunities. It is hardly an accident that the public concern about dropouts has mounted in the past few years, when the relatively low, postwar unemployment rates prior to 1957 (outside of recessions) have turned into persistently high rates of 5 percent or more for the last six years. Whatever the level of unemployment, the most poorly educated have come off second-best.[7] As long as this remains true, a shift to a relatively high overall unemployment rate will tend to raise the rate among the dropouts to alarming proportions, even if there is an equi-proportional increase in the unemployment rate for every educational group. Between March 1957 and March 1962 total unemployment among persons 18 years old and over rose from 4.1 to 6 percent—an increase of 46 percent; actually, the largest proportional rise was for college graduates, whose rate rose by 114 percent—from 0.7 to 1.5 percent—while the rate for high school dropouts (nine to eleven years of schooling) rose by 59 percent—from 5.1 to 8.2.[8] In a relative sense the dropouts' position improved, but obviously it worsened in absolute terms considerably more than was true for the college graduates. And a 1.4 percent unemployment rate is no cause for public consternation, but a rate of 7.8 percent is another matter.

Dropouts are unemployed with such frequency and for such durations that the effects not only on them but also on the rest of society become severe. There are both pecuniary, or transfer, effects—relief, welfare, and unemployment compensation payments—and real effects, such as the added crime and delinquency among uneducated youth. Both effects would be tempered and dropouts would constitute less of a social problem if we were successful in maintaining "full" employment.

On the other hand, one could argue that if there were fewer dropouts we would in fact be more successful in maintaining full employment. This is presumably the implicit argument of those who point to the high unemployment rates among dropouts as

[7] The question of why this is so is beyond the scope of this paper.
[8] Computed from data in Denis F. Johnston, "Educational Attainment of Workers, March 1962," *Monthly Labor Review*, Vol. 86 (May 1963), p. 507, Table 4.

evidence of the need to reduce the number of dropouts. The question is this: if there were fewer dropouts would unemployment rates among graduates be affected? In other words, to what extent, if any, would the employment of persons who would have been dropouts except for some prevention program (such as we discuss later) be at the expense of other workers? The usual argument against dropouts is presented on a micro basis and implies that any prospective dropout who changes his mind and completes school can expect the present employment and income prospects of the average high school graduate. (There is also the factor of differential ability, but this will be discussed later.) By implication, the unemployment among the rest of the labor force would be unaffected.

This represents a rather extreme, though conceivably correct, view of aggregate unemployment as essentially a "structural" phenomenon—and simply an educational structure phenomenon at that. Such a view seems to me to be distorted. This is not the place for a thorough analysis of the structural versus aggregate demand concepts of unemployment, but a few comments are in order. Consider the following data on unemployment rates by educational attainment, as of March 1962:[9]

Years of School Completed	Unemployment Rate
Less than high school graduation	8.1
High school graduation	5.1
Some college education	2.6
TOTAL	6.0

It hardly seems plausible that the nation's 6 percent unemployment rate could have been reduced to 2.6 percent if, by the wave of a wand, we could have provided everyone with at least some college education. Perhaps it could have been reduced somewhat, but surely not to such a remarkably low level. If not, then to some extent improved employment experience of the present dropouts would bring worsened experience for the present graduates. By the same token, it is unlikely that the mere completion of high school by the 46 percent of the labor force who were not

[9] U.S. Bureau of Labor Statistics, "Realities of the Job Market for the High School Dropout" (May 1963; mimeographed), p. 19.

already graduates would have brought a virtually full point reduction in the aggregate unemployment rate in March 1962, or an even greater reduction—1.4 percentage points—in the March 1959 unemployment rate (total unemployment—6.2 percent; high school graduates without additional schooling—4.8 percent[10]).

The fallacy of composition committed by assuming implicitly that a reduction in unemployment among dropouts will necessarily bring a corresponding reduction in aggregate unemployment takes various guises. In a recent government agency pamphlet imploring students not to quit school, a drawing shows a nongraduate and a graduate waiting before a door marked "JOBS"; an arm from behind the door is beckoning the pleased graduate, while the glum dropout is bypassed.[11] Moral: the graduate gets the job. But if *both* applicants were graduates, should we assume that the employer would hire both?

Even if a reduction of the number of dropouts reduced their unemployment *entirely* at the expense of others—leaving aggregate unemployment unaffected—there might still be an economic case for discouraging dropouts. One reason is that efforts at prevention partly represent the provision of information to young people, and better information regarding the consequences of decisions is certainly in the interest of efficient decision-making. A second reason is that society may try to protect the prospective dropout from himself by impressing him with the likelihood that he will have different preferences and a longer time horizon a few years later.

These reasons for attempting to prevent dropouts relate to assisting each individual to do what is in his own interest. But there may also be benefits to other people when dropouts are prevented—even if, to take the extreme case, aggregate unemployment is not affected. I submit that much of the concern about unem-

---

[10] See U.S. Bureau of Labor Statistics, "Factbook on the School Dropout in the World of Work" (no date, but probably 1963; mimeographed), p. 8; and BLS *Special Labor Force Reports*, No. 1, "Educational Attainment of Workers, 1959" (1960), p. A-7.

[11] "Handbook for Communities" (National Stay-in-School Campaign; GPO, 1958), pp. 10-11.

ployment and its consequences is really concern about the concentration of unemployment: in certain geographic areas, in certain racial groups, and in certain age groups. For example, if every member of the labor force were unemployed 6 percent of the year (three weeks), this unemployment rate would be far easier to bear than if 6 percent of the labor force were unemployed the entire year and the remaining 94 percent were never unemployed. Similarly, with respect to age and education: when unemployment is concentrated among young people, and particularly among the young dropouts, frustration and despair are likely consequences, from which individual delinquency, gang activity, and more serious crime are likely to spring. A broader distribution of unemployment by age, education, race, and geography would bring not only savings in social costs to others but also possibly savings in transfer payments, as the extreme effects of long-term unemployment tended to be replaced by the less severe financial and psychological effects of more equally distributed unemployment. The need for welfare assistance, unemployment compensation, and other payments to the needy would probably decline.

The discussion so far might seem to indicate that my concern for dropouts is only materialistic and disregards the nonmonetary significance of education. Certainly it is true that the acquisition of knowledge, development of the intellect, and the opportunity to lead the full life are products of education which have enormous worth even if not a dollar of extra income or of reduced cost of crime is involved. However, these nonmonetary considerations were presumably just as important thirty or forty years ago; thus the rising level of national concern over dropouts must be explained by other factors—either changes in attitude (preferences) or changes in circumstances. Certainly the high unemployment rates since 1957 have aggravated the dropout problem, although it might have existed in some degree in any case. Since unemployment tends to be greatest among the least educated, and since high school dropouts have recently become less frequent, the concentration of unemployment in the diminishing dropout group would become a "problem" even with aggregate unemployment below the recent 5 to 6 percent rates.

Returns from Dropout Prevention

What are the economic consequences of dropouts? How much, for example, does the failure to complete high school affect marginal labor productivity? To begin with, assume that income is a reasonable measure of marginal productivity (earnings, though still imperfect, would be better), and that high school dropouts are essentially the same as high school graduates—at least in the sense that the dropouts could be, if they remained in school, as productive as the graduates (this assumption will be considered later). Then, as a first approximation, we can estimate the incremental productivity loss attributable to dropouts by comparing the present values of lifetime income streams of dropouts and graduates. Since the incomes of graduates and nongraduates vary considerably among the eight groups for which data are available, separate estimates of lifetime income streams should be prepared for each.

Table 1 presents estimates of differential "lifetime" incomes between dropouts and graduates as of age 18 and age 16 for eight groupings—male and female, whites and nonwhites, and South and non-South. Results are presented at both 5 and 10 percent discount rates. The dollar figures in the table are simply the discounted lifetime income differentials between high school graduates who have not gone on to college and persons who are not graduates but did complete some high school. For each region-color-sex class the differential present values, V, were computed as follows:

$$(1) \qquad V_k = \sum_{n=k}^{75} \frac{(Y_n^g W_n^g - Y_n^d W_n^d) P_n}{(1 + r)^{n-k}},$$

where Y_n^g is income of graduates, g, who have income at age n, W_n^g is the proportion of graduates, age n, who have income; Y_n^d and W_n^d are the corresponding figures for dropouts, d; P_n is the probability that a person aged k will survive to age n at least; and r is the discount rate. Income is assumed to cease at age 75.

Table 1 shows, for example, that prevention of a white male

TABLE 1. Present Values of Differential Expected Lifetime Median Incomes;
High School Graduates vs. Dropouts, by Major Region, Race, and Sex,
at 5 and 10 Percent Discount Rates, 1949[a]

Region, Race, and Sex	Present Value of Differential at Age 18, at Discount Rate of		Present Value of Differential at Age 16, at Discount Rate of	
	5% (1)	10% (2)	5% (3)	10% (4)
Non-South				
White				
Male	$4,000	$1,740	$3,420	$1,180
Female	5,290	3,230	4,600	2,470
Nonwhite				
Male	2,160	1,040	1,750	650
Female	2,930	1,750	2,500	1,290
South				
White				
Male	7,160	3,060	6,240	2,280
Female	5,950	3,420	5,250	2,670
Nonwhite				
Male	2,250	1,010	1,820	610
Female	1,680	850	1,380	570

[a] Computed from data in U. S. Bureau of the Census, *1950 U. S. Census of Population*, Vol. 4, *Special Reports*, Part 5, Chap. B, "Education" (1953), Tables 12 and 13; and U. S. Public Health Service, National Office of Vital Statistics, *Vital Statistics Special Reports*, Vol. 41, No. 1, *United States Life Tables, 1949–51* (Nov. 23, 1954), pp. 10–11.

student's dropout in the non-South is "worth" $4,000 as of age 18,
if a 5 percent rate of discount is appropriate, or $1,740 if a 10
percent rate is justified. (We are assuming, for now, that if a
potential dropout did change his mind and complete school he
would have the same income experience as other graduates of
the same age, color, sex, and region.) It is striking to note that
at the 5 percent rate white dropouts of either sex are more costly
(that is, involve larger losses) in the South than in the North;
the reason is that income differentials between the North and
South narrow as the level of educational attainment rises. Hence,
while northern graduates and dropouts have larger incomes than
their southern counterparts, the expected monetary value of additional education is greater in the South.[12] It appears, therefore,

[12] This is true, not merely for high school, but for all levels of education through

that at this point in the analysis the following tentative conclusion may be put forth: if the costs of preventing dropouts of white persons are not much larger in the South than in the North, then a prevention dollar is a better financial investment in the South; this is, however, apparently not so for nonwhites.

A number of assumptions are, of course, implicit in the type of argument that uses such present-value figures. Perhaps the most important assumption is that marginal income prospects for an additional graduate will equal the average income for current graduates, even if the marginal graduate is of average quality. However, a significant increase in the relative supply of graduates—such as might accompany a massive dropout-prevention program—would tend to reduce their marginal productivity and incomes, while the opposite was happening to dropouts; thus income differentials would narrow. On the other hand, technological changes might increase the productivity of graduates more than that of dropouts; thus income differentials might widen.

The data in columns 1 and 2 of Table 1 are not suggestive of generally impressive financial returns from prevention of dropouts, even though we have not yet presented estimates of the costs of prevention and additional schooling. And the benefits from completing high school viewed as of age 16 (columns 3 and 4) are still more modest. Remaining in school involves opportunity costs; if a dropout leaves school at age 16 he *may* have two years of earnings while the other students are still in school. When the median incomes of dropouts at ages 16 and 17 are taken into account in calculating the present values of their expected lifetime income streams, the differential in favor of the graduate narrows. This is seen by comparing the figures in columns 1 and 2, which begin at age 18, with their counterparts in columns 3 and 4, which begin at age 16 and hence include the income of dropouts at 16 and 17. Still, the apparent payoff from prevention of high school dropouts was a minimum of $1,180 for white males and twice that for white females.

college, according to my computations (not shown here), from the 1950 Census income data. (It should be noted that the Census Bureau classifies people by place of residence, not by place of receipt of schooling.)

For nonwhites, however, the income payoffs were markedly lower in both regions. Thus, the high dropout rate among nonwhites is understandable, as an apparently realistic response to the lack of economic incentive.[13] Comparative white/nonwhite unemployment rates in October 1962 for persons who last attended school during 1960-62 also indicate the smaller payoff for nonwhites from completing high school. Unemployment rates among *dropouts* were very high, but fairly similar for whites and nonwhites: 19.8 and 21.3 percent, respectively. Among *graduates,* however, the picture was quite different: 10 percent for whites and 17.2 percent for nonwhites. Judging by these unemployment rates, the average nonwhite improved his position very little by completing high school, while the average white student could look forward to halving his unemployment probability by remaining in school.

The relatively large returns for females shown in Table 1, particularly in the non-South, is partly a consequence of the fact that the proportion of female graduates who received income was substantially higher than that of the female dropouts.[14] For males, this was true to a far smaller degree; for both sexes, however, the fact that larger proportions of graduates, compared to dropouts, had income tends to raise the differential income associated with the added education. We cannot be certain to what extent the differential proportions with income—which presumably reflect differential labor force participation rates—indicate (1) different preference patterns of those who graduate and those who drop out, or (2) different employment opportunities

[13] Among white male high school seniors in 1959, 15 percent did not graduate a year later, while the corresponding figure for nonwhite males was 24 percent—60 percent higher. See U.S. Bureau of Labor Statistics, "Factbook . . . ," (see footnote 10 above). That the Labor Department cites these data under the heading "Characteristics of the Dropout" implies that virtually all of the seniors who did not graduate actually left school.

[14] The data used were for 1949, and referred to proportions of persons with income. More recent data, however, indicate a continuation of the pattern with differential proportions of female graduates and dropouts being in the labor force (and receiving earnings). As of October 1962, 71.4 percent of the June 1962 female high school graduates were in the labor force, while only 34 percent of the 1962 dropouts were. See BLS, *Special Labor Force Reports,* No. 32, "Employment of High School Graduates and Dropouts in 1962" (1963), p. 4, Table 5.

for the two groups, as affected in part by the employment market in 1949.

Insofar as preferences are the explanation, the data in Table 1 would seem to give an erroneously high picture of the financial returns from completing high school, for they also include the return from additional effort of labor force participation. At the same time, this distinction is irrelevant if one is interested in the impact of dropout prevention on market output. Also, if employment opportunities are the explanation for the larger percentages of graduates, compared to dropouts, who had income —dropouts having greater difficulty finding jobs and therefore tending to leave the labor force—then the data in Table 1 are not biased by the effects of the differential percentages of persons with income.

The estimates in the table of financial returns from completing high school should be qualified in several respects.

1. They measure differentials in income rather than earnings. From the social viewpoint the effect of education on marginal product is important; this is more likely to be related to earnings than to income, some of which may be from inherited property. However, since only income *differentials* are presented they will approximate earnings differentials the more nearly the nonearnings component of income is equal for graduates and dropouts. My expectation is that the nonearnings component of income rises absolutely with education and as a result the observed income differentials tend to overstate the social value of completing high school on this ground alone.

2. The differentials, even if they were all earnings, probably measure more than the effect of education alone; we might expect some selectivity in those who complete high school, so that some of their additional earnings reflect their greater ability and perhaps other factors unconnected with education. That is, dropouts would tend to have had lower earnings than graduates, even if they had not dropped out. For example, most available evidence shows that dropouts do have lower IQ's than graduates. Although this may in part reflect different attitudes toward schooling rather than "inherent" intelligence, it suggests that dropouts would have lower incomes than graduates, even if the

two groups of students had the same amount of schooling. This is a second reason for believing that the data in Table 1 overstate the monetary value of preventing dropouts. The differences in ability nevertheless should not be exaggerated. The Bureau of Labor Statistics study of dropouts in seven communities between 1952 and 1957 disclosed the following distributions of IQ's, where "graduates" refers only to those who did not go on to college:[15]

IQ's of High School Graduates and Dropouts

	Under 85	85-89	90-109	110 and Over
Graduates	10%	11%	63%	16%
Dropouts	31%	15%	48%	6%

More than half (54 percent) of the dropouts had IQ's in the "average" (90-109) range or above. Estimating the mean of the "under 85" range at 80, and the mean of the "110 and over" class at 120 (for both graduates and dropouts), the weighted mean IQ for graduates was 99.8 and for dropouts 93.0. A study in New York State showed a larger percentage of dropouts with relatively high IQ's: 12.1 percent had over 110 and 30.4 percent below 90, compared with the 6 and 46 shown by the BLS study.[16] Moreover, IQ differentials may reflect quite imperfectly the productive potential of students, for IQ's measure achievement as well as "ability," and students who are poorly motivated to stay in school may achieve less than they would if a successful program of dropout prevention altered their motivations. Nevertheless, the differences in IQ's of graduates and dropouts do suggest that some downward revision of the data in Table 1 is called for.

3. There is still another reason for believing that the estimates in Table 1 overstate the productivity value of preventing dropouts. As pointed out earlier, a decrease in the number of dropouts would not be likely to reduce the aggregate unemployment in the economy to the full extent of the differential in employment rates of graduates and dropouts. Therefore, the social

[15] *From School to Work* (March 1960), p. 5.
[16] Lorne H. Woollatt, "Why Capable Students Drop Out of High School," *Bulletin of the National Association of Secondary School Principals*, November 1961, pp. 1-8.

value of preventing dropouts would be overstated if it were assumed that incomes (productivities) of all other workers were unchanged while those of the former dropouts was increased.

4. There is at least one reason, however, for believing that the
average monetary value of preventing a high school dropout is
actually somewhat *greater* than the estimates in Table 1. Let us
assume that the dropouts had the same ability and motivation,
on the average, as those who graduate from high school, and
that a successful dropout-prevention program led all the erstwhile dropouts to complete school. The majority of would-be
dropouts, who completed high school but who did not go on to
college, might be expected to receive income equal to that of
the other "graduates" who did not go to college; but some *would*
go on, and would tend to have still greater incomes. As a result,
the former dropouts would average larger incomes than the
"graduates."[17] The effect of this would be small, however, if we
may judge by the statistics referred to earlier—that only 6 to 12
percent of the dropouts had IQ's of 110 or more, for those with
lower IQ's are unlikely to have much success in college.[18]

Unfortunately, satisfactory data are not available to tell us how
large a net bias the above four factors introduce into the estimates in Table 1. Until our understanding of these matters increases, any adjustment made will be arbitrary; however, for the
purpose of comparing returns from dropout prevention with
costs, some deflation of the data in Table 1 seems called for, as
will be noted again later.

So far, the estimates presented here have been of only one form
of benefit from dropout prevention—incremental market productivity as reflected by income. For all groups studied, the present
values of additional income expected if high school is completed
are positive, and since the marginal private tuition cost of completing public school is negligible, the private profitability of not
dropping out is apparently clear. But when other private costs

[17] This is another application of the concept of the "value of the option," developed in my "Education and Investment in Human Capital," *Journal of Political
Economy*, Vol. 70 (October 1962), Supplement, pp. 106-123.

[18] "An IQ of 115 . . . is often used by educators in estimating ability to complete
college successfully . . ." (*From School to Work*, p. 5).

are considered—such as the dislike for school (or at least for the type of schooling offered) and the cost of school supplies, additional clothing, and school social and athletic events—even the private profitability may disappear. For example, a recent survey in Oregon found that the average high school student spent $238 a year on extracurricular activities alone, including proms, class jewelry, and transportation.[19] Even deflating this figure for the inflation since 1949, the result is not inconsequential relative to some of the figures in Table 1.

The direct social profitability will be even smaller than the private profitability, however, since the additional education of those would-be dropouts who remain in school is not costless to society. After all, we are not really considering the prevention of a single dropout. When large numbers of dropouts are prevented, significant increases in enrollments may occur, with the result that either average class size grows, or explicit costs for additional teachers, books, and perhaps even classrooms have to be incurred. The long-run marginal costs of education are difficult to measure; however, average costs per student in 1950 were $209 for current costs and $259 for total costs.[20] If long-run marginal cost is of this magnitude, completing high school would still seem to be a socially profitable investment, for the estimates above of the discounted additional returns to the graduate compared with those to the dropout generally exceed even several years of these costs. This assumes, of course, that the appropriate discount rate is not "too much" greater than the highest rate, 10 percent, used above, and that the dropouts can do as well in school as the average graduate who does not go on to college.

Nevertheless, students *are* leaving high school before graduation. Assuming rational (purposeful) behavior, this implies one or more of the following: (1) lack of information by students and parents; (2) preference patterns in which the psychic or consumption value of the types of school available is negative; (3) ability and motivational (preference) patterns such that special-

---

[19] Reported in NEA, *School Dropouts* (see footnote 6 above), p. 9. Such costs, to the extent that they are greater than they would be if the student were not in school, represent real costs of remaining in school. At the same time, there may also be benefits associated with these costs.

[20] *Health, Education, and Welfare Trends, 1962* (see footnote 1 above), p. 61.

ized training such as vocational education would be more suitable, but is unavailable; (4) high marginal rates of time preference. Dropout-prevention programs appear to attack these factors—by informing students of the employment and other difficulties dropouts are likely to confront, and of the added opportunities they would have as graduates; by attempting to alter preference patterns; by providing more vocational education and work-study programs, thereby permitting a wider variety of choices for students; and, in the process of doing these, by trying to extend the time horizon of frequently impatient youth.

The expansion of vocational education and work-study programs merits further comment. We pointed out earlier that quitting school might be an optimal private (whether or not social) decision for a person with certain preferences, income, and market opportunities (and the market opportunities include the varieties of schooling available). Thus, the effectiveness of dropout-prevention programs will depend in part on the success of the education system in adjusting its "products" to the wants of its "customers." The dropout problem may be symptomatic of a school problem, to be solved by a reorientation of school programs rather than by direct action toward the student. The student may elect to stay in school if the educational menu is more suitable to his tastes. A recent report on vocational education states:

> The scope of the typical high school program is narrow in relation to needs of the present day:
> (a) Rural schools have given little attention to the needs of students who migrate to urban centers.
> (b) Large high schools do not offer vocational programs in relation to probable need; only one-fifth of the students attend a school where trade and industrial education are offered.[21]

In this connection it should be remembered that the students and their parents are "buying" a complex commodity—a three-in-one commodity, including short-run consumption (enjoyment

[21] U.S. Office of Education, *Education for a Changing World of Work* (Report of the Panel of Consultants on Vocational Education, 1963), p. 157.

while in school), long-run consumption (enjoyment of the new quality of life which education permits), and long-run investment (employment and income opportunities). Presumably, the latter two factors have positive values, while the first factor may have either a positive or negative value.

But consideration only of the returns which the student obtains is incomplete in an analysis of the social dropout problem. The discussion which follows deals with the stake that other people have in dropout prevention.

External Returns

We have asserted that a dropout might be said to constitute a social problem if (1) dropping out resulted from incomplete information—e.g., on income, employment, and other prospects —which was available but not known to the student; or (2) there were external diseconomies from dropouts. In case 1, society might feel an obligation to provide students and parents with additional information. This may include the information that a student's marginal rate of time preference may be greater now than it will be in the future; as a result, he may later regret having left school, even if he was correct in forecasting the results. Compulsory school attendance reduces the likelihood of such "errors." Dissemination of estimates of the present value of the lifetime income apparently sacrificed by a decision to leave high school is also part of a full information program (but the provision of data on undiscounted lifetime income differentials is not, although it may be justified on other grounds). In case 2, even if the individual has complete information, society may wish to constrain his exercise of free choice, since a student, to the extent that he is a maximizer of his own welfare, will tend to disregard the external benefits of completing high school.

Real Returns

As noted earlier, the presence and significance of external benefits depend partly on whether or not the level of education of the population affects the aggregate level of unemployment in the economy. However, the magnitude of social benefits from dropout prevention may also depend on the distribution of the unemployment. There may be external social benefits—e.g., less crime and delinquency—even if the unemployment now concentrated among those with less than a high school education were simply redistributed among the very large number of high school graduates. The returns in the form of reduced crime and delinquency are extremely difficult to quantify. In principle, the required information is (1) the additional crime and delinquency resulting from additional dropouts, and (2) the social cost of the additional crime and delinquency.

There are other real external benefits from dropout prevention which deserve mention, even though quantitative analysis is not feasible. Greater educational attainment appears to encourage greater participation in civic and philanthropic activities, as well as in democratic political processes, including voting.[22] Insofar as education is the cause of this additional participation and insofar as the results are considered to be socially desirable, education brings returns in these nonmonetary forms. To what extent completing high school has these effects is conjectural, but we should not lose sight of program benefits simply because we are unable to place a monetary measure on their value.

When a student drops out of school he is taking a step that will affect the education decisions of other students. There are both long-run and short-run time dimensions to this type of externality.

In the long-run dimension, a student who drops out is likely to pass on to his or her children attitudes (not to mention financial problems) which discourage them from educational pursuits. In a 1960-61 statewide study in Maryland, it was found that 70 percent of the mothers and 80 percent of the fathers of dropouts were themselves dropouts.[23] In recent studies in New York and in Louisiana, the findings were that two thirds of the parents

---

[22] On the first point see Charles A. Benson, *The Economics of Public Education* (Houghton Mifflin, 1961), p. 349. On the latter points, see V. O. Key, Jr., *Public Opinion and American Democracy* (Knopf, 1961), pp. 324-325 and 564-565.

[23] Daniel Schreiber, "The Dropout and the Delinquent: Promising Practices Gleaned from a Year of Study," *Phi Delta Kappan*, February 1963, p. 217.

of dropouts held "negative or indifferent attitudes toward the value of education"—feeling that the lack of a high school education constituted no serious obstacle to their children's later adjustment or success.[24] If the prevention of even one dropout changes such attitudes, the effects on the next generation may be beneficial.

The short-run dimension may involve a number of students in a given class who are marginal about leaving school. The course of action each one of these selects may depend on the behavior of other students. If one drops out, he may trigger others to do so. In this sense, too, dropout prevention brings external returns.

Another external benefit of dropout prevention is the saving in costs of administering the various programs of transfer payments, made necessary partly by the low incomes accompanying low educational attainments. Last, as a result of the improvement in the quality of the labor supply that dropout prevention would bring, we would expect not only the marginal productivity and income of the labor directly involved to rise, but also the marginal productivity of cooperating resources. Thus, the increase in aggregate output would tend to exceed the increase in income of workers directly affected.

All of these real external benefits of prevention are properly additive to the productivity benefits previously estimated, in determining the social returns from prevention.

Pecuniary Returns—Distributional Effects

It is common in benefit-cost analyses to consider only the real effects of the project, while disregarding pecuniary or distributional effects. Evaluation of the latter involves normative judgments, which, *qua* economists, we may not be willing to make. Moreover, the distributional effects can be dealt with separately in the real world. Unwanted side effects of a socially profitable expenditure can be overcome through appropriate taxes and subsidies —in principle. But the practical difficulties in so dealing with them may be too severe.

Thus I suggest that society, to make intelligent expenditure decisions, needs to be informed, not only on the allocative effi-

---

[24] *Ibid.*

ciency of its expenditures, but also on the distributive effects which constitute another dimension of a project's impact. Redistributed sums, of course, are *not* additive to real returns.

There are two forms of external distributive effects of dropout prevention which I shall consider: reduced welfare payments to dropouts, and increased revenue from them.

If unemployment can be reduced by raising educational attainments—or even if the additional education merely redistributes unemployment so that long-term unemployment of some workers is replaced by short-term unemployment for a larger number—welfare requirements are likely to fall as the severe financial effects of chronic unemployment diminish. With the average length of unemployment per unemployed person falling in both cases, the demands upon unemployment compensation funds will also tend to fall. If unemployment were distributed more equally, the usual one-week waiting periods would tend to reduce compensation costs even if aggregate unemployment were constant. On the other hand, unemployment compensation costs could conceivably increase if there were no reduction in unemployment but solely a diminution of the average term, for unemployment benefits terminate after specified numbers of weeks. If, as seems likely, a reduction in the number of dropouts would somewhat diminish both total unemployment and its mean duration, then persons other than students—employers, employees, and consumers—would benefit from the diminished transfer payments to which they were contributing.

A first approximation of the reduction in unemployment compensation transfer payments resulting from a successful dropout-prevention program could be made as follows:

Assume that if a successful program can be launched, the unemployment rate for the students affected would fall by 3 percentage points (the difference shown in the tabulation on p. 122 between the unemployment rates for graduates and dropouts, 5.1 and 8.1 percent), while unemployment rates for other persons would not be affected. (I stated earlier that this probably overstates the ability of education to reduce aggregate unemployment, but it should prove instructive to estimate the impact on unemployment compensation payments of this sanguine possibility.) Assume also that unemployment insurance coverage is complete, and that compensation payments are a linear function of the number of weeks of unemployment (thus disregarding the cut-off point and the waiting period). Then, for each dropout, i, that is prevented the "expected" annual saving in unemployment compensation, S_i, may be estimated as

$$(2) \qquad\qquad S_i = W_i(U^d - U^g)C,$$

where U^d and U^g are unemployment rates of dropouts and graduates, respectively, W is the number of weeks annually that the person is in the labor force, and C is the weekly compensation payment. Thus, for example, if a male student is encouraged not to drop out and if he may be expected to be in the labor force for the entire year once he leaves school, $W_i = 52$. The U^d, U^g, and C have no subscripts, as the unemployment rates and weekly compensation payment are assumed to apply to all persons.[25] In our concluding section, below, this very rough and biased (high) estimation procedure will be used in valuing the pecuniary external returns from a specific program of dropout prevention. While this estimation overstates the reduction in unemployment compensation payments, it neglects other forms of transfer payments associated with low educational attainment and low income.

A successful dropout-prevention program will bring another sort of income transfer. Through the tax system, society as a whole will share with the student some of the added income he will receive as a result of completing high school. As a person's income rises, his tax liabilities rise, but the cost of the public services he consumes may well rise more slowly, if it rises at all— and it may even fall. It is not at all clear that higher-income persons consume any more public services—e.g., they tend to consume *less* public school services because they more frequently use private schools. If it is assumed that there is no increase in consumption of public services with respect to additional education and income, then all of the extra taxes paid out of the additional income which high school graduates obtain is a transfer

[25] In reality U^d and U^g do vary with respect to age of worker, and so a rigorous formulation would take this into account.

benefit to others who can, as a result, either pay lower taxes or have more public services provided.

It is true that all of these income transfers exist because of welfare and tax laws; consequently, the amounts of transfer can be affected by altering the laws. Nevertheless, as a society, we have chosen to develop a fiscal system under which the financial well-being of each of us is interdependent with the well-being of the rest. In such a context, the case for public attention to dropouts is different from what it might otherwise be.

Let us say that when a student drops out of high school he sacrifices future productivity having a present value of $3,000, of which T is the present value of the additional taxes he would pay. Furthermore, as a result of his dropping out, he is more prone to unemployment and low income, so that society will make additional welfare and unemployment compensation payments to him having a present value of W. To be sure, the direct social benefit from preventing this dropout is only $3,000. Yet I suggest that it is not entirely irrelevant to our evaluation of dropout-prevention programs to recognize that of that sum, $3,000-$T$-$W$ will accrue to the direct beneficiary of the expenditure (the student), while $T + W$ will accrue to some or all of the rest of us—to other members of society.[26] With further study, we might even be able to say something about who these "transfer beneficiaries" are. And this would represent a step toward saying how satisfied we are with the distribution of benefits from this public expenditure.

Costs and Benefits of Dropout Prevention: A Case Study

In this section I shall try to evaluate the returns from and costs of a specific program designed to reduce the number of high school dropouts. The goal is to determine whether, on the basis of available information and knowledge of measurement techniques, we can say that this program is or is not justified on allocative-efficiency grounds.

[26] Robert Dorfman suggested this useful formulation of the argument.

Determining the average cost, let alone the marginal cost, of preventing a dropout is complicated. Even when expenditure data are available, the question of how many dropouts were actually prevented by the expenditures is generally speculative. The specific program to be considered here simplified this problem by including a controlled experiment from which differences between the numbers of dropouts in a control group and an experimental group could be determined.[27] The study was conducted as follows:

1. A group of "potential dropouts" was selected, and randomly divided into two groups. All of the students were at least 16 years of age, and had an IQ of at least 80.

2. The control group received normal school services.

3. The experimental group received normal school services and also "special counseling services, assistance in getting placed on jobs and remaining on jobs, and special assistance on the job from employer and school personnel." For the 429 students originally in this work-study group, 7 additional full-time counselors-coordinators were employed; in addition, other counselors gave added attention to those students.

4. School personnel received assistance from labor groups, employers, the Chamber of Commerce, community agencies, and the State Office of Employment Security.

Relevant characteristics of the two groups, including age (17.32 years for the "work," or experimental, group, and 17.28 for the control group) and IQ (93.99 versus 93.57), did not vary to a statistically significant degree.[28]

By the end of two years, 189 of the original 429 in the experimental group, or 44.1 percent, had dropped out, while the dropout percentage for the control group was 52.0 (200 out of 385). This 7.9 point difference is significant at better than the 5 percent level ($z=2.3$). Thus, it would seem reasonable to assume

[27] *The School and Community Work-Related Education Program: A Ford Foundation Project, Activity and Progress Report, 1961-62* (Shaw School, St. Louis, n.d.; mimeographed), p. 30.

[28] *Ibid.*, pp. 9, 17; see also pp. 4-21.

that the extra counseling and other services provided to the experimental group brought about the better dropout experience.

Benefits

Using the concepts discussed in the previous section, I will attempt to evaluate (1) the benefits from the prevention program of the St. Louis Public Schools and (2) the costs of the program.

The distribution by sex and race of the students in the experimental group is shown in Table 2, column 1. In the absence of additional information, I shall assume that this was also the distribution of the students who were successfully "treated"—that is, prevented from dropping out (column 2). Then, referring back to Table 1 above, one can estimate the present values of additional lifetime income which the students may expect—under the assumptions (1) that the differential retention rate of 8 percent which existed between the experimental and control groups after two years, continued (thus, for the total of 429 students in the experimental group, 34 dropouts were prevented); and (2) that those 34 were, on the average, as capable, energetic, and ambitious as other students who graduate from high school but do not go on to college.

Rather than estimate the returns from preventing thirty-four dropouts, I shall assess the average return per dropout. For illustrative purposes, I use the present-value figures from Table 1 at the 5 percent discount rate. Data for the non-South would seem

TABLE 2. Estimated Distribution of Dropouts Prevented, by Sex and Race, St. Louis, 1960–62[a]

Sex and Race	Percent (1)	Number of Students (2)
Male, White	30.0	10
Male, Nonwhite	33.8	11
Female, White	10.5	4
Female, Nonwhite	25.7	9
	100.0	34

[a] Percentages in column 1 from *The School and Community Work-Related Education Program* (Shaw School, St. Louis n.d.; mimeographed), p. 6.

TABLE 3. Computation of Present Value of Additional Lifetime Income per Average Dropout Prevented[a]

Race and Sex	Percentage of Dropouts Prevented	Present Value per Dropout Prevented (5% Discount Rate)	Weighted Present Value
White Male	30.0	$3,420	$1,026
White Female	10.5	4,600	483
Nonwhite Male	33.8	1,750	592
Nonwhite Female	25.7	2,500	642
Average	100.0		$2,743

[a] *Source: Data in Tables 1 and 2.*

to be most appropriate for St. Louis. The data for age 16 will be used, the modal age at which dropouts occur. Table 3 combines data from Tables 1 and 2 to produce the estimate for the average return in the form of capitalized additional lifetime income per dropout prevented. That return exceeds $2,700.

In the first instance, at least, this return from dropout prevention will accrue to the individual student. What financial stake does the remainder of society have? As discussed above, we can identify returns to the rest of society in a number of forms, some of which represent real social benefits, while others are transfers. Unable to estimate the quantitative impact of dropout prevention on crime and delinquency, and on the other forms of real benefits, I shall move on to the estimation of additional tax revenue and of savings in transfer payments.

In 1950, general revenues of federal, state, and local governments were roughly 25 percent of total personal income.[29] Continuation of this rate would imply that (insofar as the aggregate tax structure is approximately proportional with respect to income so that the marginal and average tax rates are equal) some 25 percent of the additional lifetime income generated by education will go for taxes. Thus, the prevention of a dropout would produce additional tax revenue of 25 percent of $2,743 (Table 3), or $684, in present-value terms. This is an estimate of mone-

[29] Computed from *Federal Reserve Bulletin*, June 1963, p. 853; and *Health, Education, and Welfare Trends (1962)*, p. 100 (see footnote 1 above).

tary returns to the rest of society under the assumption that, as income and education increase, the consumption of public services remains constant.

With respect to reductions in transfer payments, only the reductions in unemployment compensation will be estimated. In this illustration, it will be assumed that prospective dropouts who can be turned into graduates will have the currently observed lower unemployment rates of graduates, and thus will be less of a burden on the unemployment compensation fund.

According to the BLS data in the tabulation on page 122, the unemployment rate for high school graduates was 3 percentage points below that for dropouts. Let us assume that the average prospective dropout would spend only thirty-five weeks per year in the labor force for forty years of working life. The 3 percent differential unemployment rate for those weeks per year equals 1.05 weeks less unemployment per worker-year as a result of completing high school. In 1950, average weekly unemployment compensation payments were $20.76.[30] Thus the additional week of unemployment would mean an additional $22 per year in unemployment compensation, if coverage were complete. For a forty-year working life period this would amount to a capitalized value of $380 at a 5 percent discount rate. In addition, there would probably be reductions in other forms of transfer payments—such as welfare assistance—if dropouts could be prevented and especially if unemployment were reduced.

If the profitability of devoting resources to dropout prevention were assessed from the viewpoint of the rest of society, then the present value of all returns would probably be at least as great as $684 (additional tax payments) plus $380 (unemployment compensation savings), or a "total" of $1,064 per case.

Other returns discussed above are also important. They may defy quantification, but they should not be omitted from a full analysis of the real "productivity" and distributional effects of expenditures on dropout prevention or on other education programs.

[30] *Health, Education, and Welfare Indicators* (September 1963), p. 39.

Costs

In evaluating the success of a prevention program, it should be noted that comparisons of costs with the total number of potential dropouts who return to school—say, after a summer—grossly understate costs of prevention by taking credit for retaining many students who would have returned even without special treatment. The data presented above show that 48 percent of the potential dropouts in the control group were still in school two years later, *without* the special attention accorded the experimental group; 56 percent of the potential dropouts in the experimental group, *with* the special attention, were still in school. Thus 48/56, or 86 percent of the potential dropouts in the experimental group who remained in school would presumably have done so anyway. Moreover, many students dropped out even with the special attention. In short, only 8 percent behaved differently by virtue of the prevention program.

The actual resource cost per dropout *prevented* was by no means trivial. Over the two-year period, the average cost (expenditure) per student in the experimental group was $580.[31] This seems modest by comparison with the returns previously estimated in Table 1. But when account is taken of the vast majority of students (92 percent) who either dropped out in spite of the program, or seemingly would have remained in school without it, the expenditure *per dropout prevented* soars to over $7,300. And even this is not the full cost of obtaining the returns in Table 1: there are also the "normal" average costs of educating students who have been retained—for the two-year period in the St. Louis high schools this cost was $910. Adding this to the prevention cost gives a total cost estimate of more than $8,200 per "success." The figure is, however, in 1960-62 prices; thus it will be necessary to deflate it for comparability with the estimate of returns presented above. Since an index of prices of education services is not available, I shall arbitrarily use the Consumer Price Index (CPI) without attempting any real justification. Between 1949 and 1960 this index rose from 83.0 to 104.2.[32] Thus

[31] Private correspondence of March 1, 1963.
[32] *Federal Reserve Bulletin*, June 1963, p. 844.

the $8,200 cost in 1960-62 prices is equivalent to about $6,500 in 1949 prices. Still, with a cost of $6,500 per dropout actually prevented, the program is seen to cost more than the *measured* social benefits of $2,740.

What can be concluded from this rather startling finding regarding the unprofitability, in narrow financial terms, of one program of dropout prevention?

First, I would certainly not dismiss the St. Louis experience summarily as a freak. The student body is fairly similar to that of many large cities. The IQ's of the students in the program were very close to those reported in the Bureau of Labor Statistics' 1960 study in seven communities.[33] The St. Louis school system is generally regarded highly among other urban-center school systems. Furthermore, the experiment appears to have been organized carefully. The substantial financial support from the Ford Foundation implies that the Foundation found the program well worthy of attention.

Second, results of this or any other experiment should not be generalized to all efforts at dropout prevention. A variety of approaches are being tried around the nation. Thus, additional economic analyses are required of the costs and benefits of other programs before any sweeping conclusion that dropout prevention is "unprofitable," even in the narrow financial sense, is justified. We need to know much more about the production function for dropout prevention. Some of the production techniques of the other programs may be more effective, but it is extremely lamentable that most of the other programs have not been designed as controlled experiments which permit estimates of the number of dropouts actually prevented. Of more than fifty programs studied, the St. Louis program was the only one which met this qualification.[34]

Third, the program did indicate that prevention can work. And it should be noted again that all of its benefits could not be

[33] *From School to Work* (see footnote 5 above), p. 5.

[34] But even this case illustrates the difficulties of performing a strictly controlled experiment. The students knew whether they were in the control group or the experimental group, as did their parents. This knowledge could have affected the results—in either direction: for example, there might have been a "Hawthorne effect."

submitted to evaluation. The St. Louis program actually had, not
merely the one goal of prevention, but two other goals: to improve the achievement of potential dropouts, and to build their
self-esteem. I have cited no evidence concerning these two latter
goals and have made no evaluation of their importance, but the
value of the program in regard to them was certainly not zero.
Moreover, because education both shapes students' social attitudes (which tend to be transmitted from one generation to another) and contributes to effective democracy and the good life,
society may wish to prevent dropouts even if no monetary returns at all can be found. However, a complete analysis would
still include information on productivity effects. This is particularly true in the context of the public-sector resource allocation
problem, which may involve competing projects that all have
important nonmonetary return components.

Fourth, even the monetary returns of the St. Louis program
would have been larger if the fraction of the group which was
nonwhite had been smaller. Table 3 shows that 59.5 percent were
nonwhite. By comparison, only 26.3 percent of the national dropouts in 1962 were nonwhite; at the same time, however, the percentage of the dropouts who were females, 36.2, was lower than the
national average of 55.8 percent.[35] And since, according to Table 1,
the returns in the non-South are higher for females than for males,
the estimate of average returns in the St. Louis group would have
been higher if the proportion that was female had been larger.
Neither the finding of the "unusually" high proportion of nonwhites nor the low proportion of females is a reason to consider
the St. Louis dropout program grossly atypical of large city experience. In this connection, it should be noted that 21 percent
of the nation's 1962 dropouts were in farm areas, where the percentage of nonwhites is below that in urban areas.[36]

Fifth, there are factors leading us to believe that the returns
estimated above may well be too high. (1) Returns would have
been lower had it not been assumed that prevention of a dropout

[35] Jacob Schiffman, "Employment of High School Graduates and Dropouts in
1962," *Monthly Labor Review*, Vol. 86 (July 1963), p. 775, Table 5.
[36] *Ibid.*

would cause the unemployment rate to fall to the full extent of the differential rates for dropouts and graduates. (This assumption was seriously questioned in the first section of this paper.) (2) Returns would also have been lower had it not been assumed implicitly that the 8 percent dropout-rate advantage that existed for the experimental group would continue after two years. Actually, the advantage was substantially greater in the first year of the program than in the second, because extra services were diminished in the second year. This suggests that, had more time been allowed before the success of the two-year extra expenditures was evaluated, the advantage might well have been less than 8 percent.

Sixth, program *costs* may have been underestimated. This is true, for example, insofar as the retention of basically maladjusted students has adverse effects on other students.[37]

SUBJECT TO THE MANY QUALIFICATIONS which have been called for, Table 4 summarizes the qualitative and quantitative findings of this paper and indicates the many quantitative gaps in knowledge which remain. The particular prevention program studied here was found to be "unprofitable"—in terms of *measured* benefits and costs—even before benefits were deflated for the effect of such noneducation factors as ability and ambition, which affect income differentials between dropouts and graduates. For this as well as other reasons previously discussed, any net bias in the estimates of monetary benefits is likely to be upward.

The data presented here do not permit strong conclusions to be drawn, but they do suggest that any program which does not cope with the dropout hazard before students are 16 years old may be "too late." Prevention seems to be difficult at that stage; even when extensive counseling and work-study programs are

---

[37] This point came out in a conversation with Myron Joseph. Two recent newspaper articles argued that such students are actually bringing, in effect, external diseconomies for their fellow students. See Bill Gold, "The District Line," *Washington Post*, Sept. 18, 1963, p. D12; and Maurine Hoffman, "Dropout Drive Returns Unruly," *Washington Post*, Sept. 17, 1963, p. B1.

TABLE 4. Summary of Benefits and Costs per High School Dropout Prevented: St. Louis Program[a]

(+ *indicates an unmeasured quantity that enters positively;* * *indicates an unmeasured quantity that enters negatively;* ** *indicates an unmeasured quantity that is not commensurable with the preceding costs and benefits.*)

Resource Costs per Dropout Prevented		
Direct Prevention Costs..		$5,815
Additional Instruction Costs...		725
Total Resource Costs...		$6,540
Internal Benefits per Dropout Prevented		
Increased Present Value of Lifetime Income (unadjusted).............	$2,750	
Minus Adjustment for Effects of Noneducational Factor..............	*	2,750[b]
Improved Self-Esteem of Student..		+
External Benefits per Dropout Prevented		
Increased Productivity of Cooperating Resources...........................		+
Increased Social and Political Consciousness and Participation................		+
Decreased Social Costs (e.g., of crime and delinquency).....................		+
Decreased Social Costs of Administering Transfer-Payment Programs (e.g., of public assistance)..		+
Intergeneration Benefits...		+
Total Costs (per dropout prevented) Not Covered by Measured Benefits..........		$3,800
Distributional Effects..		**

[a] This table is patterned basically on one in Jerome Rothenberg's paper, which appears later in this volume.
[b] Overstated to the extent that present value of lifetime incomes was not adjusted for effects of the non-educational factors.

tried, attitudes and motivations may be too solidified. Dropping out of school is, after all, symptomatic of other problems.

I do not intend to suggest that the dropout problem is unimportant. However, existing treatment techniques are very costly. Unless more efficient methods can be devised for dealing at the high school level with dropouts (e.g., through expanded vocational education opportunities) we ought very probably to place greater reliance on other programs—programs that could be devised for younger students or that would concentrate on ameliorating the problems of dropouts (for example, by attracting them into night-school courses) rather than only on preventing the act of leaving school.[38]

[38] This paper has not explored the case for simply extending the compulsory school-attendance age.

The relatively high cost of dropout prevention observed in our case study raises questions about launching a nationwide program. Three points may be made briefly here:

1. In time, as more is learned through experience and experiment about effective dropout prevention techniques, costs per dropout prevented should tend to fall.

2. On the other hand, attempts to reduce the dropout rate further and further will lead to rising costs, insofar as increasingly difficult cases will be encountered. (Note that the St. Louis program was limited to students with IQ's of 80 or more.)

3. Expanded efforts at prevention will increase demands for certain types of personnel, with the possible result of rising factor prices.

One thing is clear. We are now essentially ignorant about the production function for dropout prevention, and we will remain so until there occurs a sharp increase in the number of prevention programs with experimental designs that shed light on the factors determining their success or failure. It is amazing how many attempts are being made around the nation to prevent dropouts, and how few are designed so as to permit assessment of their effectiveness. Without such information, benefit-cost analysis in this area is seriously handicapped.

Comments

FRITZ MACHLUP, *Department of Economics, Princeton University*

Before I address myself to the topic of school dropouts, I want first to make some generalizations about benefit-cost comparisons. For over a century and a half, people have debated about the meaning of the pleasure-and-pain calculus of individual decision-making. I hope it will not take quite as long for us to make up our minds what we mean by the benefit-and-cost calculus of governmental decision-making.

Five Types of Benefit-Cost Comparisons

Since there is a full spectrum of meanings, ranging from an implicit sentiment about the worthwhileness of certain favorite projects to an explicit calculation based on statistical data, we must do what is usually done when a spectrum is analyzed or examined: we must separate it arbitrarily into distinct bands. I propose, then, quite arbitrarily, five different types of what may be meant by benefit-cost comparisons.

Type No. 1 is the implicit sentiment, sans analysis, in favor of the particular project, the worthwhileness of the project or program being taken entirely for granted. Everyone with a favorite project will, when asked, assert that "of course" the benefits from it are greater than the costs. If you put the question to the legislators in Congress who vote *for* certain river projects and *against* increased aid to education, they will certainly say that these river projects come out much better in a benefit-cost comparison than further aid to education—even if they have neither theoretical arguments nor empirical data to support such contentions.

Type No. 2 may be seen in explicit propositions about the *value* of the program, in which you impose your own value judgments upon the people of the commonwealth. You "know" what is best for society, though you do not care to say why, or to defer or refer to the opinions of others.

Type No. 3 is more than an imposition of your individual value judgments; it is an assumption that you and others know what society wants, what kinds of benefit it may derive from the program, and how big these benefits are. There is still no analysis, but the assumption appears to be moving closer to matters that deserve to be taken seriously.

Type No. 4 is based on theoretical analysis, where you at least state what kind of empirical information you would need to make a calculation or an estimate of the benefits and costs: data on prices and incomes, and on all sorts of actual and potential outputs, individual valuations, willingness to pay, demand and supply elasticities, future increases in the labor force, future changes in technology, and so on. If you knew all these, then you would be able to tell us just how the relevant benefits and costs could be calculated.

Type No. 5 is the actual calculation of benefits and costs in which all or much of the required information is available and utilized. Since there are always some relevant items for the estimation of benefits or costs that are not measurable, due allowance for these items must somehow be made. Some opportunity costs, for example, may not be available in any usable statistical form, and you will have to rely on rather speculative chains of reasoning leading to some estimates, however vague. After you make a good case for the numerical values chosen, you will put them into your calculation.

Society and Special-Interest Groups

In the light of the above, I will classify a research project in which I was engaged about ten years ago, involving benefit-and-cost analysis, as Type 4. I had no actual data and merely attempted to state just what data I would have to have for an evaluation of benefits and costs. In any case, I wrote some 200 pages of a draft on the benefits and costs of the patent system. The draft is sitting in my bookcase, and I do not know whether I shall ever use it. At the time I certainly lacked the courage to publish, feeling sure that everybody would say: "What of it? You don't have the data. Why make all that fuss?"

In that project I did one thing which I am glad to find has also been done by Burton Weisbrod: I distinguished "society as a whole" from "the rest of society apart from a particular group." If a program or measure is definitely to the benefit of a special group within society, then it is perfectly proper to ask whether the rest of society shall accept this program or measure though the entire benefit from it accrues only to that group, only to that particular part of society.

You may say that this consideration violates the accepted standards of welfare economics; that we should consider only allocative efficiency, and not questions of distributive justice or expedience. On the other hand, in democratic decision-making it is perfectly proper to ask why society should accept a system (say, the patent system) if it helps chiefly one group (e.g., the inventors and patent owners). To be sure, the one group is part of society. But society has a right to ascertain how much of the total benefit goes to a particular group and how much goes to the rest.

I am glad to find that this point is made and the issue pursued in at least a part of Weisbrod's paper. He argues that, if most of the entire benefit, or even all, goes to the prevented dropouts and not to the rest of society, there is a debatable question whether society wants to engage in such a program. It may be held that the rest of society should want at least to have some "cut" of the benefits—that, in other words, there should be truly external benefits accruing to the rest of society.

Assume, for example, that a large fraction of the population prefers leisure and wants to work only fifteen or twenty hours a week. Needless to say, if the rest of society could take some action to induce them to work more, there would be more production. Perhaps, however, those who prefer leisure would consume all the additional production themselves. In this case, would the rest of society find it desirable to take such a measure and bear its cost? I submit that the problem that Mr. Weisbrod is dealing with has some aspects of this example.

Measurable and Nonmeasurable Items

Concerning the types of benefit-and-cost analysis distinguished above, Dr. Weisbrod's study, fortunately, is not confined to a merely formal statement of what we would have to know in order to make a calculation. He has statistical data, uses some real numbers. Needless to say, none of our problems ought ever to be solved on the basis of only those variables for which we have the numbers. There is some danger, moreover, in cases where some of our relevant factors are measurable and are actually measured or estimated, that other factors which are not measurable will be disregarded, even when their importance is not questioned. Whether, and to what extent, measurable and nonmeasurable items should be mixed in our analysis is a vexing problem.

Whenever we can say that on the basis of numerical estimates alone a particular project looks very good and that the non-measurable factors also point in the same direction, then we can be happy and satisfied. This seems to be the case in the particular problem studied by Weisbrod.

Reallocation of Funds

Another issue requires our attention. We must distinguish between allocation of funds within a particular program—for example, education as a whole—and allocation of funds among different programs—say, education, research, health, law enforcement, highways, and all the rest. Assume, for instance, that we come upon an educational project that would have a payoff of 50 percent per annum. Wonderful! Does this in itself indicate that we should transfer funds from other areas to education? Not necessarily, because the educational basket may contain programs that have a zero return or perhaps even a negative return, in which case it may very well be demanded that we first put our house—our educational establishment—in order, and use the funds already allocated to education for projects that have a better payoff. Reallocations within the area of education deserve priority over a transfer of funds from quite different areas.

Complementary Employment

I now come to a few minor points concerning Weisbrod's paper. There is, first, an assumption about re-employment and unemployment. Two possibilities are discussed: one, that the prevented dropouts will not be unemployed and that total employment will increase as a result; the other, that total employment will not increase by the number of prevented dropouts, for some of these may take the place of other persons, now employed. I should like to propose a third possibility: namely, that total employment will increase by more than the employment of the prevented dropouts. This possibility exists because of certain complementarities in skills.

What I have in mind can more readily be seen by illustrating a change in the opposite direction. If I were to select, from all of

the 70 million people employed in the United States, a small
fraction, say 70,000, or one tenth of 1 percent, and remove them
from the labor force, what effects upon total employment could
be expected? Assume I can make a fiendishly clever selection of
persons in key positions and ship them abroad. By removing
70,000 persons of very special talent or training, I might be able
to reduce employment by 10 million, for there are certain key
skills which must be available so that other people can be employed. Now I propose (although it is not a great likelihood in
the case of prevented dropouts) that possibly some of those induced to finish high school may discover what their real abilities
are, go on to college and even graduate school, and become great
leaders who, in the wake of their extraordinary activities, employ
millions of people.

Concentrated Schooling Before Dropout Age

When Weisbrod discusses the dropouts between ages 16 and 18,
some people may wonder why this problem is so serious. It is
serious because there is still a lot to be learned in the eleventh
and twelfth grades of high school. I come now to what has been
my long-time favorite scheme: I believe the curriculum of primary and secondary schools ought to be compressed so that those
who drop out of school at age 16 will have already learned what
pupils are now permitted to learn by age 18. I submit, moreover,
that in a concentrated course of study this could be learned
much better than it is now.

We need not fear that the dropouts want to avoid the last two
grades of school, no matter how old they are. The dropout age is
related to physiological maturity; hence there is no danger that too
many would drop out at ages before 15. Thus, if the curriculum
were compressed, we would not have the losses to productivity that
we now suffer as a result of the dropout problem.

Frederic Scherer spoke earlier today of the time-cost tradeoff. I
submit that something of that sort also exists in education, though
the relationship is probably inverse: in education, to do it faster
would, I believe, mean to do it cheaper as well as better. This holds,
of course, only over a certain range of schooling, since we surely

cannot compress the education of our young into a period, say, of only five years. But for the range between nine and twelve years, my hunch is that we could avoid considerable loss, and gain a great deal of productivity, by trying to educate people faster, less expensively, and better.

Economic and Noneconomic Values

Since this conference is chiefly concerned with the general problem of cost-benefit analysis, not just with the specific programs to which such analysis is applied, I shall conclude my remarks by returning to the general theme, and submit additional comments of a general nature.

I have noticed a tendency—less among the theorists than among the practical experts—to make a distinction between economic and noneconomic values. This distinction has been made both in the evaluation of benefits and in the evaluation of costs or opportunities forgone. I submit that this is a mistake. We must not put "economic values" to one side and other values—say, ethical, aesthetic, social, political, or "human"—to the other side. The economic valuation of benefits and costs of an institution, plan, or activity must attempt to take account of values of any sort and to apply reasoned argument and rational weighting to problems commonly approached only by visceral emoting.

We shall still have to distinguish between pecuniary and non-pecuniary advantages and disadvantages, and between judgments that rest on statistical records and others that are purely subjective evaluations without any supporting numerical data. But the point to note is that economic evaluation is not confined to the items for which price data are available. It comprises all pros and cons of the plan or activity under examination.

Social and Private Decisions

Cost-benefit analysis is not only used for governmental decision-making; it is applicable, and indeed regularly applied, to private decisions, in business as well as in personal affairs. When one of us considers changing his job—from one university to another, or from an academic to a government setting—he compares pecuniary and nonpecuniary advantages and disadvantages. In his list of pros and cons will appear not only the immediate salary difference and the probability of future salary increases, but also other pecuniary items, such as differences in living cost, especially cost of housing, and the expenses of moving. And there will be nonpecuniary items, such as his wife's miseries and gripes about losing old friends and meeting new people, the children's attitudes about changing schools and playmates, the better or worse opportunities to attend concerts and theaters, the inconvenience of moving books and placing them on new shelves, the loss of time in getting resettled, and so forth.

All these and many other things enter into the considerations; all the items, including the pleasure of a better orchestra and the pain of hurting old colleagues, are part of the economic valuation of the total package of costs and benefits. It is very helpful to spell out all the items and attach a numerical value to each. If some one tells me he will not move, because he cannot bear to hurt his department chairman or because he cannot give up his partners at bridge or chamber music, I am inclined to ask him whether this is worth $5,000 a year to him, or only $3,000 a year. Quantification of nonmeasurable values is often necessary for rational decision-making.

The Rate of Discount

The needed comparisons presuppose also the recognition of an appropriate rate of discount for future costs and benefits, or rate of capitalization for future flows of costs and benefits. What is the appropriate rate for government programs? Is it lower than the rate for industrial investments? If so, how much lower?

Several government analysts argue that the rate at which the government can borrow funds is also the appropriate rate for use in cost-benefit analyses of government projects. Since the government can borrow at a rate lower than most business firms have to pay, the proper rate for public programs would be below that for private investments. I question that this principle is sound.

The economic principle calls for an allocation of resources at nondiscriminatory prices: equalization of marginal efficiencies of investment in all activities is a precondition of an optimal use of resources. In the capital markets the rates of interest at which funds can be obtained are affected by the credit-worthiness of the borrower and the credit-worthiness of the project. It happens that the projects of most uncertain promise and involving the greatest risks —projects too risky and too uncertain for private enterprise—are undertaken by the government. This may be quite appropriate. But I doubt the appropriateness of using the lowest capitalization rates for the riskiest and most uncertain investments. The fact that the government is more credit-worthy than private industry ought not to cause us to calculate the prospects of all public undertakings at the lowest rates of capitalization.

Needless to say, many nonpecuniary benefits expected from public programs and investments ought to be taken into account, benefits which would never accrue to a private firm undertaking the same investments. The inclusion of these nonpecuniary social benefits may make the difference and establish the social productivity of the investment. But then, on top of this, to apply to these undertakings also a lower rate of capitalization, merely because the government can borrow at lower interest rates, is to calculate uneconomically. The absurdity of this principle can be readily seen if one imagines a proposal to the effect that all investment opportunities too unattractive for private industry, figuring at higher marginal rates of return, should be recalculated at the preferred government rates and, having passed this test, should be transferred to the public sector. I conclude that the case for public investment should be made by full accounting for all social benefits, but not by being satisfied with low rates of return.

MICHAEL S. MARCH, *U.S. Bureau of the Budget*

In his paper comparing the economic returns and costs of a project for preventing dropouts in St. Louis, Burton Weisbrod has made a timely and useful effort to examine a problem of national concern. As he points out, the Department of Labor has estimated that, during the 1960's, 7.5 million youths will drop out before completing high school—and that this is a decade when the postwar population explosion will bring 26 million young people into the labor force. The problem is compounded because slow economic and rapid technological progress are reducing the proportion of lower-grade jobs for the poorly educated and the untrained.

General Neglect of Economics of Human Investment

Before discussing specific points in the paper, I should like to speak of the general problem of human resource development in the United States—a problem that is sometimes overlooked because of the relatively high levels of health and education in the nation, as compared to levels in other nations. But that we are still failing by a wide margin to develop our human resources fully has been pointedly demonstrated by data on selective service rejections. In 1962, roughly half of the draftees examined were rejected as unfit, with the deficiencies about evenly divided between medical and educational. Yet the potential for improvement is indicated by the range of the figures on rejections for educational deficiencies in the various states—from less than 3 percent to 45 percent. President Kennedy, in fact, appointed a task force on Manpower Conservation to study this problem and recommend corrective action.[38]

Weisbrod implies at the end of his paper that, with respect to the assessment of efforts to prevent school dropouts, we may not need more research—just better research—and expresses his amazement at the lack of assessment of effectiveness. I, too, as an economic analyst whose purview spans a considerable range of social programs, including welfare, income maintenance, health, manpower, and education, have been amazed at how little basic and applied research is done in the United States on the techniques of education or on the economic value of education.

The total annual public and private outlay for health has recently slightly exceeded $30 billion a year—and of this total, about $1.5 billion (nearly 5 percent) is spent for research. Public expenditures from all sources for education are likewise in the $30 billion annual range, but the amount spent on research and

[38] See The President's Task Force on Manpower Conservation, *One-Third of a Nation: A Report on Young Men Found Unqualified for Military Service* (January 1964).

development, privately or publicly, is extremely limited. The U.S. Office of Education has indicated that the annual outlay for research on education from all sources is perhaps $30 million a year—only about one tenth of 1 percent of the total annual educational outlay. I certainly agree with Dr. Weisbrod that we need better research in the field of education, but I would also argue that we still need much *more* research and experimentation.

Initial, Broader Efforts to Determine Returns from Education

One of the great deficit areas is in the analysis of cost-benefit relationships and of its economic contribution to the economy. The economics profession owes a large debt to Theodore W. Schultz for making us aware once more of the large role that investment in human resources, particularly through education, plays in economic growth. The more recent pathbreaking analysis by Edward F. Denison has also helped provide more precise estimates on the contribution of education to our economic growth in a comprehensive framework and has added greatly to our perspective regarding the importance of human resource development.

In my years of experience in program analysis on government programs, I have heard many groups of experts conclude that more education and training is the ultimate solution for most of our crucial social problems. Yet, until very recently, economic literature has tended to give overwhelming attention to physical investment as the principal factor in economic growth and to minimize or disregard the analysis of the economic contribution of education and other investment in human resources. And I know of no place in the federal government where more than scattered effort by one or two individuals is being given to fundamental analysis of the economic role of education.

We have much more to do in developing methodology and hard analysis, and we must get on with the job rapidly. Because I feel this so strongly, it is a pleasure to find that Dr. Weisbrod, one of the younger economists who have been stimulated by Professor Schultz, is continuing to further the economic analysis of education. The project presented in his paper is of small proportions, but for the social scientists in our milieu the common hazard seems to be to have only modest resources at their disposal, even if the problems which they seek to analyze may be of great social importance. They have not yet learned (or organized themselves) to think in the same terms as the natural scientists who initiate vast research complexes and think nothing of asking for $100 million for a new atom smasher. Perhaps someone ought to do a study on the organization of the social science research industry and its production function in comparison to that of the physical and natural sciences.

Some Doubts About Broad Assumptions

Turning now to some specific points in Weisbrod's paper, I am chiefly impressed—indeed concerned—by the pessimistic conclusions which its analysis and findings support. Not only do the monetary returns from completing high school, as compared to dropping out from high school, appear to be small, but also the costs of preventive action during the high school years to avoid dropouts appear to be several times the assumed returns. This, of course, flies in the face of what we have been generally led to believe and should provide a stimulus for careful further checking lest one project and one paper undercut our resolve to hit this problem hard.[39]

The paper contains a large number of rather wide assumptions and improvisations which are used in the calculations and in the analysis. It is, for example, suggested rather strongly that the raising and the upward equalization of educational levels are essentially futile because aggregate demand would not be sufficient to employ a universally highly educated labor force. Without minimizing the significant role of aggregate demand, I wonder if this assumption is fully satisfying either on economic grounds or on humanitarian and equitable grounds.

In dealing with a long-run process of adjustment such as edu-

[39] It was unfortunate that data for placing values on some of the external benefits of education were not available to Weisbrod. It is my understanding that under the Manpower Development and Training Act a longitudinal project is to be launched to analyze the costs and benefits, particularly external benefits, of retraining in a much more intensive manner for actual cases. This should prove useful.

cation typically involves, it might not be amiss to remember the principle that people who can produce more can also buy more— that in the market economy supply and demand are to a large extent the opposite sides of the same economic phenomenon, the propensity to consume being what it is. Even if there should be some obstacles of a macroeconomic sort, recognition of the principle of individual worth suggests that maximum opportunity should be afforded for each individual to develop his fullest potential. In any event, I think that a longer-range view of economic development might suggest that both workers and employers in our highly complex and increasingly industrialized society would be in a happier position if all applicants for jobs were high school or college graduates. We cannot avoid seeing that, in the modern world, the correlation between educational attainment and economic achievement is strong.

Some Questions About Low Returns

I return now to those striking figures that show the very small return from the completion of high school as compared to dropping out of high school, for I feel that they may be suspect. Perhaps the problem lies in the use of 1950 Census data or in the calculation methodology or in the assumptions. Let me touch on some of the factors that may be involved.

Weisbrod alluded (in his footnote 17) to a concept which he has ably outlined in an earlier paper: that completion of education through its various stages is a series of options, so that part of the returns from completion of higher education can appropriately be reallotted to the completion of earlier stages. He dismisses the possibility that many of the potential school dropouts might have gone on to college, but I wonder whether this conclusion would hold if the whole process of education were properly approached early enough in life and in a long-range manner.

Another factor which has a heavy influence in the small size of the returns is the interest-rate assumption for discounting future income at 5 and 10 percent. Where public expenditures are involved, as is generally the case in education, what justification is there for using such high discount rates? Education is well established as an activity worthy of public support. Even if some individuals might entertain high time-preference rates such as 10 percent, I find it hard to see what relevance such discount rates should have for this analysis. Perhaps the average yield on long-term Treasury bonds might be an appropriate guide: the current rate used in cost-benefit analysis for reclamation projects is 3 percent. When the people in the reclamation field get around to using 10 percent, I will be ready to use it in the education field. (More of this later.)

What Services and When?

To stray a bit into the no-man's-land between economics and education, several other points which Weisbrod briefly notes interest me. First, does not the cost-benefit comparison in this instance, rough though it may be, raise a question regarding the appropriateness and the efficacy of the package of services which was provided to the students? Were the best available techniques of counseling to evoke favorable responses for continuation in school used for the potential school dropouts? Recent research along this line is reported to be productive of techniques which yield high retention rates. The key is motivation.

Second, is the time reference for the analysis and for the input of resources too limited? Is this a case where the society's failure to provide timely services results in the accumulation of such large problems for individuals that relatively modest corrective measures do not work? Would overly large and protracted—and perhaps uneconomic—input of resources be required if the problems are to be corrected? The President's Panel on Mental Retardation (1962) pointed out that a large portion of retardation stems from cultural deprivation and begins early in life. I wonder whether the pessimistic picture which Weisbrod has presented would not have a vastly brighter aspect if the resources were put in very early in the lives of the youngsters involved.[40] Would the coordinated provision of social services to the families

[40] Editor's note: Dr. Weisbrod actually did suggest that earlier resource programs should be considered.

of underprivileged children perhaps prevent them from becoming school dropouts with low IQ's and even perhaps turn them into individuals with high IQ's who can successfully complete college education with its higher economic returns?

I must object to two other points also. One is the suggestion that vocational education is a potential solution for a lot of these people under discussion. I want to point out that there is a good bit of reversal of opinion concerning the value of vocational education that has stemmed from the manpower training experience. There is a growing sentiment that more basic education is the significant thing.

Second is his mention of amelioration of the basic conditions as a possible way out of this pessimistic picture that he discovered in the figures he computed. Principally on the grounds of democratic participation in the benefits of society, I rebel very strongly at this, even though his economic calculus indicated that the costs of direct prevention would be large. I believe that positive preventive and rehabilitative action is always preferable to ameliorative action, when it is a question of allowing people to be thrown into the junk heap.

IN CLOSING, I WANT TO STRESS THE NEED for economic analysis of education and the importance of a multidisciplinary approach to the study of educational processes and problems. Too often in the past, this has been regarded as the narrow preserve of educators. I can find no better way of expressing the breadth of our national interest in education than the statement by President Kennedy in his message on education on January 29, 1963:

Education is the keystone in the arch of freedom and progress. . . . For the individual, the doors to the schoolhouse, to the library, and to the college lead to the richest treasures of our open society: to the power of knowledge—to the training and skills necessary for productive employment—to the wisdom, the ideals, and the culture which enrich life —and to the creative, self-disciplined understanding of society needed for good citizenship in today's changing and challenging world.

For the Nation, increasing the quality and availability of education is vital to both our national security and our domestic well-being. A free nation can rise no higher than the standard of excellence set in its schools and colleges. Ignorance and illiteracy, unskilled workers and school dropouts—these and other failures of our educational system breed failures in our social and economic system: delinquency, unemployment, chronic dependence, a waste of human resources, a loss of productive power and purchasing power, and an increase in tax-supported benefits.

I think this statement provides an excellent framework for analyzing the returns from education. It embraces considerations that are far broader than that of the economic calculus, although, as I have indicated in these comments, I think the economic analysis of education has been given far too little attention.

HERMAN P. MILLER, *U.S. Bureau of the Census*

By forcing severely limited data to a conclusion, Burton Weisbrod brings into sharp focus some of the areas of much needed research in the field of benefit-cost analysis. Despite all of the caveats, the author is impressed by his "startling finding" that the cost of preventing a school dropout is greater than the measured social benefits. Because a casual reading of his most stimulating and well-written paper will undoubtedly leave others with the same impression, I have tried to examine critically the procedures used to produce this result—and have found that a different manipulation of the same basic data could alter the findings considerably. I do not, however, consider my own findings as more valid than Weisbrod's. My aim was only to show that, until sound theoretical bases are developed for handling such technical problems as the appropriate discount rate, the impact of anticipated productivity gains and increases in life expectancy, the appropriate measure of income (marginal or average, median or mean), and the appropriate income-receiving unit, results produced by benefit-cost analyses are likely to be highly controversial.

In measuring the benefits, Weisbrod considered only the present value of the difference in lifetime income between high school dropouts and graduates at age 16 discounted at a rate of 5 percent. This figure was obtained essentially by projecting the income differentials by age observed in the 1950 Census, using survival rates obtained from a life table and a discount rate of 5 percent. No thought seems to have been given to the idea that even if the ratio of dropouts' incomes to graduates' incomes remained constant over time, the absolute difference would have a tendency to grow because of productivity gains. And, as a matter of fact, this tendency did manifest itself between 1950 and 1960. In both Census years the ratio of the incomes of dropouts to graduates was 89 percent; yet the difference in median income measured in constant dollars was 32 percent, representing an increase of a little over 3 percent per year over the decade.

There is also some evidence from figures on income by age in Census data that increases due to productivity are greater for the younger age groups than for the older ones. Allowance for this factor could increase estimated lifetime incomes considerably, because greater weight is assigned to incomes received during the younger ages in preparing these estimates. On the basis of historical experience, one may surmise that projections of lifetime incomes from cross-section studies are likely to understate the expected values because they make no allowance for increases in productivity. They also understate the expected values because they make no allowance for increases in life expectancy. The expectation of living from one year to the next in Weisbrod's formula is based on the life table for 1949. Although life expectancy at age 16 did not change much over the past decade, it might be expected to increase over a longer period. This factor would also tend to increase lifetime incomes.

Weisbrod has also been rather arbitrary in his selection of a 5 percent discount rate. Although it may seem a reasonable figure for a long-term interest rate, it is also a very critical figure and cannot be used without some justification. If it is accepted, increases due to productivity, especially if they are greater in the younger age groups, might very possibly offset the reductions in expected lifetime income due to discounting. On this basis one could argue that gains due to productivity offset losses due to discounting, and that the best estimate of expected lifetime income is the undiscounted value based on cross-section studies.

I have computed undiscounted values for some, but not all, of the groups shown in Table 2 of Weisbrod's paper. For example, his value of $4,000 for white males in the non-South becomes $15,000 according to my scheme of things; his $2,160 figure for nonwhite males in the same region is $7,000, undiscounted. His estimate of $7,160 for white males in the South becomes $26,000 and his $2,250 for nonwhite males becomes $7,000. On the basis of these estimates, expected lifetime income should be about three and one half times the value shown by Weisbrod—or about $10,000. This seems to change the entire picture. With a cost of $6,600 and a benefit of $10,000, the program would clearly seem to be "profitable."

The dollar values used for women are derived from their own incomes. This could provide a very misleading picture of the benefits of education. As Weisbrod points out, the reason for the high differential between female dropouts and graduates is that a larger proportion of the graduates have income. One of the reasons the dropouts are less likely to have income is that a larger proportion of them are married—many of them having become dropouts for that reason—and they are supported by their husbands' income. The problem of measuring returns to education for women is a very knotty one, and I find the answer provided by Weisbrod's analysis inadequate.

Weisbrod has used the median rather than the mean in making his estimates of lifetime income. This produces a big difference in the results since the undiscounted lifetime incomes of whites are about twice as high when based on means rather than medians. The issue of which average to use is even more critical when returns to persons with higher education are measured, because there the opportunities for high incomes are even greater. There is no theoretical justification for the use of the median. It is erroneous from a statistical point of view to designate the median as the most typical value of a distribution. It is only the middlemost value in an array. The expected value for any individual is the expected value for the distribution—namely the mean. It could be argued that potential school dropouts who are inclined to stay in school are below average in terms of ability and should therefore be given a lower value than the average obtained for all high school graduates. This may or may not be a valid argument, but it is still no justification for the use of the median.

The St. Louis case study cannot be regarded as anything more than a small pilot study. The fact that it depended entirely on the work performed by the seven counselor-coordinators should lead to the utmost caution in using the results for statistical purposes. One or two exceptional people (one way or the other) could severely bias the results. Weisbrod seems generally aware of this limitation, yet there are times when he treats the figures as though they were "hard facts."

If the St. Louis test is a controlled experiment, Weisbrod should know better than to try to judge the results before the experiment is completed, and that will take a long, long time. At times he is aware of the fact that preventing a net of 34 dropouts in two years is only the first return on an investment in counseling; but, at other times, particularly when he speaks about the "failure" of the experiment, he seems to forget this qualification. There is a danger that others who read this paper will also forget the qualifications, and this might be one of the hazards of benefit-cost analysis based on faulty data.

Concluding Statement

BURTON A. WEISBROD

A principal goal of the conference was movement toward a consistent approach to benefit-cost analyses of government investment expenditures. Thus, it is appropriate that my brief concluding remarks concentrate on some of the methodological issues which have been raised about my foray into the economics of dropout prevention.

Perhaps the most significant of these issues involved the use of cross-section data, especially on productivity (earnings), but also on mortality rates. Herman Miller, for example, argues that cross-section data understate returns: ". . . even if the ratio of dropouts' incomes to graduates' incomes remained constant over time, the absolute difference would have a tendency to grow because of productivity gains."

But the case is a good deal more complex than Miller suggests. First, *a priori*, there is no particular reason to expect this ratio to remain constant or even to fall, as Miller implies. The general growth of productivity certainly does not insure this. The supply of high school graduates relative to dropouts is rising constantly, a factor tending to narrow differentials in earnings. In fact, recent data from Survey Research Center studies show that the difference between mean annual earnings of spending-unit heads aged 35-64 who had completed twelve grades of schooling and those who had completed nine to eleven grades (the dropouts) actually fell between 1956-57 and 1961-62. The difference averaged $530 in the first period, but only $90 in the latter.[41]

Second, whatever has actually happened to earnings of dropouts and graduates, the issue with respect to the wisdom of a dropout prevention program is: what *would* happen to the differential if there were a large and successful dropout prevention program? (After all, we are not really talking about preventing one dropout.) We can be reasonably sure that such a program —by increasing still further the supply of graduates and reducing still further the supply of dropouts—would bring smaller differentials than were observed. Miller's implicit assumption that marginal productivities, and hence earnings of graduates and dropouts, would not be affected by the shifts in supply accompanying a successful prevention program seems unfounded.

Third, regardless of what would happen to observed differentials, the question would remain: how much of the differentials should be attributed to education as distinguished from such other factors as student ability or growth of the capital stock? The

[41] See James Morgan and Charles Lininger, in *Quarterly Journal of Economics*, Vol. 78 (May 1964), p. 346.

fraction of the differentials attributable to education may grow, decline, or remain constant through time.

To be sure, none of these points necessarily argues for the use of cross-section earnings differentials. But neither does any argue for the assumption that the portions of those differentials which are attributable to education would grow through time even if a large-scale dropout prevention program—and its attendant labor supply shifts—were initiated.

When it comes to forecasting changes in future age-specific mortality rates, at least in terms of the direction of change, we are on safer ground. Undoubtedly the rates will continue to decline, although it is difficult to say how much. On the other hand, earlier retirement and longer vacations in the future will tend to offset the effects of increased life expectancies, for the real point is not how long one *lives* but how long one *works*. In addition, improved "maintenance" techniques in the future will extend not only the length of human life but also the length of physical capital life. Thus, insofar as the object of our analysis is to compare the profitability of, and choose among, alternative real investments—public or private—it may make little practical difference whether we do or do not account for increased life expectancy, provided we are consistent with respect to both human and physical capital. In principle, however, this factor should be considered.

The question of whether differentials in median or mean incomes (or neither) should be used to estimate productivity benefits, also merits attention. Miller's statement that *un*discounted incomes of *all* whites "are about twice as high when based on means rather than medians" has limited significance in this context, since we are talking about *discounted* amounts, and about income differentials between high school graduates and dropouts only, not about income levels for all people. Nevertheless, Miller is right in noting that from a narrow conceptual viewpoint there is little justification for the use of medians. I also agree with him that "potential school dropouts who are inclined to stay in school are below average in terms of ability and should therefore be given a lower value than the average obtained for all high school graduates," and that this does not imply that medians should be used. But it does suggest that some differentials lower than the means should be used; as a practical matter, the medians are convenient and useful, although not ideal.

The choice of an appropriate discount rate is another—and frequently critical—unresolved issue in benefit-cost analyses. My paper used, alternatively, a 5 and a 10 percent rate, since most rates which have been proposed fall within, or close to, this range. Michael March urged the use of the long-term government bond rate or even less. While I see little conceptual justification for such a low rate I share March's view that it seems appropriate to use the same rate (whatever it is) for government investment expenditures in *all* fields. At the same time it should be noted that for the case discussed in my paper even a 3 percent discount rate would not show a productivity benefit (even before downward adjustment for noneducational factors) as great as the cost.

In any case there is no conceptual justification for Miller's position that "the best estimate of expected lifetime income is the *un*discounted value based on cross-section studies." (Emphasis added.) He argues for this on the pragmatic ground (a ground which he seemed to reject in connection with the use of median income differentials) that the failure to inflate cross-section figures for "increases due to productivity . . . might . . . offset the reductions in expected lifetime income due to discounting." I have commented above on his "productivity" point, but in any event he provides no reason whatsoever for using a discount rate which is as low as the long-run annual increase in "productivity" (per man-year?)—3 percent or less. I urge that we resolve separately and explicitly the issues of the use of cross-section data and the appropriate discount rate(s), and that we not confound the two.

I hope that the absence of critical comment, and the presence of some favorable comment, on my proposal to take into account the distributive effects as well as allocative effects of government expenditure programs imply acceptance of this position. Economists have talked for too long as though distributive effects could be disregarded because other government programs could (and presumably would) provide the adjustments needed to bring about any desired income distribution.

Finally, of course, grand policy conclusions should not be drawn from the limited quantitative findings I presented about one program to prevent high school dropouts—however, those results appear to turn out. A great deal more R&D is needed on dropout prevention. The principal purpose of the case study—and, indeed, of this conference—was to expose the issues we confront in performing benefit-cost analyses in ways that permit results for various government investment programs to be compared.

GARY FROMM*

Civil Aviation Expenditures

ON NEW YEAR'S DAY, 1914, the pilot of a single-engine biplane, with one passenger in the second seat of the open cockpit, took off from St. Petersburg, Florida, to initiate the first regularly scheduled air-passenger service in the United States. Destination was Tampa, twenty-one miles across the bay—the flying time, twenty-three minutes, one-way subsidized fare, $5. The Airboat Line's schedule called for two round trips a day, but in good weather as many as five might be made; by dint of lapsitting, a second passenger was sometimes carried. The company soon added a sister craft, similar in its fabric-covered wood construction and top per hour speed of 70 miles, but modified to carry two passengers. Bad flying weather, cold days, or engine trouble frequently interrupted service. Nevertheless, when the line gave up in April (because the Chamber of Commerce subsidy had not covered costs) the two flying boats had logged some 11,000 miles and carried 1,205 passengers.[1]

During the year 1962, over 60 million passengers were flown a total of about 44 billion miles on U.S. scheduled airlines in domestic (trunk and local) service and territorial and international service

* Economic Studies Division, The Brookings Institution.
[1] See Civil Aeronautics Board, *Handbook of Airline Statistics, 1962 Edition* (October 1962), p. 453.

172

—in all, approximately 46 lines. Three fourths of all traffic was turbine-powered. A typical long-range turbo-jet could carry from 96 to 180 passengers, climb to 30,000 feet to cruise at more than 600 miles per hour, and fly nonstop (depending on the payload) from 2,500 to 5,000 miles.[2]

This contrast with the little flying-boat operation of 1914 graphically represents the prodigious growth of the American aviation industry—but only after a very slow start. After the Wright brothers' triumph (at ten miles an hour) above the Kill Devil Hill sand dunes, ten years had elapsed before the inauguration and four-months' existence of the Florida airline. Eleven more years passed before another regularly scheduled air-passenger service came into being and even then the comment could be made that nothing in the United States could "honestly be called an . . . airline" and that air traffic control and safety regulation were nonexistent.[3]

Then in May 1926 the first of three important aviation events occurred: Congress passed the Air Commerce Act, which relieved private carriers of making large investments in ground facilities and provided rigid safety regulations. In December 1926 a three-engine aircraft, with great safety advantages over earlier types, was first used in regular passenger service. And in May 1927, Charles Lindbergh's nonstop solo flight across the Atlantic in his tiny (but carefully planned and constructed) plane firmly entrapped the imagination of the American public; for the first time, airmindedness began to infect people of all ages and walks of life. To a large degree, these three events in combination were responsible for the aviation boom that began later in 1927, starting U.S. commercial transport aircraft on the way to the worldwide dominance attained by 1938.

Since the beginning of World War II, and especially after 1945, generation after generation of new aircraft types have been introduced by the industry—each one greater in speed, capacity, comfort, and safety. Although many people still fear air travel, the number of habitual passengers has steadily increased, both for domestic and international trips. Freight, express, and U.S. mail ton-

---

[2] *Ibid.*, p. 475; and Chamber of Commerce of the U.S., *Transport Review and Outlook—Year-End 1963*, pp. 4-6.

[3] Civil Aeronautics Board, *op. cit.*, p. 456.

miles have been in the high millions for a number of years, usually with a substantial percentage increase each year. "General aviation" (the term that covers all civil aviation except commercial airlines—for example, business, instructional, and personal flying, and the use of aircraft for aerial photography, as agricultural aids, and so on) included, as of 1962, more than 84,000 planes. These accounted for over half of all takeoffs and landings at the some 270 airports with FAA control towers and for nearly all operations at the 7,800 other airports without towers.[4]

These indications of the phenomenal growth of American aviation in the latter part of its first half century are of course also indications of the industry's impact on the nation's economy. To estimate that impact, we need to know the value of all civil aviation activities (airline and other flying, airport services, civil aircraft manufacturing, and so on) in terms of their contribution to the gross national product.

The Value of Aviation Activity

In theory, the establishment of any industry's value within the economy is simple, but in practice it can be extremely complex (if not impossible). Theoretically, at least, we may say that an industry's value is equal to the net total of all the losses resulting from its elimination. Thus, in trying to establish the value of the aviation industry, the basic idea is to measure the total value existing in the economy prior to, and subsequent to, that elimination. All substitution effects (for example, the greater utilization of nonaviation transportation services) must, of course, be given an opportunity to work themselves out.

The attempt to measure the reduction in total value, whether denoted by GNP or some other quantity, can be undertaken in steps. First, the total output accruing from the industry under existing economic conditions can be calculated. This includes not only the civil aviation industry's output but also the additional production resulting from its purchases in supplying sectors and its

[4] Federal Aviation Agency, *Statistical Handbook of Aviation: 1963* (1964), pp. 6, 25, 32, 60.

sales to using sectors. Second, the effects of substitutability must be determined. In other words, assuming that unemployment rates remained at approximately the same level "without" and "with" the industry, what would be the costs of transferring the aviation industry's output, purchases, and sales to other sectors?

The first of these steps underlies the results shown in Table 1. Unduplicated gross domestic output is the sale of aviation services (excluding intra-industry transactions) by air carriers and commercial (and other "for hire") general aviation plus the costs of operating business and personal general aviation aircraft. Stated alternatively, it is the sum of values added (employee compensation, profits, depreciation, etc.) in the civil aviation industry and values added generated by purchases of noninvestment inputs from

TABLE 1. Contribution of the Civil Aviation Industry to GNP, 1960

(in millions of dollars)

Unduplicated Gross Domestic Output[a]	
Commercial Aviation[b]	$ 3,063
General Aviation[c]	555
Indirect Output Generation	
Commercial Aviation	2,645
General Aviation	133
Gross Investment	
Commercial Aviation	683
General Aviation	227
Indirect Investment Generation	
Aircraft Manufacturers	336
Other	1,498
Net Exports of Goods	
Exports	551
Imports	61
Net Exports	490
Government Expenditures for Goods and Services	
Federal	481
State and Local	174
Gross Aviation Contribution to Gross National Product	$10,285

Source: Gary Fromm, *Economic Criteria for Federal Aviation Agency Expenditures* (Federal Aviation Agency, 1962), Chap. III.

  [a] Includes adjustments for domestic port expenditures of foreign airlines and foreign port expenditures of domestic airlines. Excludes U.S. purchases of foreign passenger and freight air transportation services.

  [b] Includes commercial general aviation.

  [c] Business and personal flying general aviation (all other general aviation is encompassed in commercial aviation).

supplying industries.[5] Indirect output generation is the net value of business travel-time savings made possible by the use of air instead of other modes of transportation. To measure this gain several assumptions were required, among them (1) that the trips made by air were necessary (in the sense that they would have been made by some other mode had air facilities not been available) and (2) that rail would be the alternative mode.[6]

The first two items in Table 1—unduplicated gross domestic output and indirect output generation—represent the value of the current inputs and sales of the industry's production. Domestic and foreign aviation investment, the latter being net exports of aircraft and parts, also serve to increase GNP to levels which would not otherwise be attained. Gross investment and net exports of goods are capital outlays by providers of aviation services. Indirect investment generation is the nondefense investment of airframe manufacturers and the investment by aviation suppliers. (The former quantity can be observed directly; the latter can only be estimated by the use of input-output techniques.)

Finally, government expenditures in support of civil aviation not only further the growth of the industry but also contribute to the national output. Federal outlays (net of subsidies and grants-in-aid for airports) are largely Federal Aviation Agency expenditures for equipping, maintaining, and operating air-traffic control facilities. State and local outlays are purchases of goods and services, less the current surplus of government enterprises (i.e., revenues from landing fees, hangar rentals, concession space rentals, etc.).

Summing all the components, the gross aviation contribution to GNP totaled approximately $10 billion in 1960. This quantity,

[5] Obviously, the principles of input-output analysis are utilized here. For a justification of these computations, see the source cited in Table 1.

[6] While neither of these assumptions is entirely correct, the error is not likely to be sufficiently great to invalidate the calculations; the general order of magnitude of the figures is probably reasonable. Any failure to travel (and the use of modes with higher direct costs than rail) is assumed to be compensated for by noninclusion of the cost savings of not employing additional sales and executive personnel necessitated by slower means of travel—that is, the unavailability of air transportation would probably create a requirement for additional sales and management staff.

however, overstates the true value of the industry to the economy
—since, if the aviation industry ceased to exist, most of these outlays would be diverted to other uses. Therefore, the net contribution to GNP is much smaller. I assume that it is equal to the value
of time saved by business passengers in being able to travel by air
instead of rail (a substitution measure) plus 10 to 15 percent of
the remaining items, or $4 billion. (This approach to value deter-
mination does, of course, ignore benefits which fall outside the
scope of national income, e.g., consumer surplus and military pre-
paredness, but these need not be discussed here.)

All of the items concerning aviation's impact on the economy
can also, one assumes, be forecast into the future. There is, however, one possible deterrent to such forecasts: this industry's prospects of growth, both domestic and international, are highly dependent on the actions and support of government—federal, state,
and local. Federal control, for example, is exercised:

1. Over sales—through the medium of its own demand and
supply of transport, rate and promotional fare control, route
allocation, and entry limitation.
2. Over certain costs—through flight crew proficiency, comple-
ment minima and hours worked maxima, maintenance and
aircraft utilization standards, navigational aid requirements,
aircraft design and airworthiness specifications, and provision of
and charges for certain necessary facilities and information (e.g.,
air traffic control equipment and weather data).
3. Over the pattern of operation—through determination of
aircraft paths and altitudes under certain conditions.

At state and local levels there are few direct controls, but zoning,
taxation, and provision of and charges for airport services radically
influence the feasibility and profitability of operation.

Given this link between government intervention and the
viability and scope of aviation activities, as well as the large role
the industry plays in the movement of persons and of highly
"perishable" freight, it is important to ascertain the desired degree
and manner of implementation and execution of governmental
regulation and financial support. This paper's goal is to indicate
a segment of the research necessary for the attainment of this ob-

jective—namely, methods of determining and interpreting the benefits which arise from government aviation investment expenditures for air traffic control facilities and airports. Before examining these benefits we obviously need to explore certain relationships involved in government interventions.

Government Intervention, User Charges, and Benefit Estimation

Economic intervention and expenditures of the federal government are not usually undertaken arbitrarily; rather, they are expected to produce benefits for particular groups within the society or for the nation as a whole.[7] The same is true for state and local governments, although in these cases, the field of concern is only the welfare of their respective constituents. The areas for such action may be characterized by a threefold prospectus:

1. The provision of social wants—either public or private goods or services (e.g., justice, defense, post office).
2. Adjustments in the distribution of income.
3. Economic stabilization, growth, and development.

These objectives alone, however, are insufficient to justify governmental intrusion into the private sphere. For this intrusion to be condoned (at least in nonsocialist societies), the workings of the market mechanism must prove significantly inadequate to produce the desired ends. Specifically, intervention may be desirable (1) if there are external effects of an activity (either benefits or costs that arise exterior to an industry and not directly susceptible to its internal control—e.g., as in the case of education, disease prevention, police-protection benefits, smog and pollution control, depletion of resource costs, etc.); (2) when a natural or legal monopoly

[7] This does not imply that these groups should not reimburse the government for the costs incurred in their behalf.

I beg indulgence from readers expert in the theory of public finance for repeating a few caveats of received theory, since what may be commonplace to them in this section has, to date, seemingly not been understood by certain other personnel in government agencies who are responsible for formulating and evaluating expenditure programs.

can best produce a service (government action then being required to prevent costly duplication—e.g., to eliminate parallel power lines); or (3) when government, due to its size and power, can most efficiently provide an essential service (e.g., highways and initial development of atomic energy).

On several counts, government aviation activities fall well within the scope of the areas cited. There is a definite social want for the services provided; there are both external effects (primarily community gains resulting from income generation and air travel safety) and requirements for a legal monopoly; and there is the authority of government to create built-in enforcement powers for such bodies as the Federal Aviation Agency, the Civil Aeronautics Board, and certain state and municipal aviation agencies—making them probably best able to handle most of the tasks they now undertake.

Nevertheless, even if we decide that the promotion and control of civil aviation is a proper sphere for government activity, it need not follow that the only correct method of effecting this support is through direct action and expenditures. The ends sought may be attained by other means—such as subsidies, or the establishment of contingent liabilities, or supervision and regulation, or any combination of these. (For example, contingent liabilities in the form of potentially heavy penalties for mechanical failures might be instituted instead of the provision of government preventive inspection.) The final choice depends on the effectiveness of the alternatives (i.e., their relative efficiency in terms of the difference between benefits and costs) in achieving the fulfillment of the necessary tasks; the relative desirability (measured by benefit-cost criteria) of other government programs; the constraint of the total amount of national (or state or local) income which can be devoted to nondefense public outlays; and any political factors (such as the encroachment of government on the freedom of the private sphere, foreign relations, etc.) which—although they cannot be evaluated in economic terms—nevertheless influence the method and extent of government intervention.

It is desirable to consider the determination of government expenditure policy as a problem in the rational allocation of resources.[8] This presumes that at every juncture a comparison is made between the productivity of using scarce resources in the private as opposed to the government sector. In other words, the marginally acceptable government expenditure program must have greater total social value (as measured by economic impact, political judgments, and public responses) than merely permitting its funds to be spent by the private sphere.[9] This would seem to lead to the conclusion that, for every potential area of government intervention (or expenditure), the complete set of social benefits and costs which would result from its implementation must be delineated and estimated.[10]

This conclusion would be untrue. In certain instances and under particular conditions, the field of inquiry (and the benefits to be estimated) can be delimited to a narrower spectrum. In general, this is valid (1) when the social wants desired are for private goods or services; (2) when matters of equity and redistribution of income are only indirectly involved or adjustments might be more efficiently transacted through taxation and transfer payments; (3) when the industry's economic stabilization, growth, and development aspects are not essentially different from any other activity in the economy; and (4) when other social external effects are not relatively large. All private and government actions have social external effects; intervention of any form can therefore only be justified if potential losses or gains are great, or if particular groups are especially disadvantaged.

[8] In actual fact, the interplay of pressure groups and political forces exerts a considerable influence on the final decisions, so that the outcome may not be strictly equitable or an optimum allocation. Political factors, however, to the extent that they are directed toward making rational choices among alternative programs, are consistent with proper allocation in the sense that they serve to indicate a collective preference function.

[9] Ideally, funds are requisitioned and allocated among alternative expenditure programs such that the marginal social utility of the outlays is equal in private or public use. This requires that the costs, quantifiable benefits, and other effects of any program be identified faithfully and measured to the extent practicable. This information must then be communicated to decision-making representatives of the people, and ultimately to the voters themselves.

[10] It should be noted that the total value of aviation activity is irrelevant to the expenditure decision, except in the unlikely instance that consideration were to be given to eliminating aviation. Generally, only marginal changes in expenditures and value are of concern.

Government Intervention and Aviation Activities

The special circumstances that prevail for aviation activities will now be examined. First, we must recognize that air travel is not a public (collective) good, but a service of particular benefit to the user and directly competitive with other goods and services in the private sector. Air transportation users have several choices. They can use other modes of travel and shipping; they can use substitutes for transportation (telephone, telegraph, mail, or relocation of plants and warehouses); and they can spend their resources elsewhere. Second, except for occasional regional aberrations, equity and income distribution effects are not of great consequence. Third, differential income stabilization and growth effects (as compared to those of other sectors) resulting from greater or lesser government support of the aviation industry when it has attained a reasonable degree of maturity are probably not large. (The economic development aspect of aviation will be analyzed below.) Finally, further external effects arising from government aid to a mature civil aviation industry are neither unique nor visibly superior to those found in other areas. For example, national defense external effects are often claimed for aviation, but such benefits may be found in manufacturing, construction, communications, and almost any other industry (all of which would be essential to an effective military operation in the event of a large-scale, conventional-type war which required mobilization of civil resources) and cannot be cited to justify special treatment for civil ayiation.

Another external effect—the prestige which supposedly attaches to the operation of an advanced civil aviation industry—has often been the cause of large public investments. It would appear that many emerging countries today believe that they must, for prestige reasons, have their own airlines, even though air service might be provided more economically by carriers of other nations. Individual communities have succumbed to prestige desires, too, constructing magnificent terminal facilities with substantial excess capacity.

The United States government is not immune to such prestige

lures—as evidenced by its recent announcement of a development program (with an estimated cost of $1 billion, some of which will be recovered from royalties on aircraft sales) to produce a supersonic transport superior to the British-French, Mach 2.2, Concorde. Although the program will have some favorable employment and balance of payments effects, its obvious primary purpose is to maintain United States leadership and prestige in the aircraft manufacturing field. The true wisdom of this type of expenditure decision and the political value of the "show-pieces" purchased is impossible to ascertain, but the substantial public sums spent for prestige (total cost less the value of direct benefits) might be evaluated on an opportunity-cost basis. Perhaps expenditures to help eradicate poverty, slums, disease, and depressed areas, or to spur community development and raise productivity in various sectors of the economy might yield larger direct and indirect benefits (and possibly even greater indirect prestige external effects) than would outlays toward avowed prestige ends. In any case, it is important to quantify or otherwise appraise benefits in these diverse areas so that resources can wisely be allocated among them.

Government Assistance and Industrial Development

Some government assistance, however, might be considered for economic development reasons. In the past, federal aid to particular industries has taken several forms: import quotas, tariffs, operating subsidies, and investment in research and development and fixed assets. The mode of assistance has varied from one industry to another, depending upon the nature, maturity, degree of self-reliance, and requirements of the industry. Primarily, and especially where large capital investments are needed, fledgling industries have been the recipients of direct or indirect outlays of the subsidy or investment form.[11]

The philosophy underlying all such help is twofold. First, government action may be required to initiate an industry and raise it to a stage of development at which it is self-supporting, where-

[11] Facility operating expenses can be viewed in terms of subsidies and investment. Subsidy aid has also been extended to industries in which foreign or domestic competition has greatly reduced their probability of survival. Yet, in general, this type of help more properly falls under the category of hardship rather than economic development.

upon the aid is withdrawn; second, there may be increasing returns to scale which, owing to the size of the required investment
in facilities and the risks involved, would not be realized without
government help. This aid may be regarded as based on future
benefits in the form of cost and price reductions and income generation effects whose discounted value is contributed by the government to increase and equalize the stream of benefits over time.
It can also be considered as corresponding, in part, to the consumer surplus provided by the industry.

In the case of the air transportation industry, consumer surplus
represents the additional value received by air travelers and shippers in excess of the amounts paid for such services. Thus, if prices
were any higher, some customers would cease to use those air
transportation services that are currently subsidized.[12] But many
would be willing to pay somewhat more, and a few much more,
than current air fares. Nonetheless, it is difficult to justify government subsidies and investment for aviation's consumer surplus,
since numerous other unsubsidized industries also give rise to such
values.

Moreover, although the evidence is limited, any economies of
scale in the aviation industry appear to be minor. One study (employing 1958 data) found, for example, that beyond 100-200 million available ton-miles annually domestic trunk carriers operate
with constant average costs.[13] Local service airlines, however, would
achieve lower average costs with an increase in volume; some
assistance on their behalf might be contemplated. This absence of
potential economies of scale (except for the local service carriers)
leads to the conclusion that the aviation industry, by and large, is
mature and should be capable of bearing a significant share of the
costs of airports and the federal airways systems.[14] These burdens

[12] The subsidies may be either direct, via federal grants, or indirect, through
the failure of user charges (of all types) to equal FAA investment and operating
costs, municipal airport costs, etc.

[13] See Richard E. Caves, *Air Transport and Its Regulators* (Harvard University
Press, 1962), pp. 55-83.

[14] The maturity is indicated by the large volume of annual sales of certified
carriers ($3 billion in 1961), the percentage of total trans-Atlantic travel by air
(74.6 percent in 1961—since 1955 the percentage has increased by 5 points per
year), and the large and rising percentage of domestic intercity travel by air.
See CAB, *Handbook of Airline Statistics, 1962 Edition*, pp. 5, 531, and 536.

should not, however, be imposed without due regard for several factors—the need for a gradual transition to a system of charges so as not to disrupt the industry, the relative requirements and degree of utilization of facilities by various classes of aviation service producers, the desirability of further growth of the industry, and questions of equity and income distribution. Certainly on grounds of equity and efficient resource allocation, no segment of aviation ought to pay more than its proportionate share of airport and airway costs as determined by some valid measure of utilization. Nevertheless, some segments might be permitted to pay less than their share on the basis of benefits which accrue to the community by such a subsidy. Local service airlines, for example, might be aided because they further the economic development of areas of the country which otherwise would not prosper as rapidly, and because they also help major trunk carriers to reach constant-cost scale by providing convenient feeder service to outlying regions. Nevertheless, public subsidies are justified only if they induce substantial economies that are passed on to consumers in lower prices or better service.

Equity and efficient resource allocation also require that commercial aviation not be given an unjustified advantage vis-à-vis other common carriers in the competition for passenger and freight revenues. Presumably, the availability and price of transportation services by mode should be such that total social welfare is maximized after regional reallocation of real income on equity grounds has taken place. Generally this may be accomplished in the field of transportation by establishing equitable, efficient, and workable competitive conditions. Marginal adjustments in price or level of service for desired equity and income redistribution effects can then be imposed by a regulatory authority. The competitive conditions should be such that all carriers compete on a uniform cost basis.

Subsidies, User Charges, and Economic Efficiency

Students of transportation economics have frequently been misled in their examinations of subsidy and aid programs for alternative modes of transport by paying attention only to explicit outlays and not to the implicit help arising from differential taxes and hidden subsidies. Rail and truck traffic is an excellent case in point. While trucks may contribute their share of the construction and maintenance costs of highways and streets through fuel charges and tolls, they certainly pay no taxes for their extremely valuable passage rights between and through cities. Railroads, on the other hand, in many sections of the country, bear the burden of high property taxes on their rights-of-way.

The past heavy subsidization of railroads is actually irrelevant to current allocations of traffic. Past costs (or gifts of land or equipment) are sunk, and can be neglected. What must be considered is the long-run marginal cost of transportation services; if taxes are not to bias the distribution of traffic, they should be neutral (i.e., nondiscriminatory) between modes. Tax discrimination in favor of trucking alters the relative costs in the two industries, raises the demand for trucking services relative to rail, and, with inflation in property values and substantial taxes, causes profits of trucking to rise and those of railroads to fall.[15] Perhaps this stimulation of the rate of growth of one mode of transport in relation to another can be condoned, but, if so, only on the grounds of social welfare external effects.

User charges, too, should be nondiscriminatory and employed to increase economic efficiency, and not, except in unusual circumstances, to enrich the general fund. Wherever possible, they should primarily serve the purpose of confronting the beneficiaries of a service with its costs for three reasons:

1. Certainly this is equitable, since individual payments should be proportional to benefits and use (revenues raised should be devoted to maintaining and operating the facilities rather than being diverted).

[15] This is only one of the inequities and presumed misallocations of traffic that might be cited. Other aberrations are common (which is not surprising, given the nature and fragmentation of transport regulation as it exists in the United States today). Consideration might well be given to coordination, if not combination, of the regulatory functions of the CAB, ICC, and FMB within a single agency. In this regard, see the conclusions of the Special Study Group on Transportation Policies in the United States, *National Transportation Policy* (Draft Report), prepared for the U.S. Interstate and Foreign Commerce Committee, 87th Cong., 1st sess. (January 1961), pp. 107-111, 463-466.

2.   The demand response to such levies provides an indicator of the optimum level of investment and operation of the facilities on an overall basis and at particular locations.

3.   If assessed so that the costs of furnishing additional capacity are borne by those who demand it, user charges act as a rationing device, limiting congestion, encouraging use of alternative underutilized facilities, and channeling traffic into that mode which best fulfills, on economic grounds, the needs of each user. In other words, when government is cast in the role (for reasons cited previously) of being a promoter of a private good, its aim should be to duplicate a free-market mechanism—goods and services being produced only if consumers are willing to pay the full supply price.[16]

When this is not done, government programs that support high-cost facilities tend to overexpand, since there is no effective constraint on the demand for their services.[17] This represents a misallocation of resources in that a reallocation might yield a higher degree of social welfare. Nonetheless, failing to adhere to a strict user benefit-charge relationship may be desirable—if, by so doing, greater overall public satisfactions can be proven to exist and to be realizable. When the users are unwilling or unable to pay the supply price of the good, for instance, and large external effects can be reaped, the government may find it desirable to provide a subsidy equal to the difference between the supply and demand prices if this is less than the value of the external effects.

[16] Although this viewpoint has generally been accepted (see James C. Nelson, "The Pricing of Highway, Waterway, and Airway Facilities," *American Economic Review, Proceedings*, Vol. 52, May 1962, pp. 426-435) a divergent thesis has been advanced. Some economists adhere to a marginal-cost concept of pricing which, in effect, would result in practically zero user charges. They maintain, for example, that, since the cost of permitting an additional auto to cross a bridge once the bridge is constructed is nil, no toll ought to be assessed for such a privilege; see Harold Hotelling, "The General Welfare in Relation to Problems of Taxation and of Railway and Utility Rates," *Econometrica*, Vol. 6 (July 1938), pp. 242-269. This argument may have some short-run merit, but it makes no long-run provision for covering the costs of the investment other than through general fund revenues.

[17] At times the reverse is true. Without user charges and any clear indication of demand, the government may be reluctant to provide high-cost facilities for which consumers are in fact willing to pay the full costs.

TO SUMMARIZE BRIEFLY. When government expenditures net of user charges are contemplated, the resources absorbed in public use must have a marginal social productivity at least equal to that which would be realized by permitting the funds in question to be spent in the private sphere.

Such evaluation must weigh not only the welfare realized in the current period as a result of present outlays in either public or private applications, but also future benefits. Thus, government subsidies in the area of economic development of an industry can be justified if the welfare consequences of the future income generation are sufficient to repay the costs, taking proper account of time discount factors. One should, therefore, continuously monitor government development programs to ascertain whether the industry being aided has attained a sufficient degree of maturity to assume part or all of the burdens of its future growth; whether government action and supervision are necessary to provide some of the requisites for growth and protection of the public interest; and whether any external effects are of a nature and significance to warrant government intervention and assistance.

This has several implications for estimating the benefits of programs that are in areas where aid to producers of private-type goods or services is necessary and is to be provided. Namely, when the industry being assisted is mature and external effects are no greater than for other competitive items in producers' or consumers' budgets, net expenditures over user charges should be nil. The principal question that must be answered is whether the direct benefits to users are sufficient to induce them to pay the costs of the government-provided services.[18] When unique social

---

[18] This does not mean that the government should ignore social costs and benefits in formulating its user-reimbursed expenditure programs. Expenditures (whether reimbursed or not) should not be made if the net social welfare consequences are negative. Furthermore, if within a program there is a choice between project alternatives of differing economic or social welfare efficiency, that project with the greatest net positive social welfare impact should be chosen. In most cases such alternatives are either directly competitive or can readily be identified.

To avoid excessive demand or use of the government-provided services (with a consequent misallocation of resources or a social congestion cost), the fully compensating user charges must actually be imposed and collected. In fact, where social congestion costs are high and indivisibilities or other factors preclude the

external effects of large magnitude exist, however, or relative adjustments are to be made in the welfare of particular groups in a society, some government subsidy aid might be justified. These external effects or benefits of equity transfers must be explicitly identified and have a social welfare value (estimated on either an economic or a political basis or both) equal to or greater than the opportunity social value of the funds if they were expended in alternative uses. The principles outlined above will be employed in the next two sections below, in trying to determine a benefit-estimation methodology for government expenditures on air-traffic control facilities and airports.

Benefit Estimation for Facilities Expenditures

As we have indicated, the aviation industry would probably decline significantly if the functions performed by various government bodies, and especially by the Federal Aviation Agency, were eliminated. The FAA supports aviation by giving general guidance to the design of aircraft equipment, by enforcing safety and operating regulations, and by equipping, maintaining, and operating the federal airways and terminal area air traffic control systems.[19]

All of these activities can be carried out at varying levels of efficiency and effectiveness. Although effectiveness and efficiency are related and can be measured in the same terms, conceptually they are not identical. (The distinction is arbitrary, but it is useful.) Efficiency is herein determined, for a fixed agency budget, by the establishment of procedures and the proper allocation of FAA expenditures among competing uses so that public benefits are maximized and no waste exists in the employment of these funds.

---

possibility of economically establishing fully reimbursed additional capacity, charges greater than the cost of the provided services may have to be levied to lessen demand. Where congestion takes the form of bunching of demand in certain peak hours, a similar decision between increasing capacity and modifying demand can be made; in this case, the charges should discriminate against peak period use so as to alter the time distribution of demand. Both situations represent a choice between sets of Pareto optimal points.

[19] It also administers the Federal-Aid Airport Program and acts in an advisory capacity to communities in the design and construction of civil airports. Benefit estimation for airport expenditures is discussed in the next section.

Thus, for any level of expenditure, the agency should be on its multidimensional production frontier in the neighborhood of the highest social welfare attainable.[20]

Altering the magnitude of the FAA budget, thereby permitting revision of use procedures and facilities of the airway and air traffic control system, causes greater or lesser public benefits, i.e., changes in the level of effectiveness.[21] These may be characterized by movements along the production frontier hyperplane and, in a dynamic context, along the expansion path. The expenditure decision that must be made (assuming that resources are used with 100 percent efficiency) is whether the increased effectiveness purchased with additional outlays and changes in the patterns of use is sufficient to induce users to pay higher charges or to justify the costs of any additional public subsidy. Thus, to make optimal investment decisions, one must ascertain the economic value of the physical ineffectiveness eliminated by the use of superior or additional procedures, equipment, or systems. Stated simply, the cost of a modification should be contrasted with the benefits of decreased ineffectiveness and the willingness of users and society to pay for direct, and any indirect, social welfare gains.

The Nature of Benefits and Their Measurement

The choice of physical criteria for measuring the performance of present or proposed air traffic control and airways systems involves two considerations. First, what measures are best suited to

[20] With an expenditure program based on user demand and reimbursement through charges, such optima perhaps cannot, due to external effects, be realized without additional intervention. It may be necessary to impose user fees in excess of the supply price of government-provided services (this should be done when it is a feasible means of altering user demand or of paying for all or part of the social costs resulting from the users' operations), or to utilize government subsidies and thus eliminate the causes of public dissatisfaction.

[21] Effectiveness is a complex concept, especially since it is not independent of consumer demand and can, therefore, be altered radically by shifts in preferences. If flights were to be staggered evenly throughout the day and an aviation system (airports, airways, and traffic control) was designed to accommodate this load—and then a change in preferences occurred such that flights were bunched at certain hours—the peak-load capacity of the system would be found inadequate. The resulting delays, although not attributable to system design, should, however, be measured and viewed as ineffectiveness.

depict the physical ineffectiveness of the systems and can these indicators be readily translated into economic terms? (That optimum mechanical measures can easily be utilized for economic analysis is not necessarily true.[22]) Second, are the measures of performance so structured that an analysis of causes and effects can be made which will permit the identification and estimation of the ineffectiveness responsibility of the systems, as opposed to that due to the actions of aviation operators or the public?

Past studies (as this one does) deemed it desirable to define ineffectiveness as the failure of the aviation system to minimize delays, diversions, cancellations, and accidents of aviation operators in relation to specifications established through engineering standards (with due regard for safety). Although other measures might also be employed (the number of near mid-air collisions probably should be used as a potential accident indicator), these four perhaps best reflect the overall operating ineffectiveness of the aviation system.[23]

Physical ineffectiveness manifest in delays, diversions, cancellations, and accidents involves costs to several groups—the FAA, aviation operators, and the users of aviation services. As ineffectiveness increases, the burden that must be endured by each of these groups, and by the public at large, rises by disproportionate amounts. The effect on the FAA and aviation operators is primarily one of greater operating costs, while that on aviation users is a diminution in consumption demand and value. It is useful, therefore, to explore the impact of physical ineffectiveness for each of the cost and demand elements separately.

[22] For example, as traffic passes from one zone to another a radar operator "hand-off" is performed. Occasionally, a missed hand-off occurs, increasing the probability of a mid-air collision. However, unless a significant number of accidents take place, assigning a value to the missed hand-off is practically impossible. To minimize such events is obviously desirable, but if one is primarily interested in the ultimate economic performance of the air-traffic control system, this type of criterion can readily be ignored.

[23] Unfortunately, space does not permit the description here of techniques that could be employed to measure current or potential system ineffectiveness on a physical basis; the interested reader is referred to the following sources: Fromm, *op. cit.*, Chaps. IV, V, XI, and also Appendix IV; and G. Lanka, "A Report on Benefit-Cost Analysis and Performance Measurement of the Air Traffic Control System" (unpublished paper, Office of Policy Development, FAA, April 1963).

Although the FAA spends large sums in improving the effectiveness of the aviation system, the persisting presence of ineffectiveness results in even greater costs. Accidents, of course, require costly investigation. Disruptions of service—delays, diversions, and cancellations—do not greatly increase the variable costs of air traffic control at the time of occurrence, but do necessitate expenditures for equipment and staff which would not be necessary if the same volume of traffic could be handled without such disruptions.

That ineffectiveness engenders increased costs for aviation operators is most readily apparent in the case of delays. Time spent in holding en route, in the terminal area, or in taxi delays is costly —involving additional fuel, maintenance, and crew expense. For example, in 1961 the variable operating cost (i.e., excluding all fixed charges) per "block-to-block hour" for a large turbo-jet was about $465; a minute of delay after leaving the gate thus cost almost $8.[24]

The Value of Delay Time

For passengers, the potential and actual losses due to accidents and the other forms of ineffectiveness mean a decline in the value of the aviation services consumed, and can result in diminution of demand for the services. If a passenger is killed, or badly hurt, in an accident, the completely negative value of the service for him is obvious. But delays, diversions, and cancellations can all be costly in differing ways, depending, for example, on the purpose of the journey, the extent of need to arrive on time, and the extent of the ineffectiveness.

Certainly the extent of a delay has some relationship to its cost to the passenger. In all likelihood, the result of brief delays is little or no loss; of somewhat longer ones, a loss nearly equivalent to passenger income per unit time; and of extremely lengthy ones, a loss more than proportional to income. That is, the loss function is nonlinear. No empirical data are available on this question, but *a priori* we might reasonably estimate the value of delay per unit

[24] See Fromm, *op. cit.*, p. VI-3. Block-to-block time is the number of minutes between pulling the blocks at the departure station and putting them in place at the destination station.

time as the annual income of the average air traveler divided by the number of annual hours worked.[25] Theoretically, a measure of the marginal value of air travelers' time should be used, but this cannot be obtained without extensive field experimentation. Therefore, an average income-per-minute statistic, if being used to justify government subsidies, should be treated with extreme caution; if viewed only as an indicator of potential willingness of passengers to pay for delay reductions, it can be used with greater confidence.

When any valuation of delays is being made, the time losses must be measured from an ideal standard, postulated on the basis that an aviation support system might be devised to permit an aircraft to proceed from origin to destination with no disruptions of any kind—no clearance or taxi delays, no holding in the terminal area or en route, and no climbing to an undesired altitude. In other words, gate-to-gate trips would be made in minimum time at normal maximum safe-cruising speed.

Traffic congestion, of course, causes actual performance to deviate from the ideal, making delays a frequent occurrence. Since airline trip-times are established on average "block-to-block" experience, these delays are built in to schedules.[26] Though passengers may be unaware of them, these built-in delays lower the value of air journeys to aviation users. The leisure and business time lost by vacationers and executives must be valued at its opportunity cost—the pleasure or output which might have been realized had the delay not occurred—or, as just described, approximated as equal to the salaries of the individuals involved. And in the case of a business trip, the value of the time lost applies not only to the passenger but also to whatever other persons at the place of destination are associated in consultations, contracts, and so on. If the proper data were available, all of these costs would have to be included in the determination of the impact of ineffectiveness.

[25] This measurement presents some difficult problems—among them how to determine income and hours worked, how to treat housewives' and children's income, and so on. Nonetheless, given the average mix of types of passengers on flights, especially in the peak hours when most delays occur, plausible figures can be derived.

[26] For consideration of seasonal factors and other complications in determining ideal standards, see Fromm, *op. cit.*, pp. V-2 to V-6.

The Value of Human Life

But of all the impacts of ineffectiveness, the losses that result from accidents are the most drastic. On both a unit and a total-cost basis, aviation accident costs far exceed those of any other service disruption. This measurement implies that a specific dollar value can be placed on injuries and on human life, a concept to which the moral principles of western society are supposedly antithetic. Yet it is true that our society is continually making economic decisions that place an implicit value on human life, even if no explicit judgments are voiced. Automobile accident fatalities, for example, would be drastically reduced if cars were more heavily reinforced, if pedestrian crossings were protected by gates or eliminated through use of moving-belt underpasses, and if police supervision of traffic were greatly increased. Yet such measures have not been implemented—apparently because their cost, in relation to the value of the lives they would save, appears prohibitive; thus, the value of a life is being implicitly downgraded.

For the aviation industry, however, an explicit set of values must be derived, since investment and operating expenditure decisions must be made regarding facilities that promote aviation safety. Once derived, the set of values may then be used to serve several purposes.

1. *As an indicator of the amount that aviation users might be willing to pay for increased safety.* We are likely to assume that individuals place an infinite value on their own lives. But even a cursory examination of the facts destroys that premise. People expose themselves to danger in their avocations, work in hazardous occupations because of high pay, or otherwise risk their lives for other presumed personal gain. Thus, many people implicitly assign a value to their lives. For two situations that are identical except for the possibility of death, this value would be the individual's willingness to pay for the safer choice divided by the increase in the probability of survival.

Alternatively, if this subjective valuation of life were given, the amount that an individual would be willing to pay to decrease the probability of a fatal aviation accident per trip (or passenger-mile) could be determined. For example, in 1962, among 54.9 million passengers carried over 34,710 million passenger-miles in certificated route, air-carrier, scheduled domestic service, there were only 158 fatalities in 5 accidents. Thus (excluding 37 deaths due to dynamite sabotage), there were .34 fatalities per 100 million passenger-miles flown—or, with an assumed median trip length of about 500 miles, the probability of a passenger being killed on a trip was approximately .00017 percent. If he valued his life at $400,000, he should be willing to pay at least 68 cents per trip to reduce that probability to zero. Note that this is not insurance to be paid in the event of death, but an expenditure to reduce the probability of death.

Many persons, of course, are willing to pay as much as 25 cents per $5,000, for specific air trip insurance. (These high charges are due, not to indemnification payments, but to the costs—including profits—of providing the insurance service.) This willingness probably stems partly from gambling propensities and lack of knowledge of accident probabilities, but mostly from a person's high life valuations, inadequate ordinary life insurance in his investment portfolio, and the desire to protect his family.

2. *As a benchmark against which the implicit human life values of projects that increase safety can be compared.* For each project, a calculation should be made of the value of its effectiveness gains in the form of reductions in delays, diversions, and cancellations. Users' willingness to pay for eliminating these losses should then be estimated and subtracted from the cost of the project. The remaining amount can then be divided by the expected number of lives to be saved (fractional units are acceptable since this number is based on expected probabilities of fatality rates) to obtain an implicit project value per human life saved. This information should prove useful in ranking the relative desirability of projects, those with the lowest implicit values being superior on a social rate-of-return basis. (Note, however, that the ultimate criterion for public subsidy is not the rate of return but the *absolute* difference between social external effects and costs; also, that users'

payments for increased safety will decrease the need for subsidy.)

3. *As a guide and justification for public subsidy, if social welfare benefits over and above users' willingness to pay are to be ascribed to a potential safety expenditure.* Since there are many fields of activity where government subsidy might save lives, the mere existence of such potential benefits in aviation does not in itself justify a subsidy. The gains from subsidizing safety in aviation should be shown to be superior (or at least equal) to those in other sectors. In other words, the cost and value of saving a life in aviation can be contrasted with the outlays technically necessary to accomplish the same end in nonaviation situations and the willingness of the community (in practice) to accomplish this purpose.

The derivation of a specific value for human life is difficult, but not intractable, and arbitrary, but not capricious. Since the definition of a social welfare function limits the component losses to be included in this value, it is a critical ingredient in this process.

Some economists argue that the relevant social welfare concept should encompass the losses of all remaining members of society, but exclude those of the victim himself. They maintain that the loss suffered is equal to the output that the individual would have produced had he lived, less the cost of his own consumption. I dispute these arguments; clearly, the welfare function must include the individual himself and should not be purely materialistic.

This obviously involves some double-counting, since the individual derives satisfactions from his total income, as does his family. (In essence, this approach which sums both sets of utilities is analogous to the treatment accorded public collective goods—the consumption of one person need not detract from that of another.) While this double-counting for some purposes would be fallacious (e.g., for calculation of the loss of future output), its use as an indicator of the external effects of aviation activity may be considered valid.

Other losses resulting from a fatal accident include the loss in contributed community service time, employers' recruiting and training costs, and accident investigation costs. On the basis of this methodology and the income and age characteristics of the average individual killed in an aviation accident in 1960, a value of $373,000 was assigned to an air-carrier fatality and $422,000 for a general-aviation fatality in that year. The $373,000 was the sum of the "value" of the individual's life to himself, $210,000, and the following economic losses: to his family, $123,000; to the community, $28,000; to his employer, $4,000; to the government, $4,000; and to airlines, $4,000.

The present value of the individual's earning stream and assets is computed from an average salary of $13,000, a yearly increase of 2½ percent in salary, assets of $25,000, 40 as the average age at death (a lower age would raise expected lifetime earnings and the present value), and a discount rate of 6 percent. These figures are in 1960 dollars; "average salary" and "average age at death" are based on the "Fortune Airlines Study" (March 1959) conducted by *Fortune* magazine for the Travel Research Association and Port of New York Authority. Calculations are made on a pretax basis, since it is assumed that benefits equivalent to the amounts paid are derived from the disposition of tax money. Of this amount, $185,000 also represents the loss to the economy in the form of decreased output. The assumption is made that individuals are paid their marginal products (if they are exploited, the loss to the economy is even greater), but it is not necessary to assume that there is full employment, since the incremental losses in efficiency at each stage after the chain of substitution of personnel has taken place would probably approximate the passenger's salary. (For further justification of these procedures, see Fromm, *op. cit.*, pp. VI-18 to VI-25.)

Estimating Overall Ineffectiveness Costs

All of these costs (for accidents and for delays, diversions, and cancellations) are direct—in the sense that they immediately diminish the value of air travel. Ineffectiveness, however, also has an indirect, detrimental impact on demand. With disruptions, demand is more inelastic and lower than if they did not exist, thereby reducing the consumer surplus and the total satisfaction derived by the air traveler. And both current and potential users of aviation services probably travel less frequently or use alternative modes of transportation, thus realizing lower consumption satisfactions than they could attain. Unfortunately, only a partial measure of the diminution in demand effects can be derived from available data. We can estimate the impact of ineffectiveness on demand for those people who continue to travel by air, but not for those who, if they cannot take an air trip without the real or imagined danger of an accident (or other inconvenience), will not fly or will use another mode to reach their destination. Nevertheless, to the extent possible, we should calculate the demand effects of disruptions and include them in the benefits which would accrue if effectiveness were improved (see Table 2, note g).[27]

An estimate of the 1960 costs of ineffectiveness of the overall airway, terminal area, and airport system is presented in Table 2.[28] The figures represent the gains (including some external effects and assuming that resources freed have an opportunity value equal to their market price) that might be realized if the aviation system were capable of operating without delays, diversions, cancellations, and accidents. Such a capability cannot, of course, be accomplished with one bold modification of the facilities; rather, an evolutionary use of improved equipment and procedures is required. Each contemplated change should be analyzed and an evaluation made of whether (1) the increased effectiveness it produces is sufficient to induce users to pay for the facilities, or (2) the external effects are of a nature and amount to justify public subsidy.

[27] While the previous benefits were of an efficiency nature, representing resource savings, the value of increases in demand is only partially a social gain, since the expenditures and some portion of the consumer surplus are a diversion from other sectors. From a social welfare standpoint, only the increase in consumer surplus is a benefit.

[28] These costs are based on a specially conducted sample survey of air carrier delays, diversions, and cancellations on several hundred thousand flights. Ineffectiveness caused by the aviation support system, as contrasted to that by aviation operators or users, was explicitly identified. This experience, per flight, was extended to general and military aviation flying. Accident data were taken from CAB statistics; unit costs were derived from airline CAB Form 41 reports, special general aviation surveys, and a myriad of other published and unpublished sources.

There was, of course, no means of ascertaining whether the FAA was operating the system at 100 percent efficiency. Therefore, these costs, although termed ineffectiveness losses, may include elements of inefficiency.

TABLE 2. Aviation Support Ineffectiveness Costs, 1960

("units" in dollars, "amounts" in thousands of dollars)

Ineffectiveness by Reason of	Air Carrier Unit	Air Carrier Amount	General Aviation Unit	General Aviation Amount	Military Aviation Unit	Military Aviation Amount	Subtotal Amount
Delays							
Outlay Costs	$186.00[a]	$ 34,650	$ 16.35[a]	$ 2,250	$188.50[a]	$16,550	$ 53,450
Passenger Costs	255.00[b]	58,950	28.50[b]	2,700			61,650
Total		93,600		4,950		16,500	115,100
Diversions							
Outlay Costs	169.00[b]	2,550	44.55[b]	150	94.25[c]	600	3,300
Passenger Costs	637.50[b]	9,550	57.00[b]	200			9,750
Total		12,100		350		600	13,050
Cancellations							
Outlay Costs	50.50[b]	3,450	15.00[b]	250			3,700
Passenger Costs	633.50[b]	43,450	71.00[b]	1,200			44,650
Total		47,900		1,450			48,350
Accidents							
Outlay Costs	770,000[c]	39,950	8,700[c]	24,700			64,650
Other Outlays[d]	12,000[e]	5,300	6,000[e]	14,000			19,300
Passenger and Other[f]	361,000[e]	181,950	416,000[e]	360,750			542,700
Total		227,200		399,450			626,650
Total							
Outlay		80,600		27,350		17,150	125,100
Other Outlay		5,300		14,000			19,300
Passenger and Other		293,900		364,850			658,750
Total		$379,800[g]		$406,200		$17,150	$803,150[g]

Source: Gary Fromm, *Economic Criteria for Federal Aviation Agency Expenditures* (Federal Aviation Agency, 1962) Chaps VI and VIII. (All unit costs in this table are based on average aircraft types, etc. Outlay costs are of an out-of-pocket expense nature—e.g. increased fuel consumption; passenger costs are imputed values for time, human life, etc.)

[a] Delay cost per aircraft block-to-block hour.
[b] Per aircraft delay hour, diversion, or cancellation, as appropriate.
[c] Depreciated value of destroyed aircraft; amounts include estimates for repairing damaged aircraft.
[d] Employer and accident investigation costs.
[e] Per fatality; injury cost estimates included in amount.
[f] Passenger, family, and community losses.
[g] Does not include $23.8 million of demand diminution air carrier revenue loss.

Benefit Estimation for Airport Expenditures

The relationship between an adequate system of airports and the potential value of aviation services is obvious—without numerous conveniently located airports, the availability of aircraft and airways is of little utility. As of January 1, 1962, there were 3,004 general use and 2,715 limited use airports for conventional type planes in the United States and its possessions on which 85,593 aircraft were based.[29] These varied greatly in their capabilities. Of all airports in the 50 states, only 2,043 were paved, 2,288 were lighted, and about 270 had FAA-operated airport traffic control towers.[30]

Unfortunately, summary statistics on the concentration of location of U.S. airports have not been published. Ideally, information on the number of airports, by class, in each Census Standard Metropolitan Statistical Area (SMSA) should be available; better still would be airport statistics by functional economic area. This data would be extremely valuable in the determination of the availability and adequacy of aviation services. We can draw some inferences, however, from statistics on aircraft ownership and the air commerce traffic pattern.

In 1960 there were 675 counties with airline service and 2,397 others in which aircraft owners resided. Of the active general aviation aircraft, 72 percent were located in the first group of counties, leaving an average of about 8 planes—normally in the smallest size classes—in each of the localities without airlines (the least populated counties, of course, had far fewer aircraft).[31]

The concentration of scheduled airline service is also marked. In 1961, the three largest U.S. cities enplaned 26.1 percent of all

[29] General use airports offer public facilities for aircraft maintenance, refueling, etc.; limited use airports are available to the public but are not equipped to offer minimum services. In addition, 1,466 airports were in restricted use (use by the general public restricted except in case of forced landing or previous arrangement) with 1,903 based aircraft. Total seaplane bases and aircraft numbered 285 and 1,038 respectively, and heliports and helicopters 245 and 438. (*FAA Statistical Handbook of Aviation, 1962 Edition*, p. 11.)

[30] *Ibid.*, pp. 6-7, 31.

[31] See George R. Borsari, "Economic Planning for General Aviation Airports" (Federal Aviation Agency, 1960; mimeographed).

U.S. passengers, 57 communities enplaned 75.6 percent, and 148 communities enplaned 95.8 percent. The 395 other communities with airline service enjoyed 19 percent of the aircraft departures but had only 4.2 percent of the enplaned passengers.[32] Comparable concentrations are found on a city-pair basis. Of a total of 38,432 airport locations linked by origin and destination of air travelers, the top 100 city-pairs (35 of which were New York or Newark with another city) enplaned 41.1 percent of the domestic passengers; the top 200 enplaned 52 percent; and the top 500, or 1.3 percent of all city-pairs, 67 percent.[33]

These figures do not, of course, reveal the pattern of available nonstop airline service, which is an important consideration in the economics and benefits of air travel. In general, aircraft characteristics and the cost of air-carrier operations are such that long hauls can be justified only for large volumes of passenger movement, i.e., there are economies of scale by length of haul. This has resulted in an airline traffic pattern like a series of planetary gears or "wheel spokes," which radiate from the largest hubs to the smallest: there is nonstop service between all major international gateways; nonstop service between most large hubs and the major gateways; nonstop service between the medium hubs and the nearest large hubs and also the closest "key" hub (whether New York, Chicago, Los Angeles, Washington); nonstop service between the small hubs and the closest medium or large hub; and multiple or nonstop service from nonhubs to the closest small, medium, or large hubs.[34] This route structure is, in part, altered

[32] Federal Aviation Agency, *Air Commerce Traffic Pattern (Scheduled Carrier): Calendar Year 1961* (June 1962), pp. 5 and 10.

[33] CAB, *Handbook—1962 Edition*, p. 426.

[34] Federal Aviation Agency, *Longest Airline Nonstop Flights* (May 1960), pp. 2-10. In *Air Commerce Traffic Pattern*, p. 5, FAA uses the following method of hub classification:

	U. S. Enplaned Passengers	
Type of Hub	Percentage	Number in 1961
Large (L)	1.0 or more	550,115 or over
Medium (M)	0.25 to 0.99	137,529 to 550,114
Small (S)	0.05 to 0.24	27,506 to 137,528
Non-hubs (N)	Less than 0.05	Under 27,506

by the number of airports, since traffic concentrations are reduced as the number of service points is increased. In turn, the number of airports influences the benefits that aviation users and operators reap from air travel. (These benefits are discussed below in sequence, and benefit estimation for community and national airport investment is then outlined.)

Benefits of Airport Expenditures to Aviation Service Users

Individuals travel (and freight is shipped) because benefits are realized in the transfer from one location to another. These benefits, on a net basis, vary with the purpose and the costs of the journey. If the costs exceed the benefits, the trip is perhaps not undertaken; if costs are small, travel becomes more frequent. Basically, the costs are the monetary outlays for travel, the time spent in getting from origin to destination (portal-to-portal), and the risk, convenience, and comfort of the overall journey. For air travel, the provision of airports influences these cost factors and thus, as noted above, also affects the benefits of using aviation services. Primarily, these benefits are reductions of travel time via the use of air (the risk, convenience, and comfort aspects will be discussed subsequently).

Three situations can be distinguished in determining the benefits which may accrue from air-carrier airport investments:[35]

1. Communities in which aviation services are generally unavailable or (except for long trips) not very valuable because of exceptionally long ground times or high costs of reaching an airport. In the United States, these are usually small, low-population, and sometimes relatively isolated communities.

2. Communities in which air travel is feasible and utilized extensively, but where relocation or provision of additional airports would substantially reduce ground time for many passengers and induce many other individuals to travel by air rather than not traveling at all or using an alternative mode.

[35] General aviation airports and general aviation use of air-carrier airports will be treated separately.

3. Communities where airport capacity is inadequate or saturated, resulting in congestion and ineffectiveness.[36]

While benefit determination in each of these three instances (the second and third of which might occur in combination) involves many of the same aspects, the cases are sufficiently dissimilar to be analyzed separately.[37]

For inbound and outbound passengers, the benefits of a community airport with airline service might be approximated by the time and cost savings of utilizing air directly from the community rather than combinations of other modes of transport. The calculation should encompass the total time and all costs involved between the ultimate air origins and destinations. Although this procedure is theoretically correct for identical trips made between the community and other cities—identical except for the lack or existence of an airport—it is invalid for benefit evaluation of induced or diverted air traffic. The measurement technique overstates the gains of providing an airport when trips are lengthened, air origins altered, or individuals who would not otherwise have traveled do so when air service is instituted. Since the air travel expenditures made by induced traffic passengers are substitutes for other outlays, the only additional benefits they realize are increases in consumer surplus from spending the funds for more or different aviation services instead of for something else.

Thus, to ascertain the net benefits of providing an airport, we must know the "with and without" travel patterns of inbound and outbound air passengers and their demand functions for aviation and alternative expenditures. But such information is unavailable, and therefore any attempt to measure "true" benefits is doomed to failure. All that can be done is to make what, *ex*

[36] Saturation may occur and cause ineffectiveness only during certain hours of the day on specific days of the week. To the extent that congestion losses were experienced in 1960, they are reflected in the costs shown in Table 2.

[37] The analysis that follows assumes that the airport investment decisions to be made are concerned with marginal additions to an extensive pre-existing system, rather than creation of a basic aviation network. In other words, it is possible to consider the effects of establishing any two marginal nodes as being independent. However, when large changes are to be undertaken or an entire route pattern is to be constructed (as is likely in the provision of transportation facilities in emerging nations), simultaneous equation techniques must be utilized to estimate system and individual airport benefits. Local interactions must also be taken into account when more than one airport is planned in a given area.

ante, seem to be reasonable assumptions: in this particular case it may be justifiable to evaluate the benefits for airline passengers who enplane or deplane in the community as being equal to the time and cost savings of not having to employ automobile ground travel to reach the closest appropriate small, medium, or large hub where relatively convenient or frequent air service can be obtained.[38]

Elements similar to those described in the preceding assumed case are found in the analysis of airport investment decisions when consideration is being given to improving airport access. Many of today's airport sites have been inherited from the era of the "flying Jenny" or were chosen without regard for the convenience of air travelers; as a result of this wide divergence between airport locations and the air-traffic generating areas of communities, the demand for and value of air travel are reduced. Thus it may be desirable to relocate many of the existing airports, to provide additional facilities more strategically situated, or to improve the means and lower the costs of access to existing facilities. The passenger benefits from this type of airport investment are, primarily, the reductions in travel time and cost between the airport and the air travelers' origins and destinations in the community.[39]

[38] This is a general guideline, and should not be applied uncritically in all situations—e.g., when small, medium, or large hubs are more than two hours driving time from the community and a nearby nonhub has feeder service. (It also involves some further assumptions, particularly about hub locations and travel patterns, which we do not delineate here.)

The savings are calculated on a net basis: automobile ground-time less air-time and ground-travel costs less air-costs to the appropriate hub airport. In some instances, only one hub must be considered, in others, several. The unit value of time may be determined by the techniques described in the preceding section— annual income of the average air traveler from the community divided by hours worked.

A similar calculation can be made for the speed gain of freight shipments; here, the relevant benefits are for truck vs. air costs to the appropriate hubs, the interest-cost savings on the difference in the value of goods in transit, and some additional arbitrary premium for the reduction in other costs (customer dissatisfaction, "down" machine time, etc.) made possible by more rapid deliveries.

[39] Because holding for several hours at airport freight terminals is frequent and city to airport trucking charges are based on tons or ton-miles, the benefits of reduced freight hauling costs from increases in accessibility are probably not very significant and generally can be ignored.

But, once again, an estimation of such gains is made exceedingly difficult, if not impossible, by the need to evaluate induced traffic. In the absence of data on local origins and destinations and demand functions for aviation and other expenditures, a logical approximation to aggregate passenger benefits of improved airport access is the difference in the weighted (by proportion of use of a ground-travel mode) sum of the time and outlay costs of reaching the more conveniently located (as opposed to the more distant) airport from the community's central business district (CBD).[40] Information on the local originations and destinations of air travelers, to the extent that it is available or can logically be assumed, should be used to provide additional foci to the CBD in this calculation.

The analysis and determination of benefits in the third case cited earlier—situations where airport capacity is saturated—are essentially identical to those described in the previous section. The inadequacy of capacity may result in delays, diversions, cancellations, and accidents—and the losses attendant on all of these things. The benefits of, and the willingness to pay for, reducing this ineffectiveness can thus be compared with the costs of providing additional capacity—either in use, by rescheduling flights, or physically, by adding runways, taxiways, other equipment, or another airport.[41] If an airport to relieve congestion is constructed,

[40] The formula for this calculation might appear as follows:

$$B = P_j\left[\sum_k a_{ik}(Y_j\,t_{ik} + C_{ik}) - \sum_k a_{jk}(Y_j\,t_{jk} + C_{jk}) \right],$$

where,
i = the less accessible airport
j = the more accessible airport
k = mode of reaching an airport
B = passenger benefits
P_j = passengers who utilize the more accessible airport
a_{ik} = proportion of passengers who use mode k to reach airport i
Y_j = average income per hour of P_j
t_{ik} = hours spent in using mode k from CBD to airport
C_{ik} = outlay cost in using mode k from CBD to airport i

[41] That the large transport capacity of air-carrier airports might be increased by prohibiting use of congested fields by general aviation aircraft has frequently been suggested. Because the attendant equity and efficiency consequences of this prescription differ in each case, it should be rigorously analyzed in every instance. Account should be taken of the air carrier's feeder role for general aviation; of the

any additional losses or gains due to the greater or lesser difficulty of ground access should also be taken into account.

The availability, accessibility, and congestion of airports also affect the benefits which aviation users can derive from general aviation flying. When an airport in a community will not accept general aviation aircraft, other modes or combinations of modes (including air) must be used to complete journeys to and from that location. Thus, for business and personal general aviation itinerant (intercity) flying, the benefits of providing an airport might be approximated by the time and cost savings of passengers and crew in reaching the closest general aviation or air-carrier airport which has a comparable aircraft service capability. If flights are undertaken with equal frequency in all directions, the savings can be readily shown to be the value of time and the monetary outlay of driving to the alternative airport; when particular directions of flight are more frequent, an adjustment increasing or lowering this saving may have to be made. Although not exactly equivalent, the business, personal, and instructional local flying benefits of the airport from an aviation user standpoint might be calculated in identical fashion to the equi-frequent directional case.

The user benefits (for these classes of general aviation) from more accessible airports that accept general aviation traffic can be estimated by calculating the differences in time and cost of reaching alternative airport sites. Data on the local origins and destinations of general aviation users (to the extent that it is available) should be used in these computations to serve as additional foci to the central business district. Normally, it may be reasonable to assume that 75 percent of the business traffic originates or terminates in the CBD and 25 percent at the location of aircraft owner-

---

proximity and adequacy of other general aviation airports; of the time distribution of congestion and the possibility of shifting traffic to other hours of the day; of the high, clear weather (VFR), general aviation capacity of a single runway (rates of 480 aircraft per hour have been realized); of possibilities of altering the mixing pattern (e.g., by short takeoffs from runway intersections); and of the marginal costs and gains of adding special general aviation facilities. See Airborne Instruments Laboratory (M. A. Warskow, H. C. Burns, T. Dayton, W. Guidi and P. H. Stafford), *Airport Facilities for General Aviation* (Federal Aviation Agency, November 1962).

ship. For personal flying, these percentages are probably reversed, and for instructional flying, 100 percent of the traffic may be assumed to originate in the proximate residential and plant areas of personal and business small aircraft ownership.

In considering the last of the three airport situations—the alleviation of congestion—the benefits should be calculated with the same methodology as employed for airline users: that is, the impact of reduced ineffectiveness on the level and value of demand for general aviation services should be ascertained.

Benefits of Airport Expenditures to Aviation Operators

The availability and general location of airports in relation to the size of the potential air-travel market greatly alters the profitability of air-carrier operations. Two influences are involved here. On the demand side, the greater the number of highly accessible airports, the larger the airline revenues will tend to be; on the supply side, however, the greater the number of stops for any specified level of passenger-mile demand, the higher the operating, maintenance, and station costs of furnishing the service—and the lower the profits. Moreover, the fewer the number of stops and marginally profitable flight stages, the greater the frequency of flights and the realization of potential passenger demand on the longer routes.

In a static situation, the number of airports which a monopoly airline would choose to serve depends on the charges imposed for use of airport facilities, air fares and the geographic pattern of demand, and the marginal cost of adding additional service. With lower landing and other airport fees, demand is greater and more evenly distributed; with smaller marginal costs, the number of airports which will be included in the route structure is larger. A particular city-pair (A-B) will be serviced if the revenue from total seat occupancy equals or exceeds the airport and operating costs between the points. That not all passengers, however, must originate their trips at A and terminate them at B is illustrated in Figure 1.

Obviously, the profitability of operations between A and B depends on the route structure before and after these points and

FIGURE 1. Aircraft City-Pair Revenues

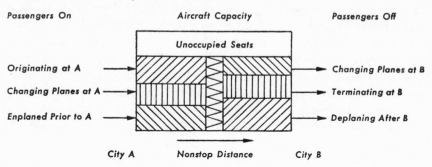

Passengers On   *Aircraft Capacity*   *Passengers Off*

Unoccupied Seats

Originating at A ⟶   ⟶ *Changing Planes at B*

Changing Planes at A ⟶   ⟶ *Terminating at B*

Enplaned Prior to A ⟶   ⟶ *Deplaning After B*

City A   *Nonstop Distance*   *City B*

the willingness of some passengers to undergo the time loss and inconvenience of changing planes and multiple stops. In other words, the additional cost from adding another service junction includes, not only the increase in airport and operating charges, but also a loss in marginal revenue from passengers prior to and beyond that city. Summing these effects over all aircraft and city-pairs which might use a particular terminal reveals whether airline service is feasible at that node.

Nonetheless, two other factors must be taken into account: the number of competitors, and the dynamic elements of supply and demand. Competition may induce a carrier either to service a particular point or to abandon such an attempt. When demand is growing and is expected to continue to grow, service may be provided at an earlier date than short-run profitability would dictate, owing to an airline's desire to insure itself a share of the future market and perhaps preclude entry by competitors. But with static demand, rising costs, or extensive competition causing low load factors in a particular city-pair market, service may be abandoned even when revenues equal costs. In both of these cases (or their variants), an airline's decision will be affected by its overall sales posture in relation to its competitors.

Such factors have implications for the desired degree of airport availability—namely, that for any level and geographical distribution of potential demand for airline service, there is an optimum number of airports that should be provided from an air-carrier standpoint. The number will vary with the charges imposed for airport use, being greater as landing fees and other airport levies are reduced. Given dynamic elements in demand and competitive conditions, the tendency is to service points even when this is not justified by profitability standards. Thus, if the supply of airports is excessive, the number of points served will tend to be non-optimal, resulting in higher marginal costs and lower marginal revenues and profitability than should ideally be experienced. Therefore, unless the external effects of serving the excess locations are of a nature and magnitude to justify these lower profits and any public subsidy, social welfare has been diminished. From an airline standpoint, the benefits of an additional airport, whether provided for availability or accessibility reasons, must be determined on the basis of the impact on overall profitability of serving the affected portion of its route structure.

Relocation (or original location) of airports to increase accessibility also has an effect on airline revenues and costs. Because sites closer to the urban areas they serve are more expensive than those further removed, airport user charges tend to rise as location accessibility is enhanced. These costs, from a carrier viewpoint, must be compared with the benefits of higher revenues which accrue from greater exposure and convenience to the air travel market. (On occasion, of course, accessibility and revenues are increased without cost to airlines: for example, when a locality improves the highway network serving the airport area.)

Finally, the results of airport expenditures to relieve congestion hardly require further exposition. The gains that airlines achieve are simply the savings in ineffectiveness costs and greater revenues from induced demand. Nonetheless, when greater capacity is created through the medium of an additional airport, all the complexities cited earlier regarding route structure impinge.[42]

The community benefits of airports and aviation activities are, of course, related to the gains of passengers, shippers, and airlines from these facilities and services. A number of studies on the

---

[42] Similar considerations (with some variations) also affect the profitability of supplemental air carriers, commercial general aviation, and firms that service the maintenance, communications, and other needs of aviation operators. Although these are not explored here, they should be taken into account when evaluating the consequences of particular airport investments.

impact of such activities on individual local areas have been conducted by city and regional planning agencies and consulting firms.[43] Normally the list of benefits in these analyses has included: (1) the employment, payroll, profits, and local purchases of the aviation service producers; (2) the general employment effect of these local purchases; (3) the expenditures of visitors for hotels, restaurants, taxis, and other goods and services; (4) the employment of persons in travel bureaus and firms to handle aviation travel arrangements; (5) the employment and income effects of local aviation construction investment; and (6) the employment and income from consumer services sold in the airport area and air freight oriented industries and commerce.

Such listings (which are usually incomplete) of the gross impact of airports and aviation services are useful, but by not taking into account the interdependence of economic activity, the possibility of substitution of one type of expenditure and employment for another, and the consequences of attempting to increase local income under conditions of local or national full employment, they are not true analyses of benefits. The existence of interdependence precludes the possibility of assigning income-generating effects on a gross basis to any industry. For example, aviation delivers convention visitors—but without hotels there would be no conventions; in this case, hotels and airlines are obviously needed in conjunction, and neither can be considered more necessary than the other (that is, they are complementary).

Potential and actual substitution effects also appreciably affect the benefits arising from any activity. If local residents who had usually spent their vacation outlays within the community are induced by the availability of aviation services to travel abroad in numbers that are not reciprocated by tourists from abroad, the community welfare will have declined from a local-income standpoint. To the extent that skills are transferable from the local vacation industry to the local aviation industry, some unemployment will be avoided; nevertheless, because net outlays are spent

[43] See, for example, Seattle Port Commission, *Seattle-Tacoma International Airport and Its Impact Upon the Economy of King County, Washington* (1962), and C-E-I-R, Inc., *The Economic Relationship of Air Transportation to the Economy of the New Jersey-New York Metropolitan Area* (August 1960).

abroad and the aviation industry is less labor-intensive than the local vacation industry, total employment will have declined. Many similar illustrations of substitution possibilities can be presented.

Finally, the benefits from attempts to increase local income via aviation facility investments are not independent of local or national resource employment conditions. When the local labor force, for instance, and especially the construction industry, is fully employed, attempts to build an airport (assuming an inelastic construction supply function and either full employment elsewhere or local entry barriers) will mainly lead to higher construction prices and little greater real investment. Similarly, the consumption expenditures of primary aviation employees may not result in greater real income and employment locally if full-employment conditions already exist.

It is evident that the interdependence, substitution, and resource employment factors greatly complicate the evaluation of community benefits from aviation investments. The ultimate measure of these benefits is the contribution made to the community's real income and employment after all substitution effects have had an opportunity to "shake themselves out." Thus, a forecast of benefits from the outlays must provide for a review of the complicating factors in the present and in the future.

It is probably best, therefore, to consider only three classes of benefits as being germane to a community's evaluation of a potential investment in aviation facilities. These are (1) the efficiency gains of aviation users and operators; (2) production activities brought to the community when persistent underemployment of community resources is expected and when the availability of aviation facilities to be provided is an essential requirement of the location action; and (3) other external effects of the aviation services.

The efficiency gains of aviation users and operators are those cited above: the value of reductions in travel time, lower freight inventory costs, and increased profitability or lower fares (if they accrue to the community) due to modifications of route structures, decreased ineffectiveness, etc. If these savings are to become benefits, however, it is necessary that the resources released have an opportunity value. Business-time savings, for instance, should result in increased output—and that they so result is a reasonable assumption, since otherwise individuals and firms would not generally avail themselves of air travel (which, on an out-of-pocket basis is normally more expensive than other modes). However, a user or operator willingness-to-pay standard is probably the best measure of these benefits for all classes (availability, accessibility, and congestion) of airport investments.[44]

In the second category of community benefits—the location of manufacturing plants or other types of establishments on the basis of advantages of the area (including as a primary precondition scheduled airline service or a general aviation airport)—the income generated by these production activities should be counted as a gain only when the community resources utilized in the process would otherwise have been unemployed. The projected aviation activities themselves, of course, should be considered in the same regard. The relevant benefits here are the local income created, including any real income multiplier effects from the use of other available resources, and the income (if taxable) of other factors imported to serve as production inputs.[45] That is, the wages and salaries of personnel transferred to the community should also be incorporated in its welfare function.[46]

To identify these income gains in the normal dynamic setting of community growth is difficult. Thus, extreme care should be exercised in using these benefits as a justification for public investment or subsidy, especially since such outlays may have high opportunity costs in the form of alternative measures for reducing the unemployment of resources. The optimum procedure is probably to consider the gains as net benefits only when there are

[44] Because in the early years of any new airport's existence traffic is likely to grow slowly, financing for the investment normally should provide for an initial moratorium and gradually increasing repayment.

[45] Note that if several community actions must be taken in conjunction to satisfy the conditions imposed by prospective local industry, the costs of providing airport facilities must be incremented by the outlays for these additional requirements; alternatively, an allocation of the total expected benefits can be made on the basis of shadow prices.

[46] This inclusion as a benefit assumes that the taxes collected exceed the marginal cost of additional required community services or that there are economies of scale in the production of such services that will improve community welfare. Further there should be no net social costs to the existing community due to the expansion in population or the importation of other resources.

local resources that are idle and when specific firms will employ those factors solely on the explicit agreement of the community to make airport availability, accessibility, or improvement expenditures.[47] For the most part, in the United States such situations occur only in small, isolated cities and counties.

The final class of benefits—other external effects—are generally of an intangible nature and not susceptible to measurement, especially on a net basis. There are educational and recreational gains (e.g., the Sunday outing to the airport to show the children an airplane), prestige gains (the community is part of the air age), "insurance" gains (potential use of aviation services for disaster relief, emergency trips for specialized medical treatment, family or business needs, etc.), and gains in additional community output.

The last-named benefit arises through the second-order effects of the efficiency gains of business aviation users. Aviation services help to attract and retain business firms in the community. These in turn draw in or cause the creation of other companies with complementary activities. The interaction of all these entities tends to create substitution effects (producing higher-valued output) and also generates external economies within the industrial and business complex. All of these are clearly benefits—but they cannot be measured, nor can an objective determination be made of the share attributable to aviation. On the other hand, there are the offsetting social costs of congestion, pollution, smog, noise, accidents, etc. And, as indicated earlier, these gains should probably not be employed as a sole justification for aviation expenditures, since similar benefits may result from many other industries and activities.

To summarize, the community benefits from aviation expenditures are primarily the efficiency gains realized by aviation users and operators. Because these represent gains in individual welfare, it is appropriate and desirable that these groups be con-

[47] Certain individuals concerned with depressed area problems have at times proposed the provision of airports as a stimulus to local production. Although per se the availability of aviation facilities would be an asset, it is doubtful that this action, taken in isolation from other prerequisites, would greatly enhance the desirability of locating in such areas.

fronted with the costs incurred on their behalf and user charges imposed to recoup the community expenditures (including imputed interest) for the facilities.[48] However, under the special circumstances in which users and operators are unable or unwilling to pay these full costs, when local unemployment (or underemployment) of resources exists, and when business enterprises are willing to create employment opportunities under the explicit conditions that the locality provide aviation facilities, a comparison might be made of the income generated (from aviation and these activities) and the required public subsidy outlays. If the income-generation benefits on an opportunity basis are sufficiently great, the aviation expenditures, user charges, and subsidies should be undertaken. All other external benefit effects of aviation should generally be ignored (with the exception of safety) in determining community expenditure policy for airports and related facilities.[49]

These principles, with some slight modification, also hold true for federal assistance to civil aviation. Three additional factors must be considered.

1.   That the income-generation gains of one community's provision of aviation facilities may result in income losses (which may be less than, equal to, or greater than the gains) for another. Localities continuously, of course, attempt to improve their own welfare, whether or not this is at the expense of other

---

[48] A discussion of the appropriate level of a community's imputed interest charges is beyond the scope of this paper, but it might be noted that these charges depend on the opportunity tax loss on the airport land taken off the tax rolls and the method of financing the airport expenditures.

[49] It would seem reasonable to expect that aviation users and operators should pay the costs of ensuring a high degree of safety. These burdens might take the form of actual outlays for maintenance, landing aids, lighting, etc., or of strict imposition of such standards and restrictions to limit the extensiveness and intensity of flying as certification standards for airmen, aircraft, and airports, restricted airspace areas, required minimum separations of aircraft in the airspace, absolute limitations on the number of airborne aircraft in a given area, and scheduling restrictions. Even with all these charges and burdens, however, air safety may be impossible to combine with a desired level (as determined by efficiency and income-generation gains) of aviation activity. In this case, given the large external effects of human life values, it may prove advisable to subsidize aviation safety, but only if the funds entailed could not more fruitfully be spent for other purposes in either the public or private sectors.

regions. Certainly, as long as all play by the rules (as set forth
in national, state, and local constitutions and charters) no federal interference in this process is justified, even when a net loss
in income occurs.[50] Any desired *ex post* adjustments (based on
interpersonal-intercommunity welfare concepts) in relative well-
being can then be carried out through the medium of federal
tax, transfer, public works, or/and procurement policies. Unfortunately, in many depressed areas such readjustments have
only recently been initiated at a level and in forms that have a
significant impact on local incomes.

When federal subsidies for airports are to be given, however,
anticipated shifts in output should be reviewed and analyzed
to determine their consistency with long-term national eco-

[50] An analogy drawn between firms and communities (although they are not
strictly comparable) is useful in supporting this proposition. Under the mechanisms of free-enterprise systems, the self-interest of individuals and firms is said to
assure an allocation of resources which yields an optimal overall magnitude and
distribution of output among industries. The government intervention that is
sometimes required under certain circumstances is usually aimed at encouraging
private production, not restricting it. (The farm program, paradoxically, reflects
both influences. Whether income maintenance for farmers is accomplished best by
output reduction is moot.) For example, competitive actions—whether in the form
of inconsequential differentiation or of technological change—that result in the transfer of output from one sector to another are not hindered (despite the availability of
large quantities of sunk assets in the outmoded industry), although short-term
hardship aid might be given to the sector suffering substantial losses. In the long run,
however (even though some government assistance may be required in the process),
it is expected that excess resources in this industry will be absorbed in other productive activities and that the value of output will be greater than if restrictions had
been imposed.

Similarly, in cases of communities' competition for productive activities, it can
be reasoned that when free-enterprise principles are permitted to operate,
they will tend in the *long run* to bring about optimum allocation and use of resources. To presume that a federal bureaucracy possesses greater wisdom than the
actions of a large number of independent communities is not inherently logical;
furthermore, there are costs of losses of freedom in interference.

Nonetheless, some federal intervention is probably desirable, especially if it tends
to perfect the market. Its aims should be (1) to inform communities of the potential loss of key local industries; (2) to provide information and guidance to
firms contemplating locational shifts to help deter unwise transfers; (3) to give
short-term assistance to ease local hardship, if necessary; and (4) to lend technical
and financial assistance (for equity and employment of idle resource reasons) to
transform and transfer potentially productive factors to useful functions and also
to help attract new industry to the region.

nomic efficiency. This is necessary because a community receiving assistance bears neither the full cost nor risk of an airport investment, and thus the potential local rate of return is effectively raised and the incentive to lure away other communities' actual or potential air travelers is strengthened. Such transfers should be aided *only* if efficiency is enhanced enough to justify federal subsidy.

2. That the efficiency gains realized by passengers and shippers through the use of local and trunk airlines may result in greater than proportional losses for other common carriers and their customers. This might occur, for example, if rail traffic is diverted to air and there are economies of scale in rail operations. Any such losses should be considered when federal assistance to a single transportation mode is proposed.

3. That intercity and interstate transportation links create external economies for widely separated industries. These gains may provide some justification for federal subsidies, but expenditure decisions on this basis still must take into account the potential diseconomies of concentrating too much activity at a single node or in a given area.

Some Final Observations

Civil aviation is now a mature industry that should be capable of assuming the burdens of sustaining its future growth. Benefit estimation, therefore, should with few exceptions take the form, not of accounting for an extensive set of direct and external effects, but of ascertaining the gains users will realize from government outlays and user willingness to pay for them. To the FAA aviation-support system (largely airway and air-traffic control operations and facilities) the gains mean decreases in the elements of ineffectiveness—delays, diversions, cancellations, and accidents; to airports they mean reduction in travel time. The difficulties in determining these benefits are almost wholly those of measurement of the demand and supply functions of aviation services. But the problems are not so insurmountable that sufficient effort could not resolve them.

Beyond this are questions of the design and financing methods of aviation systems. Within communities these revolve around airport locations and capacities; provision of rapid airport access integrated with the community's ground transportation means; land-use planning (zoning and pre-empting ideal airport sites— which also includes prevention of housing developments nearby that would be subject to the noise nuisance); proper regard for general aviation's role and requirements; and establishment of user charges. At the national level there are such considerations as neighboring cities sharing an airport when demand inadequacy exists; designing rational aviation systems (system flow over airways between airports and the optimum number of nodes, time phasing of additional capacity, and equitable decisions about the social costs of providing services for general aviation vs. air carriers); obtaining an efficient intermodal transportation network to eliminate redundant capacity and encourage coordination; formulating programs of federal aid to airports that do not distort resource allocations; and achieving full user reimbursement for government outlays to aviation. Such matters are imperative to government expenditure decisions and must eventually be more fully explored than they have been to date.

Comments

GORDON MURRAY, *U.S. Bureau of the Budget*

When commenting on a paper that is as voluminous and intricate as this one is, a road map of some sort is needed. My remarks therefore will be patterned somewhat sketchily according to the three main sections—excluding the introduction—through which Gary Fromm's argument progressed.

Benefits of Aviation Programs

To lead off his discussion of the relationships of government intervention, user charges, and benefit estimation, Fromm told us that "expenditures of the federal government . . . are expected to produce benefits for particular groups within the society or for the nation as a whole." This casual association of "particular groups" and "the nation as a whole" made me immediately apprehensive, for it seemed to imply that whether the benefits are produced for the one or the other mattered very little. In my view, the statement largely ignores a practical distinction of great importance, even though later in the discussion Fromm does make a theoretical separation between "external effects" and the direct benefits accruing to users and entrepreneurs of aviation.

The special-interest groups that have directed their efforts toward obscuring the vital distinction between themselves and the society as a whole, or toward equating the interests of the nation with their own interests, have created major problems for the government's investment policy. Indeed, such efforts constitute one of the principal reasons for funds *not* being "ideally . . . requisitioned and allocated among alternative expenditure programs such that the marginal social utility of the outlays is equal in private and public use" or among public outlays (see Fromm's footnote 9). The reality, for "the nation as a whole," is that funds are seldom requisitioned and allocated so that utility is equal for public and private use—or among alternative public uses, or even (and especially) among alternative aviation uses.

"It should be evident," Fromm says, "that government aviation activities fall well within the scope of the areas cited" and that there is "a definite social want for the service provided." To reach these conclusions, he must have facts and insights that I do not. He has, however, acknowledged that many among our population do not use aircraft as a transportation mode. But officials of some of the trunk airlines have stated repeatedly that the majority of United States citizens have never flown and possibly never will fly. I am tempted, therefore, to borrow a form of argument from Senator Wayne Morse, who suggested, during the 1963 debates on the foreign aid bill, that if foreign aid were submitted to a referendum of the American people, it would be overwhelmingly defeated. My version would be this: if the program of federal expenditures for aviation were submitted to a referendum of the American people with a detailed statement of costs by purpose, it would be, if not "overwhelmingly defeated," at least drastically curtailed.

The statement of costs would include, for example, the $12 which Uncle Sam puts up as direct operating subsidy for every traveler moving the eight to ten miles of a typical trip in a certificated helicopter, against the $8 to $10 fare put up by each of the small group of typical expense-account travelers who use the service.[51] We simply cannot presume, as Fromm apparently does, "that at every juncture a comparison is made between the productivity of using scarce resources in the private as opposed to the government sector" or that the nation as a whole rather than a particular group is benefited, or that the significance of a given type of benefit is the same, whether it goes to a particular group or whether it goes to the nation as a whole.

I believe that the questions for government economic policy that are encountered in evaluating the social marginal utility of the aviation and nonaviation alternatives which make claims on public funds are more important than any questions found when evaluating alternatives *within* the aviation category. Unwarranted assumptions as to the answers to these problems may make any refined comparison between benefits derived from inputs of public funds into various aviation facilities or services seem frivolous.

With much of Fromm's discussion of the motivations (for example, national prestige) for public investment in aviation facilities I am in agreement. The discussion is largely based on hunches —as he says, "no information on these relative benefits is available"—and in this case, my hunches happen to coincide with his. Many of his observations, however, would give little comfort to public officials attempting to decide or to guide decisions on a fair allocation of public funds. We can all agree, for example, that "if prices were . . . higher, some customers would cease to use air transportation," that some "would be willing to pay somewhat more, and a few much more"; or that "the aviation industry . . . should be capable of bearing a significant share of the costs" and "no segment of aviation ought to pay more than its proportionate share." But to what practical effect?

The general discussion of how resource allocation (private as well as public) as between railroads and trucks is distorted by inequitable public promotion and taxation is brief but good. My

[51] Certificated services in San Francisco are not subsidized.

own view is that distortion through regulation—for example, umbrella rate regulation—is probably of greater importance. In any event, we need tools for analysis, rather than general observations.

I also agree with his position that there is a constant danger of overexpansion of government programs that support high-cost facilities, when there is no effective constraint, such as user charges, on the demand for the services such facilities provide. For years federal officials have been struggling to devise practical methods of tying public expenditures to recouping charges without building inflexibilities, such as trust funds, into the government's fiscal operations.

That "the resources absorbed in public use" (in excess of recouping charges) should "have a marginal social productivity at least equal to that which would be realized by permitting the funds in question to be spent in the private sphere" may well be sound economic doctrine. And to it should certainly be added that the chosen public expenditure should have marginal social productivity at least equal to that which could be realized by permitting the funds to be spent in *any other public activity*. However, our present ability to identify, quantify, and compare benefits, to state or even order costs (especially when they result in joint products, which they so often do), and, finally, to compare costs and benefits measured in the same units is so limited that the economic doctrine, it seems to me, is of very little use to the public official who is under pressure to make investment decisions *now*. (I will make reference later to what I think is presently a more promising alternative to this dependence on quantification.)

I must disagree with Fromm's idea that, among the great run of proposed programs for public investment, the one with "the greatest net positive social welfare impact . . . can readily be identified." I am sure, at least, that this identification cannot be made with anything like the precision which I conceive to be meant by his term "marginal social productivity."

Physical Ineffectiveness of Aviation Systems

In his discussion of benefit estimation for facilities expenditures, efficiency is defined as "proper allocation of agency expenditures." Here the program of the Federal Aviation Agency is being considered. The stated problem is to "ascertain the economic value of the physical ineffectiveness *eliminated* by the use of superior or additional procedures, equipment, or systems" (my emphasis). I object very strongly to the list of manifestations of system ineffectiveness "delays, diversions, cancellations, and accidents" if the implication is that they are of the same order. The lack of safety comes first and is different from all the others. In official doctrine, it is actually not a manifestation of "ineffectiveness" at all. This means that safety will be the same for all systems, insofar as the system per se influences it. Every other part of the system must yield as much as need be to achieve maximum safety. It seems doubtful to me, accordingly, that the value of human life has any place in the calculation of benefits from reduced ineffectiveness. Whatever the system, if the game is being played according to the rules, accidents are due to extraneous causes—mostly human weaknesses—and are not the fault of the system.

I would challenge Fromm to support the statement that delays, diversions, or even cancellations (other than strike-caused) have resulted in diminution of demand, or that any businessman, because his plane was late in departing or arriving or was diverted by fog or congestion to another airport, elected to make a subsequent trip by car or train or to stay at home. The whole discussion of the value of personal time left me in a state of mild exasperation. Fromm tells us that no data are available on how businessmen occupy their time on airplanes. But whether a businessman reads reports, drinks martinis, talks to the stewardess, looks at *Sports Illustrated,* or prays that he will get down safely seems to me to be quite beyond the conjecture of the economist or the measurement techniques of the market surveyor. I simply do not agree that *a priori* estimates produce plausible figures of even potential willingness to pay for delay, and I doubt the usefulness of such figures, in any event.

Benefit Estimation

Concerning the discussion of benefit estimation for airport expenditures, I content myself with three comments:

1. In regard to data that would provide a basis for estimating the net benefits of an airport, Fromm states, "Such information is unavailable, and therefore any attempt to measure 'true' benefits is doomed to failure. All that can be done is to make what . . . seem to be reasonable assumptions." But who is to determine whether the assumptions are reasonable? In the mundane world, reasonableness is very much a matter of one's point of view and interests.

2. "As a result of wide divergences between airport locations and the air-traffic generating areas of communities," we are told, "the demand for and the value of air travel is reduced." This statement (if I understand it) seems to be pure conjecture within the limits commonly experienced in our large cities.

3. "Given dynamic elements in demand and competitive conditions, the tendency is to service points even when this is not justified by profitability standards." Maybe. But far more important for public policy, in my view, is another proposition: given the liberal subsidy policy of our government, not only is there a tendency but also sharp pressure on the part of the special beneficiaries to require service even when it is not justified by profitability standards or by any other criteria consistent with the national public interest. What then?

What I am about to say in conclusion applies not only to Fromm's paper but also to the central subject of the conference. When benefits can be readily quantified (and that, it seems to me, is not very often), by all means, let it be done. When they are large and, with respect to any given project, relatively few and direct, one need not wander far into the realm of conjecture. I doubt, however, that the infinite variety of individual utilities can be standardized or abstracted, much less aggregated.

As I suggested earlier, there is an alternative approach, which has been well described by Henry Fagin, Harold Wein, and others.[52] We might term it "an education-for-democratic-involve-

[52] See, for example, Fagin's chapter in Harvey S. Perloff, ed., *Planning and the Urban Community* (Carnegie Institute of Technology and University of Pittsburgh Press, 1961), pp. 105-120; and Harold Wein's paper delivered at the Conference on Transportation Research held at Woods Hole in August 1960.

ment" approach to public decision-making. It has its limitations, but it avoids the spurious process (or so it seems to me) of trying to quantify benefits that are indirect and very diffuse. In brief, its requirements are:

1. That money costs of all major public expenditures be described with accuracy and completeness in terms that can be understood by the decision-making representatives of the people and ultimately by the people themselves.

2. That what the expenditures are expected to do be described as completely as possible; never mind the benefits, or disbenefits (since they are seldom minded anyway).

3. That decision-makers and, wherever possible, the democracy, listen, consider, and aggregate net marginal utility by voting on proposals.

Obviously, the approach demands a long program of civic education and further development of political mechanisms (1) to bring about the necessary understanding and (2) to induce the necessary involvement of the populace. But since a great deal of time and money has been spent on theoretical refinements of cost-benefit techniques—with little practical effect—we can probably afford some exploration of possibly more practical alternatives.

WILLIAM VICKREY, *Department of Economics,*
Columbia University

I should like to comment briefly on four problems that have been raised in this discussion: (1) the tendency to base charges for transportation facilities on historical cost; (2) the problem of intangible, indirect benefits; (3) adaptation to errors in planning transportation facilities; and (4) the value of safety in transportation.

Use of Historical Cost

It is surprising that the idea of the just price persistently reappears in discussions of transportation pricing policies, even when the purpose is declared to be efficiency in the use of resources rather than the achievement of equity. Suggestions that the level of charges for airport services be set on the basis of average cost, or to recover the original cost of the facilities, are based to a large extent on equity considerations. Furthermore, it is always tempting to resort to a measure that will be certain and relatively easy to compute, even if it clearly leads to inefficiency, rather than a measure as conducive to efficiency as can be devised, even though this may require the use of estimated relationships involving a relatively wide margin of uncertainty. Administrative convenience and the striving for "equity" thus conspire to produce a policy of setting charges at a level that will recover the cost of the facility from its users.

Clearly, however, if efficiency is the goal, historical cost should not be used in the determination of charges. If an airport is constructed, it is quite irrelevant—from the standpoint of making the best use of it—that it may have been built at great expense rather than having been received as a free gift of nature. Taking cost into account suggests that one is prepared to allocate the use of the scarce air space, which cannot be added to, in terms of a price that reflects the cost of building airports, which can be added to. Moreover, to charge the aviation industry in general on the basis of gasoline use or number of passengers or gross revenues is clearly inefficient if, in fact, the problem is to alleviate congestion at specific places and specific times.

Two different types of situations may be distinguished. The first is the case of the small-town airport, where, once the decision to build has been made, the marginal cost of additional use may be negligible within any reasonable planning horizon. The second is the case of the metropolitan airport, where the important marginal cost may be the cost of economizing the air space rather than the cost of creating additional airport facilities.

Typically, one finds that there is considerable congestion at a big city airport on Friday afternoon. At other times, traffic flows smoothly. The usual solution to this problem is to build a new facility, rather than to attempt to ration the use of the existing facility. Since the new facility is more elaborate and more costly than the old, prices are raised. Actually, rational allocation of resources in such cases would require the reverse behavior: a high price should be charged *before* the new facility is completed, and the price should be lowered *after* it is built to ensure that full use will be made of it. Such an arrangement, however, would hardly meet the approval of a person who believes that construction costs must be recovered from the users of the facility.

Measurement of Intangible, Indirect Benefits

In 1844, the originator of the concept of consumers' surplus, Jules Dupuit, declared firmly that "there is no utility that nobody will pay for." This does not mean that indirect benefits should not be counted if collecting a payment for them is actually impossible. It merely means that the value of a service is the whole area under the demand curve up to the point at which its use is cut off by charging a price. Admittedly, if discriminating perfectly by charging different prices to different users is not possible, the maximum revenue that can be obtained from consumers will be something less than the full value of the service to them. Nevertheless, if there is some way of estimating what this ultimate willingness to pay might be, I would be prepared to rely on it, at least conceptually, as a basis for making "lumpy" decisions, whether as to large lumps of investment, or as to radical discontinuous changes in operating practices.

Suppose, for example, the construction of a local airport is justified in part on the basis that a community which does not have its own hospital will benefit if its residents can be taken to a neighboring hospital more quickly when emergencies occur. The value of this benefit is represented in the upper portions of the demand curve—it is part of the total value being estimated. In principle, this is no different from trying to estimate the amount that a businessman might be willing to pay if he is in a hurry. This might be reflected in the middle portion of the demand curve.

In brief, I believe that there is no sharp dichotomy between the direct measurable benefits and the indirect benefits that were discussed at this conference. In principle, the indirect benefits are represented in the demand curve, but actually they are merely further up on the curve than the direct benefits. An attempt to discriminate between them is, in my opinion, essentially arbitrary and probably futile.

Adjustments to Planning Errors

Government officials (like businessmen) may make mistakes. They may err as to whether an airport should be built, where it should be built, and when. And the doctrine that, if it is built, the users must be saddled with the cost simply compounds whatever errors were made originally. For example, if a newly built airport attracts very little traffic to it, an attempt to recover the original cost by raising charges will drive out what sparse traffic there is. Thus the dubious benefits from the mistaken expenditure are even lower than they might have been had the pricing policy been rational.

Conversely, if an inadequate airport is built and subsequent greater-than-cost charges cannot be levied, congestion is created. The community derives less benefit from the airport than it would if traffic were restrained by appropriate increases in charges because, although the volume of traffic is larger, this gain is more than offset by increased costs, including the cost of added delay and congestion for the more urgent uses.

The Value of Safety

A great many irrational evaluations of the safety element in transportation have been made—not so much in overestimating or underestimating the value of safety, but in being concerned with safety *only* when somebody can be held directly responsible. I have in mind what seems to be happening in the New York subways: the signal engineers apparently believe that the essence of their job is prevention of rear-end collisions. As a consequence, the signaling system is designed with a tightness that means, among other things, that the trains cannot be operated as frequently as somewhat lower standards would allow them to be. The net effect is more overcrowding on the subway, which impels more people to use surface transportation, whereby more people may be killed on the New York streets and the intersuburban highways than are saved by the strict signaling standards—but this is not brought home to the signal engineer.

A somewhat similar relationship exists between highway traffic accidents and aviation accidents. If an air crash occurs, it is conspicuous news, and the ensuing federal investigation, in which an attempt is made to place the blame on someone or something, is also conspicuous. Highway accidents involve fewer people per accident and are therefore far less conspicuous. In the aggregate, however, many more fatalities occur on the highways than in the air, even on a per passenger-mile basis. Yet there is virtually no attempt to fix the blame on the persons who may be more or less directly responsible for so many deaths.

Such examples indicate that our evaluation of the safety factor in transportation is probably quite irrational in significant respects. Much more work needs to be done on this problem to maximize safety on an overall, rather than a fragmented, basis.

Concluding Statement

GARY FROMM

Reply to Mr. Murray—General Criteria

Gordon Murray's comment indicates that he has long labored in the bitter vineyard of the Budget Bureau's bureaucracy and long battled the ubiquitous lobbyist. In a country as large as the United States, it cannot be expected that all government actions will benefit the nation as a whole. In fact, a good many will not. Even national security expenditures benefit some groups more than others—for instance, the exposed industrial targets that garner most of the country's strategic protection as well as its defense production reap the greatest gains from larger defense outlays. All that can be done in a large, geographically dispersed society such as ours is to design a balanced set of efficient equitable programs that *in toto* will increase national welfare. Any single program of the set need not have the support of the majority of the electorate nor need it be ideal; it may merely be a counterbalancing force for other inequities. For example, states with disproportionate areas of nontaxable federal lands frequently are accorded disproportionate shares of grants-in-aid programs. Similarly, the relative welfare of different income groups can be adjusted, not by tax and transfer measures alone, but also by differential expenditures on their behalf.

Yet Murray's caveat is well taken. The danger that some interests will gain more than others is ever present, especially since the democratic and legal mechanisms of our society place such heavy reliance on adversary proceedings. In the long run the meek may inherit the earth; in the short run, however, the weak succumb to the strong, who unfortunately are not always the righteous. Therefore, for programs that do benefit particular industries or groups, care must be taken to ensure, in whatever manner possible, that a desired distribution and advancement of national welfare is achieved.

That we do not now possess the measurement tools to accomplish the allocation of funds on the basis of marginal social utility, I agree. That I necessarily restricted myself to some general observations, I also lament. But my essay assignment to discourse on benefit estimation for civil aviation expenditures may already have strayed too far afield in even outlining the principles with which Murray takes so much issue.

Moreover, his own proposal of adherence to the Fagin, *et al.,* approach can only lead to retrogression in the efforts to upgrade the present *ad hoc,* arbitrary, and sometimes capricious techniques of federal budgeting. Certainly there can be no argument about detailing the money costs of all major public expenditures with completeness and accuracy. Nor can there be any quibble about describing as completely as possible what the expenditures are expected to do, or about representative or electorate voting on proposals. But I cannot agree that it is desirable to ignore benefit quantification that is difficult and to presuppose that approval of a vaguely defined program implies a correct allocation of funds within that program.

Government agencies are all too prone to formulate expenditure proposals along the following lines: Item: 10,000 black boxes; Cost: $5,000 each; Justification: will reduce the variability of the TVOR direction finder signal by 10 seconds of arc. Or, for a major program: Item: $50 million grants-in-aid for school construction; Cost: $50,000 per classroom; Justification: to reduce "overcrowding" in existing metropolitan schools. The relevant questions here are not what physical changes will be effected, but what benefits will accrue. We must ask whether the black boxes reduce ineffectiveness; and if so, by how much? And how much is this worth in comparison to the cost (of black boxes)? We also must ask whether reducing overcrowding improves the quality of education and attempt to get some measure of the degree of change, and its value. Assuming that governmental resources are limited, and they are, only then is it possible to choose rationally between more classrooms, higher teaching standards, additional vocational training later in life, parent education for home training, educational home television, and so forth.

Comparisons between black boxes and classrooms are even more difficult, but unless an attempt is made to quantify the benefits of both, the choice between them gets mired in the quicksand of bias, conjecture, and ignorance. As in projects that involve saving human lives, the failure to assign explicit values merely results in the choice of implicit ones. Mr. Murray alleges that we have already spent too much on cost-benefit analysis designed to aid in these decisions, but with little practical effect. In my opinion, the failure to reap the gains of applying the technique lies more with the failure of the Budget Bureau to demand its proper utilization than with the shortcomings of the analysis itself.

Some Specifics

A reply to the particular points raised in Murray's comments is also needed. First, as my essay states, accidents are the most important manifestation, both on a unit- and total-cost basis, of ineffectiveness. Murray tells us that the air traffic system and procedures are designed "to achieve *maximum* safety" (italics mine) and that "accidents are due to extraneous causes—mostly human weaknesses —and are not the fault of the system." The sooner we dispose of such myths the better. Human beings, of course, are part of the man-machine aviation system and their errors, whether they are made by controllers or pilots are subject to reduction by substitution and addition of more navigational and other aids.

Obviously, the simplest way to eliminate aviation accidents is to ban air traffic. Such a proposal is not as ridiculous as it may seem. Figures derived from a study of airline landing and approach accidents for the period 1946-1958 by the Flight Safety Foundation reveal that 77 percent of all such incidents in the United States occurred at night; additionally, 94 percent of all accidents took place under instrument flight rule (IFR) conditions.[53] As to fatalities, 64 percent of total U.S. fatalities took place at night, 97 percent of the total under IFR. Thus, it would appear that some curtailment of night IFR operations would have a beneficial effect on the aviation accident rate. Furthermore, since over 70 percent of all U.S. air carrier accidents take place in the terminal area,[54] stricter minimum landing requirements for night operations might well accomplish the desired traffic and accident diminution. (Current FAA regulations do prescribe stricter minima for night operations. Consideration could, however, be given to making these even more stringent in order to attain lower accident rates.) Another means would be to provide improved takeoff, landing, and ground and terminal area guidance equipment. Either alternative entails costs, whether they be in the form of the net value of curtailed demand or in the form of direct outlays. Only when accidents have been eliminated can it be said we possess a maximum safety system.

Finally, Murray's denial that ineffectiveness and poor airport location reduces demand disregards the available evidence. Not only have passenger surveys confirmed that individuals anticipate and react to potential ineffectiveness, but statistical analyses show significant reductions in demand when delays, cancellations, diversions, and accidents are at their peak.[55] The impact of airport location on demand has also been confirmed in numerous studies.[56]

Reply to Mr. Vickrey

As the present high priest of imposing surcharges to limit congestion, it is not surprising that William Vickrey advocated the remedy

---

[53] Otto E. Kircher, *Critical Factors in Approach and Landing Accidents, Part I, Statistics* (Flight Safety Foundation, Inc., 1960), pp. 19-21.

[54] Civil Aeronautics Board, *U.S. Air Carrier Accidents: Statistical Review* (1959).

[55] See Fromm, *op. cit.*, Chap. VII.

[56] For example, see Landrum and Brown, *Study of the Effect of Airport Accessibility on Realization of Air Passenger Potential* (Cincinnati, April 1961).

once again in his comments here. As a matter of fact, I accept the
principle in my paper. Nor is it inconsistent that at the same time
he adheres to short-run marginal cost (SMC) pricing. Yet, as I have
pointed out above, the adoption of SMC pricing under conditions
of economies of scale almost inevitably leads to unjustifiably high
demands for competitive, publicly produced goods and services in
relation to those produced by the private sector. Capital put in
place cannot simply be treated (as Vickrey is wont to treat it) as a
free good. A private firm that adopted such an approach would soon
go out of business; a government that does so soon finds itself con-
fronted with ever-increasing deficits and a distortion of resource
allocation. Full cost recovery is desirable on equity grounds but,
additionally, in free enterprise systems its economic rationale is even
more compelling. This does not require that equal repayment be
obtained in each time period; equal charges for incidence of use
would be more appropriate. However, where planning errors in
the direction of overcapacity have been made, downward adjustment in charges for airport use may be desirable.

Vickrey's second concern, the close similarity between direct
and indirect benefits, can be stated even more succinctly—"benefits
are benefits." I agree. Nevertheless, to make any indirect benefits
the justification for government subsidy for publicly produced but
privately competitive goods is fallacious. On Vickrey's grounds, *re-
ductio ad absurdum*, almost every "lumpy" productive private investment (including the local taxi-cab operator) might be entitled
to public assistance. Of course, there are cases where unique, large
external effects do justify subsidies.

I concur with Vickrey on his last point, but would go even
further. More needs to be done not only to increase safety on an
overall transport network basis, but also to increase the efficiency
of the total system. The government should regard transport not
as individual modes—air, highway, rail, and water—but as a function to satisfy movement demands. Only then can an efficient national transport system be designed.

HERBERT MOHRING*

Urban Highway Investments

IN COMPARISON TO THE ANALYSIS of many other types of public outlay, optimizing the expenditures on transportation facilities and measuring the benefits attributable to them seem almost simple tasks. Like all other public expenditures, transportation investments, in addition to yielding primary benefits, can have substantial income redistribution effects. However, unlike public housing and urban renewal expenditures, they do not appear to have technological or pecuniary externalities as their major ends.[1] Furthermore, the aesthetic, humanitarian, and other "nonmarket

* Department of Economics, University of Minnesota.

[1] Actually, the income redistributional aspects of public transportation expenditures reflect both technological and pecuniary externalities. That these externalities are not major ends of government expenditures does not, of course, make them any less real. It seems doubtful that those people on whom externalities impinge would be made appreciably happier or sadder by learning that their windfall gains or losses are accidental and do not result from explicit introduction of their utility functions into the planning process. This statement applies with equal force to farmers whose rocky pastures are converted by interchanges into motel sites and to city dwellers whose streets are made quieter by traffic diversion or who may suffer in any of a variety of ways from being located in or near the path of a new highway. In any case, the benefit measures developed in this paper ignore externalities—plus or minus, pecuniary or technological. My basic excuse for this shortcoming is the conventional one: the data required to place dollar values on externalities are lacking.

benefit" arguments that are often used to justify subsidies to such areas as education, research, and the arts seem to apply little to transportation.

Despite some few exceptions, the existence of tangible dollar benefits has been the typically argued justification for federal subsidies. And for highways (which compose the largest category of transportation investment expenditures) the principle of 100 percent—or greater—user responsibility seems firmly established in the federal government and in almost all state governments.

In brief, the vast majority of highway and, more generally, transportation investment benefits are either directly or indirectly reflected in market transactions. In sorting these transactions out to determine optimum expenditure programs and to measure net benefits, it seems reasonable (or at least more reasonable than it would in dealing with certain other public expenditure categories) to ignore the redistributions of income that may be reflected in these transactions. And contrary to what seems to be common belief, in this process of optimization and measurement one can place exclusive reliance on the tools that a competent economist would use to place a value on a dam, a steel mill, or any other public or private investment project. That movement through space is involved rather than production at a single point in space does not mean that special tools of analysis or special standards of valuation must be developed.

To be more specific, in the absence of scale economies, the procedures a highway authority ought to use in establishing and pricing an optimum highway network are formally identical to the process through which a competitive market reaches long-run equilibrium. Employment of the following two simple operating rules would lead to both a Pareto-optimal utilization of an existing (perhaps nonoptimum) transportation network and, ultimately, a long-run optimum network: (1) establish short-run marginal cost prices for the use of each link in the existing network; (2) alter the size of each link to the point where toll revenues equal the total costs to the authority of providing that link. Ways in which the validity of this perhaps surprising contention could be demonstrated are sketched in the section that follows this introduction.

In the second section, information is developed on some of the most important costs of urban highway transportation. Section three begins with brief discussions of the techniques of relevance in measuring investment benefits in competitive economies and of the "noncompetitive" elements that complicate the analysis of · highway investment benefits, and concludes by presenting rough estimates of the benefits of new expressway construction, obtained by implementing the competitive benefit estimation framework with data on the levels at which the Minneapolis-St. Paul metropolitan area highway network is currently being utilized. To avoid bogging the discussion down with statistical minutiae, the more involved computational procedures are relegated to an appendix.

Characteristics of an Optimum Transportation System in a Competitive World[2]

Since the Pigou-Knight "two-roads" controversy, if not before, economists have generally been aware that a zero price for trips on a highway network would, in general, lead to a misallocation of resources even though additional trips impose no costs on the highway authority involved.[3] This is true because the cost of a trip on a given road increases with the level of traffic on that road. On most highways these congestion effects set in at fairly low traffic levels and increase steadily as the level of traffic increases; they do not (as the engineer's use of the term "highway capacity" might seem to connote) set in only at a substantial traffic level and increase rapidly thereafter.

To put it differently, above some typically small traffic level, addition of a vehicle to the traffic stream on a highway imposes a definite if small cost on each of the other vehicles comprising that stream. Therefore, in the absence of tolls, the marginal private

[2] This section is an outgrowth of work done at the Transportation Center at Northwestern University in 1960-61 under the sponsorship of the Automobile Manufacturers Association and the U.S. Bureau of Public Roads. I am deeply indebted to Constance Schnabel, Robert Strotz, and Martin Bailey for many helpful suggestions.
[3] See A. C. Pigou, *The Economics of Welfare* (Macmillan, 1920), p. 194; and F. H. Knight, "Some Fallacies in the Interpretation of Social Cost," *Quarterly Journal of Economics,* Vol. 38 (1924), pp. 582-606.

cost of a trip on the highway would be less than its marginal social costs. Since the demand for trips is presumably not totally inelastic and since drivers can be expected to base their travel decisions on the private rather than the social costs of their trips, an optimum level of utilization of the highway could be achieved only by levying positive tolls.

To repeat, that the congestion phenomenon implies the social desirability of user charges for highways and other transportation facilities is by no means a novel conclusion. The problems involved in optimizing the use of an existing highway network through the imposition of congestion tolls have been examined in several studies.[4] However (with, to my knowledge, only two exceptions[5]) none of these studies has gone on to analyze the relations between optimum tolls and the costs of an optimum transportation network. The prevailing impression appears to be that these two quantities are quite unrelated. For example:

Since user fees limited in this way [i.e., to the difference between short-run marginal private and social costs] may not create total revenues sufficient to attract capital to highways and in limited cases may yield more revenues than could be invested efficiently in highways, the marginal-cost pricing economists tend to deny that any relation, or close relation, should exist between user fees and capital investment. The rule of self-liquidation as a general guide to efficient investment is thrust aside as unnecessary and as a substantial hindrance to efficient utilization of existing highways. What specific rule for efficient road investment is to be substituted is far from clear; presumably it would be a matter for planning according to social surplus criteria of investment, often involving subsidy expenditures.[6]

[4] Most notable among them is *Studies in the Economics of Transportation* by Martin Beckmann and others (Yale University Press, 1956.)

[5] Herbert Mohring and Mitchell Harwitz, *Highway Benefits: An Analytical Framework* (Northwestern University Press, 1962), pp. 80-87; what follows in the present paper is an extension of the reasoning developed in the 1962 study. Robert Strotz has recently developed alternative and in several respects more general proofs of these propositions in "Urban Transportation Parables," which is scheduled to appear in a forthcoming collection of papers presented at a 1964 conference sponsored by the Resources for the Future Committee on Urban Economics.

[6] James C. Nelson, "The Pricing of Highway, Waterway, and Airway Facilities," *American Economic Review*, Vol. 52 (May 1962), p. 426.

Advocacy of short-run marginal cost pricing does not necessarily imply advocacy of thrusting aside "the rule of self-liquidation as a general guide to efficient investment . . . as unnecessary and as a substantial hindrance to efficient utilization of existing highways." That this is the case can perhaps most easily be seen by establishing the relationship between the process of determining and pricing an optimum highway system and the process by which, say, a competitively organized widget industry reaches long-run equilibrium.

In the short run, a competitive widget producer can be relied upon to establish that level of output which equates his short-run marginal cost and the prevailing market price. Such an equality is a necessary condition for Pareto optimality. That is, "price equals short-run marginal cost" is a necessary condition for achieving an organization of economic activity in which it would be impossible to make some individual or group better off without simultaneously making some other individual or group worse off. The revenues resulting from a "price equals short-run marginal cost" output policy can conveniently be broken into two parts. First are those funds that go to cover the widget producer's variable costs—outlays for factors such as labor and raw materials whose use can be varied on short notice. Second is the difference (normally positive, but never negative) between his total revenues and his variable costs. This difference is commonly characterized as a quasi-rent on the widget producer's fixed capital equipment.[7] Given constant returns to scale, it can be shown to equal the market price of widgets times the marginal physical product of that equipment.

In the short run, the quasi-rents earned by the owners of widget-producing capital equipment need not equal the cost of that equipment. If quasi-rents exceed the market rate of return on the outlay needed to reproduce these assets, new productive capacity can be expected to enter the industry. As a result, output levels will increase and both prices and quasi-rents will decline. Long-run equilibrium can be defined as having been reached when

[7] Ignoring, for simplicity, the possible existence of specialized entrepreneurial skills or other attributes of the firm to which rents might be imputed.

sufficient new capital has entered the industry to equate quasi-rents with the market return on reproduction costs.[8]

Differences Between Highways and Normal Commodities

Differences do exist between highway networks and the typical competitive industry of economics texts that would invalidate the statement resulting from substitution of "highway" for "widget" wherever it appears in the preceding paragraphs. Differences that can for convenience be grouped under the heading "noncompetitive elements" are discussed briefly later. Three additional differences warrant discussion at this point.

First, in textbook discussions, business firms are normally depicted as hiring or owning all of the inputs required to place their products in the hands of their customers. However, the buyers of transportation services must provide at least one input—their own time or that of the goods they ship—essential to the provision of the services they consume. Indeed, those who travel in private passenger vehicles and other highway users provide almost all of the variable inputs as well as some of the capital inputs required for their trips.

That buyers of trips play both a consuming and producing role does not eliminate the necessity of short-run marginal cost pricing to the achievement of Pareto optimality. However, the toll required to establish a marginal cost price for a trip is equal, not to the trip's short-run marginal cost, but rather to the difference between this magnitude and the variable costs borne by the trip taker in question. This toll would cover such of the variable costs of trips as are borne by the highway authority (e.g., some maintenance expenditures) plus a quasi-rent on the capital invested in highway facilities. Just as with the widget industry, in the short run this quasi-rent need not equal the market return on the reproduc-

---

[8] Also being ignored here are those costs of maintaining capital equipment that are independent of the level at which that equipment is used. Taking them into account would require modifying the end of this sentence to read something like this: " . . . the market return on reproduction costs *plus* those maintenance and depreciation costs that are independent of the rate at which capital is utilized." Those maintenance and depreciation costs that do depend on the rate of capital utilization are, of course, included in variable costs.

tion costs of these facilities.[9] Just as with the widget industry, if the quasi-rent exceeds the market return on the costs of reproducing a segment of a highway system, expanding that segment would be in order. And again, just as with the widget industry, long-run equilibrium could be defined as having been reached when the quasi-rent on each segment of the highway system equals the market return on its reproduction costs.

The second difference warranting discussion at this point is that the industry typically treated by economics texts is one faced by a demand relationship which remains fixed for a more or less substantial period of time. Discussion of shifts in industry-demand relationships is typically limited to shifts that give rise to the entry or the exit of productive capacity. However, many—perhaps most—industries face demand relationships that vary regularly and more or less predictably from hour to hour, day to day, week to week, or season to season—periods of time too short to permit capital stock to be altered. Clearly, highway and most other forms of transportation are characterized by this sort of demand variability. For example, during the average 1958 weekday the hourly traffic on the highway network of the Twin Cities metropolitan area ranged from a low of about 14,000 vehicle miles between 3 and 4 A.M. to a high of around 1 million vehicle miles between 4 and 5 P.M.[10]

For the purposes at hand, it is convenient to distinguish two groups of variable demand commodities: (1) those for which storage costs are, and (2) those for which they are not, small enough that the effects on output levels of short-term demand variations can be moderated appreciably by inventory variations. Discussion of the behavior of the second group—to which transportation activities seem clearly to belong—is considerably simplified because consumption during each demand period must be met by production during that period. Thus, demand periods are linked only by a fixed capital plant. The existence of a variable stockpile of finished goods need not be taken into account.

Suppose, then, that widgets belong to this latter group of com-

[9] Plus those maintenance and depreciation costs that are unrelated to highway use.
[10] See section below on "Benefits of Urban Freeway Investments," and Appendix, pp. 271 ff.

modities. Suppose, to be more specific, that the widget industry is
competitively organized; that widgets cannot be stored; that the
demand for them is substantially greater during the fall and
winter than during the spring and summer; and that widget-
producing capital equipment cannot be employed in other activities and cannot be varied over the demand cycle. Under such
circumstances, in each demand period each widget producer could
be expected to establish an output level which equates his short-
run marginal costs with the market price. As in the invariant demand case, the revenues a widget producer receives in each period
can be broken into two parts—variable costs and quasi-rents. If,
as is normally the case, widget short-run marginal costs are increasing, market price, output, average and total variable costs, and
average and total quasi-rents during the peak period will all be
greater than during the off-peak period.

In the short run, the sum of a widget producer's peak and off-
peak period quasi-rents need not equal the market return on the
outlay required to reproduce his capital plant. Should the sum of
these quasi-rents exceed the market return, it would pay the
widget producer and others to invest in new widget-producing
equipment. Doing so would lead to increased output, a lower
price, and reduced quasi-rents in both demand periods. Similar
to the invariant demand case, long-run equilibrium can be defined as existing when the sum of peak and off-peak quasi-rents
equals the market return on the cost of replacing widget-producing capital.

In brief, the only appreciable difference between the conventional invariant demand case of economics texts and that dealt
with here is that "quasi-rent on invested capital" must be replaced by the "sum of quasi-rents over the course of the demand
cycle" in defining long-run equilibrium.[11] In turn, the only for-

[11] It may be worth noting that this conclusion seems to be at variance with that
reached by Steiner, Hirshleifer, and others in discussing the so-called "peak load
problem" faced in electric power generation and other activities. See Peter O.
Steiner, "Peak Loads and Efficient Pricing," *Quarterly Journal of Economics*, Vol.
71 (November 1957), pp. 585-610; Jack Hirshleifer, "Peak Loads and Efficient Pricing: Comment," *ibid.*, Vol. 72 (August 1958), pp. 451-462; and Steiner, "Peak Loads
and Efficient Pricing: Reply," *ibid.*, pp. 465-468. Hirshleifer, for example, con-

mal difference between widget production and highway transportation is that transportation consumers provide some of the
variable costs of their trips. This fact has been found to yield
characteristics of an optimum transportation system that do not
differ fundamentally from those of the conventional competitive
industry. It thus seems to follow that, in the absence of scale
economies, short-run demand variations also imply characteristics
of an optimum highway system not differing fundamentally from
those of the conventional competitive industry.

One final difference between highway transportation and the
competitive industry of economics texts warrants consideration at
this point. As an example of what seems at first glance at least to
be the most plausible of the common arguments for subsidies to
mass transit systems, consider:

To make people pay what it costs is self-defeating for the reason that
one of the broad social justifications of a new investment in rapid
transit facilities is to relieve the urban traffic dilemma by inducing
people to give up the use of private motor vehicles or to remain on
public transportation if they are being discouraged by poor service.[12]

Such statements seem to amount to the contention that some
auto drivers (or the public at large) ought to be willing to subsidize (i.e., to bribe) other auto drivers to switch to transit, because
the remaining auto drivers would benefit from the resulting reduction in highway congestion. Put in these terms, the argument
loses some of its initial appeal; after all, an analogous argument
could be applied to any other pair of substitute commodities. A
shift from orange to apple consumption, for example, would (at
least in the short run) reduce orange prices and increase apple

---

cluded (p. 462): "It is clear that any pricing principle based upon allocating . . .
total capacity costs will not, except in special cases, be consistent with efficient
prices in terms of the marginal conditions." This conclusion may very well be correct. However, the culprit is not short-run variation in demand, but a level of
capital investment unequal to the long-run optimum level or something other
than constant returns to scale.

[12] William Miller, *Metropolitan Rapid Transit Financing*, A Report to the
Metropolitan Rapid Transit Survey of New York and New Jersey (1957), p. 62.

prices—thereby benefiting both apple producers and the remaining orange consumers. Both groups therefore ought to be willing to offer bribes to encourage such a shift. On the other hand, both orange producers and apple consumers would suffer from the shift and therefore should be willing to offer at least partially offsetting bribes. The two sets of potential bribes would, in fact, exactly offset each other if both orange and apple markets were initially in long-run competitive equilibrium.

Clearly, at least some of the same considerations apply to the markets for auto and transit trips. While auto drivers and transit operators would benefit from a shift away from auto transportation, present users of public conveyances would likely suffer. Public conveyances would become more crowded and the duration and number of stops would likely increase. Thus, just as in competitive apple and orange markets, the two sets of bribes might cancel each other out.

However, at least one peculiarity of transportation activities suggests that the orange-apple analogy is inapplicable.[13] As was noted earlier, users of transportation facilities supply at least one variable input required for the provision of trips—their own time or that of the goods they ship. Contrary to one of the implicit assumptions on which the preceding analysis was based, the value of this scarce resource quite likely varies substantially among users; this would pose no problems if neither indivisibilities nor scale economies were involved in the provision of transportation facilities. Under such circumstances, each origin-destination pair could be provided with separate facilities for each travel-time valuation class. The preceding analysis would then apply to each of these facilities.

But indivisibilities do exist in the provision of transportation facilities. Each railroad track must have two rails, and each highway or country road must be at least as wide as the vehicles that use it. Producers of trips having widely different costs therefore normally utilize any given transportation facility. That the occupant of one vehicle values his time at $50 an hour and the oc-

[13] Another peculiarity, the existence of substantial indivisibilities in the provision of mass transit services, is discussed briefly in a later section.

cupant of another at 10 cents might well seem to make the orange-apple analogy inapplicable. That is, this presumed fact might seem to open the way to mutually beneficial and socially desirable bribes even if both public and private transportation were in the equivalent of long-run competitive equilibrium.

Unfortunately, I have been unable to devise a rigorous non-mathematical way of evaluating this possibility. Therefore, the following assertion must suffice: recognition of differences in the travel-time valuations of individual travelers does not lead to alterations of the above conclusions. It can be shown that, in the absence of scale economies, for a long-run optimum highway network (i.e., one for which the quasi-rents generated by marginal cost tolls equal the market return on invested capital) the benefit-maximizing toll system would be one requiring each member of each user class to pay exactly the costs his trip imposes on others. It would be undesirable, from the viewpoint of either society or auto drivers, to have each auto driver pay *more* than the costs he imposes on other travelers, thereby encouraging still more people to travel by bus or subway.

This contention can perhaps be made more plausible if it is put in a broader context. Individual tastes in various commodities —furniture, clothing, housing, and autos, for example—do differ. This being the case, if a variety of choices was available in these product groups (and no cost penalty is involved), society as a whole would seem better off than if alternatives were few in number. The more alternatives available, the more likely is each consumer to find a combination of specifications that conforms closely to his tastes. So, too, with tastes in transportation routes; most importantly, the rates at which individuals would be willing to exchange dollars for time vary considerably. Thus (again, if no cost penalty is involved), the availability of routes possessing a wide variety of toll and time combinations would clearly give each traveler a better chance of finding a personally optimum travel mode than if no choice was available. However, the fact that tastes differ does not in itself justify subsidizing either a particular product or those who buy it.

The Costs of Urban Transportation

It is useful to distinguish four categories of urban transportation costs: the opportunity costs of vehicles; the opportunity costs of the people and goods that occupy the vehicles; vehicle operating costs; and the capital and maintenance costs of the roadways on which the vehicles travel. The cost items that fall within these categories are so many that, given the space limitations of this paper, they cannot possibly all be dealt with here. Only the following components of urban transportation costs will therefore be treated:

1. The opportunity costs of private passenger vehicle occupants—that is, the value they place on their travel time.[14]
2. The relationships between travel time and vehicle operating costs on the one hand and the rates at which freeway and arterial street capacity is utilized on the other.
3. The capital costs of freeways.

Operating Speed, Operating Costs, and the Value of Travel Time

Travel time is bought and sold in no existing market; assigning a value to it is therefore rather difficult. Indeed, the authors of the AASHO "Red Book" (1960), apparently regarding objective determination of its value as impossible, proceeded to pick a number out of the air: "A value of travel time for passenger cars of

[14] In dealing with the costs of private passenger trips, transportation studies commonly assume that the opportunity costs of the vehicles themselves are zero— that the only relevant costs are outlays for gasoline, oil, tires, vehicle maintenance, and the like. This assumption has been adopted here, although not without considerable misgivings.

The opportunity costs of a car are quite likely zero when it is parked at its owner's home. It does not follow, however, that its opportunity costs are zero when its owner or some member of his family is driving it. Clearly, the opportunity costs of a car used, for example, to drive to work are greater than zero if someone else would have used the car had it been left at home. Warner found the presence of other drivers in a family to be among the most important variables determining the choice of mode for a work trip. See Stanley Warner, *Stochastic Choice of Mode in Urban Travel* (Northwestern University Press, 1962).

$1.55 per hour, or 2.59 cents per minute, is used herein as representative of current opinion for a logical and practical value."[15]

Even accepting this rather conservative value, travel time turns out to be by far the most important cost of urban travel. It thus seems worthwhile to point out that a value of travel time is implicit in a driver's selection of a target speed. An increase in his speed reduces the time costs of a trip. Also true, however, is that an increase in speed definitely increases vehicle operating costs, and probably increases accident costs and the psychic wear and tear of driving. A utility-maximizing driver would presumably select that operating speed which minimizes the total costs of his travel. Sufficient data are available to provide a rough estimate of the distribution of values implicit in choices of operating speeds.

To be more specific, it seems reasonable to write a representative vehicle operator's total trip costs per mile, C, as

(1) $$C = F(S, N, \overline{Z}) + V/S^*(S, N, \overline{Z}),$$

where F denotes all trip costs other than that of travel time; V is the value the driver and his passengers collectively place on an hour's travel time; and S^* is the speed at which they actually travel. V/S^*, then, is the time cost of traveling a mile. Both F and S^* are presumed to be functions of the driver's *desired* speed, S; traffic volume, N; and such highway characteristics as the number and width of lanes, curvature and grade standards, and access controls (the vector $\overline{Z}$).

On a very lightly traveled road, a driver can attain his desired speed. That is, on such a road, S^* equals S and equation (1) therefore becomes

(1′) $$C = F(S, O, \overline{Z}) + V/S.$$

To repeat, a utility-maximizing driver would attempt to travel at the speed which would minimize the total costs of his trip. Differ-

[15] American Association of State Highway Officials, Committee on Highway Planning and Design Policies, *Road User Benefit Analysis for Highway Improvements* (1960), pp. 103-104. The 1960 value of $1.55 an hour appears to have been obtained by adjusting a 1949 estimate of $1 an hour to reflect price and income increases during the subsequent eleven years.

entiating equation (1') with respect to S and setting the resulting expression equal to zero yields

(2) $V = S^2 \partial F/\partial S.$

Sufficient data are available on the effects of speed on gasoline and oil consumption and tire wear—the major components of vehicle operating costs—to enable estimation of V at alternative values of S.[16] On straight, level rural roads and on urban expressways where traffic signals and stop signs do not interrupt traffic flows, data in the "Red Book" suggest that, regardless of traffic level, approximately the following costs prevail for representative vehicles:[17]

(3a)   gasoline: $\$0.30/(13.2 + 0.40S - 0.0076S^2)$

(3b)   oil: $\$0.45/(1600 - 21S)$

(3c)   tires: $\$0.0010 + 1.5 \times 10^{-8}S^{3.2}$

Regarding $F(S, O, \overline{Z})$ as equal to the sum of these equations, Table 1 lists the travel-time values implied by equation (2) for representative values of S. If (as the *Highway Capacity Manual* indicates[18]) desired speeds on high-quality, straight, level rural highways are approximately normally distributed with mean and standard deviation of 48.5 and 8 miles per hour respectively, $\overline{V}$, the mean travel-time value for the occupants of *all* vehicles can be obtained by evaluating

(4) $\overline{V} = \int_{-\infty}^{\infty} S^2 \; \dfrac{\partial F}{\partial S} \; n(S; 48.5, 8)dS,$

which works out to be approximately $\$2.80$.

The analysis and the data underlying these estimates are rather

____

[16] Few if any data are available on such important subjects as the effects of highway characteristics, traffic volume, and desired speeds on the frequency and severity of accidents and on driver comfort and convenience. It seems reasonable to expect that both accident and comfort and convenience costs increase with both traffic density and desired speed. Since no data are available on these costs, however, it is assumed in what follows that both of them are unrelated to both traffic density and desired speed. This assumption, it should be noted, imparts a downward bias to the estimated distribution of travel-time values.

[17] AASHO, *op. cit.*, pp. 100-126. Gasoline, oil, and tires are respectively assumed to cost 30 cents a gallon, 45 cents a quart, and $100 a set.

[18] U.S. Bureau of Public Roads, *Highway Capacity Manual* (1950), p. 32, Fig. 6.

TABLE 1. Travel-Time Values of Vehicle Occupants Implied by Alternative Desired Operating Speeds

Speed (S)	$V = S^2 \dfrac{\partial F}{\partial S}$
(m.p.h.)	*(dollars per hour)*
20	−0.02[a]
30	0.13
40	0.62
50	2.06
60	7.38
70	7.82

[a] Minus sign indicates that travel time (at speeds less than about 25 m.p.h.) has a negative value for travelers.

rough, to say the least. In addition, one might argue that they are of dubious validity both because the average driver is only dimly aware of the relationship between speed and operating costs, and because few people drive average cars. These two sources of variability imply that the true standard deviation of travel-time values is larger than that which would be computed on the assumption that all people drive average cars and are fully aware of the cost relationships involved. However, the mean of the distribution would be over- or underestimated by these procedures only if the average driver's estimates of the relationship between speed and operating costs were biased. No evidence of any sort exists on the nature of these possible biases.

Operating Costs and the Volume/Capacity Ratio[19]

If equations (3a, b, and c) are reasonably accurate in specifying the form of the vehicle operating-cost relationships involved in taking trips on rural roads and urban freeways, then travel time and perhaps comfort and convenience would appear to be the only user costs that vary with traffic density. That is, if the hourly output of trips between two points on a road is conceived of as a function, $N = f(T, C, \bar{Z})$, of three inputs—travel time (and com-

[19] The highway production functions and related data on which the operating-cost estimates presented in this section are based are developed in greater detail in the first section of the Appendix.

fort and convenience), T, operating costs, C, and highway characteristics, $\bar{Z}$—equations (3a, b, c) suggest that the partial derivative
of this function with respect to C is a constant.

Several studies have been undertaken that provide information
on the relationship between N and T. While their specific forms
differ, all are alike in an important respect: all imply that the trip
production function is one for which the marginal product of
travel time both decreases for all positive travel-time values and
becomes negative beyond some finite travel-time input. That is,
the trip production function has a "Stage III" and is one for
which the boundary between stages I and II occurs at either negative or zero travel-time inputs. Hence, there is a maximum rate at
which any given highway can produce trips. If the rate at which
people attempt to make trips exceeds this capacity level, the output of trips will actually fall.

Table 2 lists the average travel-time costs per vehicle mile that
are implied by these production functions at various volume/
capacity ratios and at alternative values of $1.55 and $2.80 per
hour. The specific values for any given ratio do differ somewhat;
however, all of the relationships are alike in implying that travel-

**TABLE 2. Travel-Time Costs per Vehicle Mile Implied by the Greenshields,
Norman, and Underwood Travel-Time Relationships**

(cents per mile)

Volume/ Capacity Ratio	Travel-Time Value = $1.55/veh. hr.			Travel-Time Value = $2.80/veh. hr.		
	"Green-shields"	"Norman"	"Underwood"	"Green-shields"	"Norman"	"Underwood"
0.1	3.3	3.3	3.3	5.9	6.0	6.0
0.2	3.4	3.5	3.5	6.1	6.3	6.3
0.3	3.5	3.6	3.6	6.4	6.5	6.5
0.4	3.6	3.8	3.8	6.6	6.8	6.9
0.5	3.8	4.0	4.0	6.8	7.2	7.2
0.6	3.9	4.1	4.3	7.1	7.5	7.8
0.7	4.1	4.4	4.6	7.3	7.9	8.4
0.8	4.4	4.6	5.1	7.9	8.3	9.3
0.9	4.9	4.9	5.8	8.9	8.8	10.6

Source: See Appendix, section on "Operating Costs and the Volume/Capacity Ratio."

time costs do increase substantially with increases in the rate at which highway capacity is utilized.[20]

The preceding discussion has been based on uninterrupted traffic flows of the sort that prevail on high quality rural highways and expressways. Analysis of such flows is quite simple, at least in comparison to the problems involved in dealing with urban arterial streets. The characteristics of traffic flows on arterials vary from complete congestion in the cues that form behind stop signs and red traffic lights, through an intermediate stage as vehicles accelerate to normal operating speeds on leaving these impediments, to the free-flow characteristics that correspond to the existing traffic volume as modified by whatever speed limits may be enforced. Traffic flow characteristics are also affected by, *inter alia*, the frequency of right and left turn maneuvers, the amount of pedestrian traffic at intersections, the amount of curb parking, the number and spacing of bus stops and signals, and (but only for low traffic volumes) the sequencing of red and green cycles at successive traffic lights.

Perhaps because of their complexity, the interrelationships among volume, density, and travel time on urban arterial streets have seldom been studied systematically. Of the few existing studies, only one provides information that can be analyzed in a fashion similar to that employed with the relationships developed for rural roads and urban freeways. This study suggests that, just as with freeways and rural roads, travel time per mile and hence travel-time costs on urban arterials vary substantially with the volume/capacity ratio (see Appendix, page 269).

An at least tentatively plausible case can be made for ignoring vehicle operating costs in relating travel costs to traffic volume on urban expressways and rural highways, but no such case can be made in respect to urban arterial streets. In the Chicago Area Transportation Study, both vehicle operating and accident costs for a trip segment were found to be negatively related to the

---

[20] These relationships could easily be made to yield relationships giving *marginal* travel times per vehicle mile and hence, given an estimated travel-time value, optimum tolls. To take the most extreme case, the Underwood relationship at $2.80 a vehicle hour implies optimum tolls of 0.2, 2.1, and 16.0 cents per vehicle mile at volume/capacity ratios of 0.1, 0.5, and 0.9, respectively.

TABLE 3. Time, Operating, and Accident Costs per Vehicle Mile on Two-Way Urban Arterial Streets at Various Volume/Capacity Ratios

(cents per mile)

Volume/ Capacity Ratio	Operating and Accident Costs (1)	Time Costs at:		Total Costs at:	
		$1.55/hr. (2)	$2.80/hr. (3)	$1.55/hr. (Col. 1+Col. 2)	$2.80/hr. (Col. 1+Col. 3)
0.1	3.9¢	6.5¢	11.7¢	10.4¢	15.6¢
0.2	4.2	7.0	12.6	11.2	16.8
0.3	4.4	7.5	13.5	11.9	17.9
0.4	4.6	8.0	14.5	12.6	19.1
0.5	4.9	8.6	15.5	13.5	20.4
0.6	5.3	9.2	16.6	14.5	21.9
0.7	5.7	9.9	17.9	15.6	23.6
0.8	6.0	10.8	19.5	16.8	25.5
0.9	6.4	11.9	21.5	18.3	27.9

Source: See Table 2.

average speed for that segment.[21] (The operating and accident cost factors determined in this study are summarized in the Appendix.) By combining estimates of travel-time, operating, and accident costs, total travel costs at various traffic levels can be estimated. Table 3 contains such estimates for various volume/capacity ratios on two-lane arterial streets.

The Capital Costs of Freeways

Currently, the only technique being employed to an appreciable extent to alleviate urban traffic congestion is investment in additional highway capacity. Some of these additions to capacity have involved widening or otherwise altering existing arterial streets, but most of them have involved the construction of entirely new, high-speed, limited-access expressways. Table 4 summarizes the estimated capital costs of the urban freeway improvements in Minnesota that are planned for completion during the next decade as part of the interstate highway system. For comparison, data on the costs of rural portions of the system are also included.

[21] George Haikelis and Hyman Joseph, "Economic Evaluation of Traffic Networks," *Bulletin 306*, Highway Research Board (1961), p. 55.

TABLE 4. Estimated Capital Costs of the Interstate Highway Systems in Minnesota[a]

(in thousands of dollars)

Costs	Urban						Rural		
	Six Lane			Four Lane			Four Lane		
	Average	Approximate Range		Average	Approximate Range		Average	Approximate Range	
		Low	High		Low	High		Low	High
Per Mile									
Right-of-Way	$1,882	$400	$4,900	$233	$ 30	$4,000	$ 24	$ 4	$ 50
Clearing and Grubbing, Utility Relocation, Earth Moving, etc.	539	200	2,800	288	60	650	85	9	150
Paving	304	120	440	222	170	280	197	170	280
Per Structure									
Overpasses	299	120	950	219	120	510	140	15	235
Interchanges	549	190	1,670	425	210	2,020	259	62	694

Sources: State of Minnesota, Department of Highways, Estimate of the Cost of Completing the National System of Interstate and Defense Highways in the State of Minnesota (July 1, 1957); Ibid., First Revised Estimate of the Cost of Completing the National System of Interstate and Defense Highways in the State of Minnesota (August 1, 1960).

[a] Add, in addition, approximately 15 percent to cover engineering expenses and contingencies.

A few words are perhaps in order on the factors underlying the substantial variations that exist in these costs. Right-of-way costs are very high near the central business districts of Minneapolis and St. Paul, but drop sharply outside them. The $30,000 per mile urban four lane low occurs in a small town. Variations in rural right-of-way costs appear primarily to reflect differences in the quality of farm land. The costs of the clearing and other activities required to prepare the right of way for pavement are also highest near the two largest central business districts and decline rather slowly with distance from them. The sources used do not provide detailed information on the character of individual overpasses and interchanges. For example, full clover leafs are not distinguished from lower cost "diamond" interchanges— access roads parallel to the freeway that require some traffic on the intersecting road to cut across opposing traffic streams. The highest interchange costs appear to occur where two or more freeways or a freeway and two or more arterial streets intersect near the two major central business districts.

The Benefits of Urban Freeway Investments

That the "benefit/cost ratio" of urban freeway investments is probably greater than 1, at least in the Twin Cities metropolitan area, can be demonstrated fairly easily. As was noted earlier above, if the cost and other characteristics of real world transportation activities are "competitive," and if optimum toll collections on any given road-link exceed its maintenance and capital costs, expanding the link's capacity would be in order. Federal and state excises on gasoline comprise the only appreciable source of tax revenue that varies directly with vehicle mileage. In Minnesota these taxes total approximately 10 cents a gallon. The driver of an average car (according to the "Red Book" definitions) who desires to travel 48.5 miles per hour on a rural road or urban expressway averages 14.7 miles per gallon of gasoline. Such a driver therefore pays a toll of approximately 0.7 cents per mile. Analysis of the relationships between travel time and volume/capacity indicated in Table 2 suggests that such a toll, at alternative average travel-time values of $1.55 and $2.80 per vehicle hour, would be optimum at respective volume/capacity ratios of 40 and 30 percent.

The average speeds underlying the estimated city street costs in Table 3 range from 13-24 miles per hour at volume/capacity ratios of 90 and 10 percent respectively. One study determined that, at these speeds in city traffic, average gasoline consumption for a 1951, six-cylinder car would be on the order of 14 and 18 miles per gallon respectively.[22] At average speeds of 18 and 24 miles per hour, equation (3a) suggests gasoline consumption rates of 17 and 18 miles per gallon respectively. Again, assuming gasoline excises of 10 cents a gallon, these fuel consumption rates imply an actual tax burden range of 0.5 to 0.7 cents per mile. Even assuming that the "Red Book" estimate of $1.55 per hour for the average value of travel time is correct, analysis of the relationships underlying Table 3 suggests that such tax rates would be optimum only for a volume/capacity ratio of less than 10 percent. Taking yearly registration fees of roughly $20 per passenger vehicle into account would add 0.4 and 0.1 cents per vehicle mile to the respective tax costs of 5,000 and 20,000 mile per year drivers and would raise total tax payments to the range of 0.6 to 1.2 cent per vehicle mile. Again assuming $1.55 per hour to be the correct travel-time valuation, the upper bound of this range would be the appropriate toll for a volume/capacity ratio of less than 20 percent.

Using the procedures described in the Appendix, data generated by the Twin Cities Area Transportation Study (TCATS) in 1958 were used to estimate the average volume/capacity ratio for the area's highway network for each half hour during the average 1958 weekday. These ratios ranged from lows of 2 to 5 percent between 1:30 and 5:30 A.M. to a high of approximately 81 percent between 4:30 and 5 P.M. Between 7 A.M. and 6:30 P.M., 53 percent was the lowest ratio experienced on an average weekday.

The inferences to be drawn from these statistics are, however, subject to two (at least partially) offsetting biases. On the one hand, volume/capacity ratios vary from road to road and, indeed, from one side of a road to another in the metropolitan area. The

[22] A. J. Bone, "Travel Time and Gasoline Consumption Studies in Boston," *Proceedings of the 31st Annual Meeting of the Highway Research Board* (1952), pp. 440-456.

relationship between volume/capacity ratios and vehicle operating costs is nonlinear. Therefore, the implicit assumption in the procedures used that all roads have the same volume/capacity ratios leads to underestimates of average optimum toll levels, as well as of average operating costs. On the other hand, not all travel in the area is on city streets; some expressways exist, primarily in outlying areas. As comparison between Tables 2 and 3 suggests, city street vehicle operating costs and optimum tolls are greater than those on freeways at all volume/capacity ratios.

Be this as it may, current highway user charges appear to be substantially below those that would be required to equate the marginal private costs and the marginal social costs of private passenger vehicle trips on urban streets. To repeat, the gasoline and vehicle license taxes paid by the average urban driver range between 0.5 and 1.2 cents per vehicle mile—amounts equal to optimum city street tolls at volume/capacity ratios in the 10 to 20 percent range. For most of the day, however, actual volume/capacity ratios are in excess of 50 percent. The current level of highway user excises is at least sufficient—indeed, probably more than sufficient, given the rural bias that seems to exist in the highway expenditure programs of most states—to support the current highway maintenance and expansion program in the Twin Cities metropolitan area. Both an increase in user taxes and an expanded highway investment program would therefore seem clearly in order.

The above conclusion is based on an analysis that assumes constant returns to scale and short-run marginal cost pricing. A number of "noncompetitive" elements in the provision of transportation services are of considerable importance in urban transportation planning. While none of them appears to affect my conclusions to any appreciable degree, a few words about the most important of them nevertheless seem desirable.

The competitive assumptions discussed earlier do not hold in at least three important respects. First, the provision of highway facilities clearly involves indivisibilities and, as the data on the capital costs of freeways (in the preceding section) suggest, also appears to involve scale economies. The costs of paving six-lane expressways are less than 150 percent of those for four-lane expressways. The same consideration applies to overpasses and to the amount of land required for rights-of-way.[23]

The provision of mass transit services also involves substantial scale economies. A mass transit patron, in addition to whatever fare he pays, must supply the valuable resource of his own time when he takes a trip. Two sorts of time costs can usefully be distinguished: (1) time en route and (2) waiting time—time spent walking from his origin to a bus stop, waiting there for a bus to come, perhaps waiting at a transfer point for a second bus to come and, finally, walking from a bus stop to his final destination. It is with respect to waiting-time costs that scale economies arise in the provision of mass transportation services.

Once a rider is aboard a bus or a subway car, an increase in the number of passengers very likely increases his time en route by increasing both the number and the duration of stops. At the same time, however, an increase in the demand for trips on a route would likely result in increased service frequencies and therefore in reduced waiting time. Further increases in demand may provide additional reductions in waiting time by allowing a more dense route network and may actually reduce time in transit by enabling the introduction of skip-stop or express service.

A complete analysis of these scale economies would require information not presently available on, for example, the operating costs and speeds associated with vehicles both larger and smaller than the fifty-passenger bus currently operated by most transit companies, the producer and user costs associated with various route densities and service frequencies, and the degree to which users are familiar with transit schedules. It is possible, however, to develop rough estimates of the orders of magnitude of the subsidies associated with the time savings that would result if the service frequencies on existing routes were increased.

---

[23] The design standards for both four- and six-lane portions of the Interstate System in urban areas call for an 8 to 60 foot median strip between opposing traffic streams and a total of 13 feet for paved shoulders for each of the two roadways. In addition, a variable but substantial amount of land is provided as a buffer between shoulders and adjacent access roads or property. As a result, only 40 to 50 percent of the right-of-way for a typical four-lane urban expressway is devoted to traffic lanes. Expanding such a highway to six lanes therefore increases right-of-way land requirements by only 10 to 15 percent.

Unfortunately, to describe the procedures involved in making these estimates would require a lengthy digression. It must therefore suffice to assert that the scale economies associated with increasing service frequencies imply an optimum mass transit subsidy in the Twin Cities area equivalent to roughly 10 to 20 percent of current fare revenues.

The second noncompetitive element of importance concerns the fact that the total money outlay for almost any transportation facility increases with the number of stages in which that facility is constructed. The magnitude of these "stage construction diseconomies" depends on the nature of the facility involved. At one extreme, for example, for three lightly traveled sections of the Interstate System, the total outlays involved in initially building two-lane facilities with later expansion to four lanes were found to be on the order of 20 percent more than those required for initial four-lane construction.[24] At the other extreme, a four-lane bridge over a four-lane highway must be almost completely rebuilt when the highway is expanded to six lanes. Similarly, when the grade and curvature standards of a highway in rolling terrain are improved, a very large part of the initial highway investment is obliterated.

The existence of stage construction diseconomies means that it would be highly inefficient to adopt a policy of continually expanding each lane in a highway network for which the demand is increasing. Rather, an optimum investment policy would involve building each individual link to standards higher than those required to minimize the costs associated with the initial demand for its services, and improving it only when the rate at which it is used exceeds its optimum utilization rate by a perhaps substantial amount.

As for the third "noncompetitive" element, the marginal costs of providing transportation services in peak and off-peak hours and hence the optimum toll for these services differ substantially. A variety of systems have been proposed for varying tolls on

[24] Robley Winfrey, "Economy of Constructing Four Lane Highways of Interstate Highway Immediately, Compared to Constructing Two Lanes Immediately and Two Lanes in the Future," *Bulletin 306*, Highway Research Board (1961), pp. 64-80.

urban transportation facilities through the day—ranging from extremely complex mechanical and electronic devices to variable taxes on parking facilities in highly congested areas. All of the proposals present substantial—perhaps insuperable—legal, political, and/or cost obstacles. This is particularly true of those that deal with private highway transportation. The possibility must therefore be faced that it may prove infeasible to establish marginal cost prices for some or all forms of metropolitan transportation.

If marginal cost prices in one line of economic activity cannot be established, the best alternative system would not necessarily be one for which marginal cost pricing prevailed in all other lines of activity. In particular, if it proves politically, legally, or financially impossible to set marginal cost prices for the use of highways by private vehicles, these "second best" considerations may well make it desirable to provide mass transit subsidies even if transit facilities have constant returns cost characteristics.

As noted before, each of these noncompetitive elements is of considerable importance in planning transportation improvements, but none of them appears to affect significantly either the basic conclusion that an expanded highway investment program would be desirable or the orders of magnitude of the benefits associated with such a program. The only important effect of stage construction diseconomies would seem to be that of making the investment planning process much more difficult than it would be if no penalties were attached to expanding highway capacity by small increments. The existence of economies of scale in highway construction implies that, in an otherwise competitive system, a transportation plant of less than optimum size would result if investments were undertaken only when the total costs fell short of the total revenue derived from short run marginal cost prices. It also implies that basing benefit estimates on the *average* costs of new highway capacity (as is done below) would lead to underestimates of the *marginal* benefits associated with increments of freeway capacity. Finally, "second best" considerations coupled with the existence of scale economies in the provision of mass transit services do not appear to affect highway investment benefits as such. Rather, they simply suggest that even greater benefits could be realized by using some part of highway user taxes to subsidize mass transit activities rather than to expand highway capacity.

Turning to the actual measurement of highway investment benefits, there is, to repeat, good reason to ignore so-called "nonmarket benefits" and the redistributions of income that highway improvements may produce. The aesthetic and humanitarian contributions of transportation expenditures seem small, and it is difficult to find clearly defined groups of worthy individuals that are either seriously harmed or greatly benefited by these expenditures.[25]

In dealing with commodities for which neither "nonmarket benefits" nor income redistribution are of importance (the vast majority of private market goods and a large number of government provided goods), consumers' surplus benefit measures seem appropriate. Accordingly, the metropolitan freeway benefit estimates developed below are essentially measures of the effect of changes in the price of the commodity "weekday passenger vehicle miles (or their equivalents) in the Twin Cities metropolitan area" on the area under the demand curve for that commodity.[26]

To denigrate consumers' surplus as a measure of much of anything seems to be a current fashion, at least partially because of the fact that consumers' surplus measures tend to yield somewhat less dramatic magnitudes than do some of the more imaginative devices that have been used to measure public expenditure benefits. But consumers' surplus measures have also been abused because their relevance is misunderstood when the marginal utility of money is not constant and when the effects of changes in the price of one commodity on the demand for other commodities cannot be ignored. A few words in explanation and

---

[25] But again, see footnote 1 above.

[26] The data used distinguished three classes of vehicle—passenger cars, truck-taxis, and buses. Truck-taxis and buses were converted into private passenger vehicle equivalents by multiplying the average number of these vehicles on the road during each half hour by 2 and 5.5, respectively. The former factor appears to be commonly employed in traffic analysis work. The reason for using the latter is discussed in the Appendix.

defense of consumers' surplus measures therefore seem in order.

Consider, then, the consumer depicted by Figure 1. He can spend all of his money income, OA, on other goods, or use some of OA to purchase commodity X. The line ABC, determined by the price of X, indicates combinations of X and the money value of other goods that can be purchased with OA dollars. If the consumer is a utility maximizer, he will, of course, reach equilibrium at B in Figure 1—the point at which his budget line, ABC, is tangent to an indifference curve.

Now suppose that a technological innovation reduces the price of X and hence makes a new budget line, ADF, of relevance. The consumer will once more move to a point of tangency, this time at D. Two alternative money measures of the benefit derived by the consumer as the result of this price change are commonly employed: AL, the amount of money that would have to be taken from him to leave him just as well off with the new set of prices and new money income as he was at point B before the price of X fell; and PA, the amount of money he would need to be given to make him as well off as he was at point D if the price of X returned to its former level.

Suppose that OH and OK in Figure 2 refer respectively to the old and new prices of X. Suppose also that X is a superior good—a good for which the income elasticity of demand is greater than zero—and that DAB is the "normal" demand schedule for the good, i.e., the demand curve traced out by points of tangency such as B and D in Figure 1 when the consumer's dollar income *and* the price of all other goods remain unchanged. Several things can easily be demonstrated:[27]

1. The "income compensated" demand curve traced out by points of tangency such as B and M in Figure 1 (i.e., points corresponding to altering the price of X while simultaneously extracting just sufficient money from the consumer to make him indifferent between the new price and income on the one hand and the price and money income associated with the original tangency at B on the other) will, in general, be shaped like $D'AC$ in Figure 2.

[27] See, for example, J. R. Hicks, *A Revision of Demand Theory* (Oxford University Press, 1956), Chap. 8.

FIGURE 1

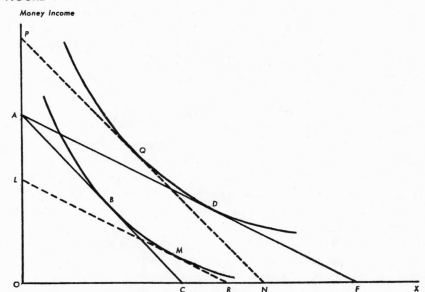

Money Income

FIGURE 2

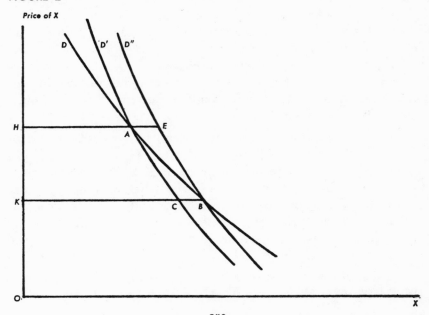

Price of X

2. The demand curve associated with points of tangency such as Q and D in Figure 1 will have a configuration like $D''EB$ in Figure 2.

3. The area $HACK$ in Figure 2 has a value equal to that of the line AL in Figure 1.

4. The area $HEBK$ in Figure 2 has a value equal to that of the line PA in Figure 1.

Hence, if X is a superior good, $HABK$, the area under the "normal" demand schedule between the new and old prices over-estimates the amount of money that would have to be taken from the consumer to make him indifferent between the old, higher price and money income on the one hand and the new, lower price and money income on the other. Similarly, $HABK$ under-estimates the amount of money that would have to be given the consumer to make him indifferent between the new price and his old money income on the one hand and the old price and new income on the other. The magnitudes of the differences between $HACK$, $HABK$, and $HEBK$ depend on two things: the income elasticity of demand for X and the proportion of the consumer's income that is spent on X. In general, the greater the income elasticity and the greater the proportion of the consumer's income that is spent on X, the greater the differences among these magnitudes will be. For most goods, however, the differences will be small.

The validity of the foregoing truths is in no way affected either by how much additional satisfaction results from the change in the price of X or what effect the change has on the consumer's demand for other goods. Whether the additional utils resulting from the price change are proportionally greater than, less than, or equal to the ratio of AL to AO, AL still represents the income that could be extracted from the consumer while leaving him indifferent between the old price and money income and the new price and money income. Similarly, if commodity X is, for example, margarine, AL is just as valid a benefit measure if butter is among the commodities on which OA could be spent as it would be if butter did not exist. That is, in computing the consumers' surplus benefit measure, it is *not* necessary to take account of the fact that the demand schedule for any commodity will normally shift to the left when the price of one of its close substitutes falls.

One important proviso must be attached to this last point, however: in drawing a "normal" demand schedule, money income and all other prices are assumed to remain unchanged. A reduction in the price of one commodity will typically lead to leftward shifts in the demand curves for substitute commodities. If the resources producing these commodities are to some degree immobile, such demand shifts will lead to falls in their prices. As a result, the nonproducing consumer of these products will, in general, be able to reap benefits greater than those implied by AL in Figure 1 or $HACK$ in Figure 2. In aggregating over society as a whole, however, these benefits do not represent net social gains. Rather, they result from reductions in the quasi-rents received by producers of the substitute products. As a further result, to the extent that product X has close substitutes (or close complements, for that matter) with less than completely elastic supply schedules, the observed demand relationship for X will be steeper than DAB in Figure 2. Thus, under these conditions, the observed demand relationships will definitely lead to underestimates of $HABK$ and $HEBK$ and may even lead to underestimates of $HACK$.

In summary, consumers' surplus computations based on observed price and quantity combinations will not, in general, provide exact estimates of the income compensating magnitudes $HACK$ or $HEBK$. Such computations will typically lead to underestimates of the latter magnitude and may or may not lead to underestimates of the former. However, the resulting inaccuracies will tend to be small unless the income elasticity of demand for the commodity being studied is great; and/or the commodity comprises a quite substantial share of the average consumer's budget; and/or the commodity has quite close substitutes with quite inelastic supply schedules. These characteristics seem to apply to no appreciable degree to the product under consideration here—passenger vehicle equivalent miles in the Twin Cities metropolitan area.

Presuming the area's current highway network to be of less than optimum size, the construction of urban freeways would reduce the cost of passenger vehicle miles in two ways. First, it would add to highway capacity, thereby reducing volume/capacity ratios and vehicle operating costs by an amount exceeding the opportunity costs of the capital employed. Second, urban freeways would provide "better" vehicle miles than do arterial streets; at any given volume/capacity ratio, the speeds attainable on freeways are roughly double those attainable on city streets. Thus, a driver starting from a point adjacent to an expressway interchange and going to a destination adjacent to another interchange on the same freeway could cut the time required for his trip roughly in half by using the freeway rather than city streets.

Most trips, however, do not have such fortunate combinations of origin and destination. In general, a driver must go somewhat out of his way to avail himself of the higher speeds attainable on freeways. Taking the freeway will reduce the total time and money costs of his trips, but will also typically increase the total number of miles he travels.

I have no direct information on the degree to which use of freeways increases the circuity and hence the mileage of trips, but some inferences on the matter can be drawn from a highly artificial model. Consider a city consisting of a central business district (CBD) surrounded by a circular residential area in which population densities are the same regardless of distance from the CBD. Suppose that travel takes place only from residences to the CBD; that it is possible to travel in a straight line between any two points in the city at a cost proportional to the distance involved; and that the number of trips any household takes per time period is the same regardless of either the cost of traveling a unit of distance or the distance of its residence from the CBD.

And suppose finally that an *un*limited access expressway is constructed radially from the CBD along the X axis to the periphery of the city and that the time and money costs of traveling a mile on it are a fraction β of those prevailing in the remainder of the city. The freeway will be used, not only by households living adjacent to it, but also by households at some distance from it. It can, in fact, be shown that all residents living in a pie-shaped wedge—the "impact area"—bounded by the line $\beta = |\cos^{-1} \Theta|$ will go to the artery along a route parallel to the boundary of the

FIGURE 3

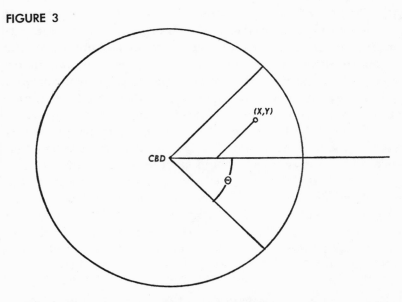

impact area and then follow it to the CBD.[28] For a resident of (X, Y), doing so reduces the time required for a trip from

$$k (X^2 + Y^2)^{1/2} \text{ to } k [\beta X + (1 - \beta^2)^{1/2} Y].$$

By integrating these two expressions over the impact area, aggregate travel time for its residents can be obtained before and after introduction of the expressway. For $\beta = \frac{1}{2}$, the ratio of "after" to "before" travel times for the area as a whole turns out to be 82.7 percent. Similar integrations yield before-and-after distances covered. Again for $\beta = \frac{1}{2}$, "after" distance is 110 percent of "before" distance. Furthermore, the distance covered by impact-zone residents is split exactly 50/50 between the expressway and the surrounding area.

On the basis of this admittedly meager evidence, it was assumed that new freeway capacity equal to A percent of arterial street capacity would be utilized by a fraction $A/(100 + A)$ of total miles. The circuity involved in freeway trips was assumed to add 5 percent of this fraction to both freeway and off-freeway vehicle

[28] See Mohring, "Land Values and the Measurement of Highway Benefits," *Journal of Political Economy*, Vol. 69 (June 1961), pp. 236-249.

miles.[29] The benefits of the first hypothetical increment in freeway capacity were measured by the difference between estimated total vehicle operating and time costs involved (1) in current travel on the existing arterial street network and (2) in the somewhat larger number of vehicle miles associated with the expanded network. A similar procedure was followed in dealing with succeeding increments to capacity.

This estimating procedure involves the assumption that the demand for *trips* in the area is totally inelastic—that is, freeways are assumed to generate additional vehicle miles only by virtue of the greater circuity of trips on them. No account is taken of the benefits and costs entailed in the additional trips that would almost certainly be generated as a result of a freeway-associated reduction in travel costs. That this zero demand elasticity assumption would impart a downward bias to benefit estimates *if* optimum user charges were levied can be shown fairly easily. Given the current pricing system, however, the assumption might or might not lead to underestimated benefits. On the one hand, the benefits derived by those who take the additional trips are ignored; on the other, generated trips would increase volume/ capacity ratios and hence costs above the level that would have prevailed had they not been made. Whether the unmeasured benefits of generated trips are more than offset by the additional costs to which they give rise would appear to depend on the elasticities of the trip demand and the marginal vehicle-operating and time-cost functions. The more elastic the former function and the less elastic the latter, the more likely that the procedures used would overestimate true benefits. Unfortunately, no reliable evidence is available on the elasticity of demand for trips.

To be somewhat more specific about the computational procedures, TCATS data imply the capacity of the metropolitan area's streets to be approximately 22,600 vehicle miles per minute or 1.36 million vehicle miles per hour. The basic and practical capacities of an urban freeway have been stated to be respectively

---

[29] That is, during a period in which N passenger vehicle-mile equivalents would have been traveled on city streets if no freeway capacity had existed, freeway and off-freeway travel of $n = 1.05\ NA/(100 + A)$ and $N - 0.954\ n$ respectively was assumed.

2,000 and 1,500 vehicles per hour per lane.[30] Assuming a *possible* capacity of 1,800 vehicles per lane-hour implies that 1.88 miles of four-lane freeway would provide a vehicle mile capacity equal to one percent of that of the area's arterial streets. Table 4 indicates the average cost of 1.88 miles of four-lane freeway, with an average number of interchanges and overpasses, to be $2.25 million.

For computational ease, weekday travel was aggregated into three groups: during the 10 hours in which densities of less than 30,000 passenger vehicle equivalents prevailed; during the 7.5 hours with densities of 30,000 to 48,000; and during the 6.5 hours with densities of more than 48,000. These average densities were used to determine volume/capacity ratios, and hence average travel times and aggregate passenger-vehicle equivalent miles, for each of the three periods studied. Assuming no freeway capacity, 2.11 million vehicle miles at an average speed of 23.4 mph., 5.45 million vehicle miles at 17.6 mph., and 5.93 million miles at 15.8 mph. during the first, second, and third time periods, respectively, were estimated—a total of 13.5 million passenger-vehicle equivalent miles per day.[31]

To obtain total pre-expressway variable travel costs, the Chicago Area Transportation Study cost factors (see Table A-2 in the Appendix) and travel-time values of $1.55 per hour and $2.80 per hour were applied to the estimated average speeds during each of the three time periods studied. Freeway capacity equal to 10 percent of nonfreeway capacity was then introduced into the system. Traffic was assigned according to the procedures outlined above, and freeway and nonfreeway volume/capacity ratios and average speeds were recomputed for each of the three time periods.[32] In turn, the CATS cost factors and the alternative travel-time values were used to compute new total variable costs. The results of these

[30] *Highway Capacity Manual*, pp. 36, 47.

[31] If anything, these figures are probably a bit high. The Twin Cities Area Transportation Study estimated a total of 2.4 million passenger-vehicle and truck-taxi trips during an average weekday and a median passenger-vehicle trip length of 2.2 miles. Dividing passenger-vehicle plus two times truck-taxi trips into the estimated total of 13.5 million passenger-vehicle equivalent miles implies an average trip length of 4.4 miles.

[32] In computing freeway average speeds, the "Norman" volume/capacity travel-time relationship was used (see Appendix, pp. 268-269).

TABLE 5. Weekday Gross Benefits of Freeway Investments
to Present Travelers on Twin Cities Metropolitan Area Streets

Ratio of Freeway Capacity to Street Capacity (percent)	Variable Costs ($1,000)			Total Variable Costs ($1,000)		Incremental Daily Benefit per $1,000,000 of Investment	
	Operating & Accident	Travel Time at:		Travel time at:		Travel time at:	
		$2.80/hr.	$1.55/hr.	$2.80/hr.	$1.55/hr.	$2.80/hr.	$1.55/hr.
0%	$696	$2,170	$1,201	$2,886	$1,897		
						$9,900	$6,400
10	648	1,995	1,104	2,643	1,752		
						8,000	5,200
20	610	1,853	1,026	2,463	1,636		
						5,400	3,500
40	561	1,659	918	2,220	1,479		
						3,600	2,300
60	531	1,526	845	2,057	1,376		
						2,800	1,700
80	508	1,424	788	1,932	1,296		
						2,000	1,200
100	493	1,350	748	1,843	1,241		

computations together with similar computations for freeway
capacities equal to 20, 40, 60, 80, and 100 percent of total nonfreeway capacity appear in Table 5.

For the first 10 percent increment in expressway capacity, these
estimates indicate a saving in vehicle-operating, accident, and time
costs of $9,900 and $6,400 per day per million dollars of capital
outlay at respective travel-time valuations of $2.80 and $1.55 per
vehicle hour. That is, at these respective valuations, pay-out
periods of 101 and 156 weekdays would be associated with the
first 10 percent increment in capacity. Assuming 300 weekday
equivalents per year, rates of return of 180 to 300 percent per
year (less depreciation and maintenance) are implied. Expanding
freeway capacity from 80 percent to 100 percent of arterial street
capacity would yield an incremental benefit per $1,000,000 in-
vested of $1,200 per weekday even if travel time is valued at only
$1.55 per hour. At this rate, incremental benefits would aggregate
to incremental costs in 833 weekdays, implying a marginal rate
of return on invested capital of more than 35 percent. Finally,

even if no value were placed on travel-time savings, the operating cost savings provided by this last increment of freeway capacity would aggregate to its capital costs in approximately ten years.

Summary and Conclusions

The main tasks of this paper were (1) to place highway and, more generally, transportation activities in a broader framework, and (2) to develop rough estimates of the benefits to be derived from expanding urban freeway capacity. Regarding the first task, I have tried to demonstrate that an optimum highway system can be characterized in essentially the same terms as can long-run equilibrium in a competitive industry. In the absence of scale economies, a highway authority could maximize the benefits to be derived from an existing, perhaps nonoptimal, highway network *and* ultimately provide an optimum network by adopting two quite simple decision rules: setting the price of a trip on each link of the network equal to its short-run marginal costs; and expanding or contracting the capacity of each link to equate the quasi-rents earned by the link with its capital costs. In brief, the paper has demonstrated that, again in the absence of scale economies, equality of user charges and system costs would characterize an optimum highway network just as it characterizes long-run competitive equilibrium. Subsidies to transportation systems would seem to be justifiable only for the same reason that theory tells us justifies subsidies to any other line of activity—the existence of scale economies. On the other hand, given constant or increasing returns to scale *and* a state of affairs in which the quasi-rents on highway facilities exceed their capital costs, investment in new facilities would be in order.

As for its second basic task, the paper has first developed estimates of the costs of urban freeway capacity and of the value of time-in-transit to vehicle occupants. Mathematical relationships have also been developed showing the effects on travel time and vehicle-operating costs of changes in the rates at which urban streets and freeways are utilized. Data on current traffic levels, street capacity, and user charges were then introduced. Comparison of present user charges with those appropriate to prevailing volume/capacity ratios suggest that the present Twin Cities metropolitan area high-way network and, quite likely, those of most other large urban areas are of substantially less than optimum size.

The paper concludes by developing rough estimates of the bene-fits that would result from expanding urban transportation systems through construction of new freeways. The magnitudes of estimated benefits are critically dependent on both the value attached to savings in travel time and, of course, how much additional free-way capacity is assumed to be built. At an assumed travel-time value of $3 an hour, *marginal* rates of return of approximately 300 and 60 percent respectively appear to be associated with additional freeway capacity equal to 10 and 100 percent of present arterial street capacity. If travel time is valued at $1.55 a vehicle hour, the respective marginal rates of return are about 180 and 35 percent. Finally, even if no value is placed on travel time, vehicle-operating cost savings alone would provide respective marginal rates of return of 60 and 10 percent.

Appendix: Computational Details

Operating Costs and the Volume/Capacity Ratio

It seems reasonable to conceive of the hourly output of trips on a road as a function, $N = f(T, V, \overline{Z})$, of three inputs: travel time (and comfort and convenience), T, vehicle-operating costs, V, and a vector of highway characteristics such as lane widths, curvature and gradient standards, and the like, $\overline{Z}$. Equations (3a, b, and c) suggest that, for rural roads and urban expressways, the partial derivative of this function with respect to V is a constant.

Several studies have been undertaken that provide information on the relation between N and T for reasonably wide, straight, level rural roads. Greenshields and Huber, for example, found an approximately linear relationship to exist between average speed and traffic density (D)—the number of vehicles occupying a mile of road at any instant of time.[33] For what can be referred to as

[33] Bruce D. Greenshields, "A Study of Traffic Capacity," *Proceedings of the 14th Annual Meeting of the Highway Research Board* (1935), pp. 448-474; and M. J. Huber, "Effect of Temporary Bridge on Parkway Performance," *Bulletin 167*, High-way Research Board (1957), pp. 63-74.

conditions of "free flow," Norman found speed and volume to be linearly related up to the "capacity" of the highway.[34] Greenberg and Underwood respectively found that relationships of the form $D = a\ e^{-bS}$ and $S = a\ e^{-bD}$ provided better fits to the data they studied than did linear relationships between S and D.[35]

By relying on the identities volume *equals* density *times* "space mean speed" and density *equals* volume *times* "space mean travel time," these relationships can be shown to imply relationships between traffic volume and travel time of the following sorts:[36]

Greenshields: $N = (aT - 1)/bT^2$
Norman: $N = (aT - b)/T$
Greenberg: $N = ae^{b/T}/T$
Underwood: $N = (a + b\ \ln T)/T$

Relying again on the fact that the mean desired operating speed on high-quality, straight, level rural highways is 48.5 miles per

[34] See O. K. Norman, "Results of Highway Capacity Studies," *Public Roads,* 1942, pp. 57-81; and *Highway Capacity Manual,* pp. 36-43. The *Manual* utilizes three measures of capacity: basic, possible, and practical. The first two refer respectively to the maximum number of vehicles per hour that can pass a point on a road under ideal and actual conditions. "Practical" capacity is the maximum number that can pass "without the traffic density being so great as to cause unreasonable delay, hazard or restriction to drivers' freedom to maneuver under the prevailing roadway and traffic conditions." All references to "Capacity" in this study apply to possible capacity or, more accurately, to the maximum number of vehicles that can pass a point on a roadway under the conditions that prevail at that time. In addition to free-flow conditions, Norman also notes the possibility of what can be referred to as "saturated flow"—all vehicles traveling at the same speed and maintaining a spacing characteristic of that speed. Under saturated flow conditions, volume and speed bear a curvilinear relationship increasing from zero vehicles per hour when all cars are stopped to intersect the free-flow speed relationship at capacity volume.

[35] H. Greenberg, "A Mathematical Analysis of Traffic Flow," *Tunnel Traffic Capacity Study* (Report V, Port of New York Authority, 1959); and Robin T. Underwood, "Speed, Volume, and Density Relationships," in *Quality and Theory of Traffic Flow* (Yale University, Bureau of Highway Traffic, 1961).

[36] "Space mean speed" should be distinguished from the more normal "time mean speed." The former is defined as total distance covered by all drivers divided by total time elapsed, while the latter is the average of the speeds of individual drivers. "Space mean travel time" is the reciprocal of space mean speed; "time mean travel time" is *not* the reciprocal of "time mean speed." In the following discussion, the (perhaps erroneous) assumption is made that all of the authors quoted imply space means, not time means.

hour, the Greenshields, Norman, and Underwood relationships can be rearranged to yield the following relationships for travel time per vehicle mile:[37]

Greenshields: $\alpha T^2 = 0.0825\ T - 0.0017$

Norman: $(48.5 - 18.5\ \alpha)\ T = 1$

Underwood: $\alpha T = 3.88 + \ln T$

where α is the volume/capacity ratio—the ratio of actual trips per hour to the maximum number of trips per hour the highway can produce. It is on these relationships that the travel-time costs per vehicle mile presented in Table 2, above, were based.

Of the few systematic studies that have been undertaken of the interrelationships among volume, density, and travel time on urban arterial streets, only one provides information that can be analyzed in a fashion similar to that employed with the Greenshields, Norman, and Underwood relationships for rural roads and urban freeways. Coleman fitted quadratic relationships to data on travel-time volume/capacity ratios observed on a sample of one- and two-way streets in several Pennsylvania cities.[38] These relationships are (after making two adjustments[39]):

[37] An additional assumption—that 30 miles per hour is the average speed when a highway is used to the capacity level—is required to implement the Norman relationship. (See *Highway Capacity Manual*, p. 32.) The Underwood and Greenberg relationships, respectively, imply infinite speeds at zero traffic densities and positive speeds for indefinitely large densities. This deficiency could be eliminated by introducing constant terms into the equations. However, doing so considerably complicates the analytical manipulations required to estimate travel-time volume/capacity relationships. The additional complexity does not appear appreciably to affect the implications of the Underwood relationship for below-capacity operations. The values obtained for the Greenberg relationships, however, are critically dependent on the assumptions made about average speeds at low volumes; hence this relationship was dropped from the analysis.

[38] Robert R. Coleman, "A Study of Urban Travel Times in Pennsylvania Cities," *Bulletin 303*, Highway Research Board (1961), pp. 62-75. The measure of "capacity" used in this study was the mean of individual intersection capacities as measured by the techniques described in the *Highway Capacity Manual*, Part V.

[39] By definition, traffic volume is a maximum when the volume/capacity ratio is equal to one. The fitted relationships had maximum volumes at volume/capacity ratios of 0.904 and 1.067 for one- and two-way streets respectively. Accordingly, the volume/capacity ratios implied by the listed formulas are respectively 1/0.904 and 1/1.067 times those of the computed regression relationships. The computed relationships implied mean speeds of 25.5 and 47.4 miles per hour for the one- and

(5)   one-way streets: $\alpha = -1.98 + 1.07\,T - 0.096\,T^2$;

(6)   two-way streets: $\alpha = -1.90 + 1.02\,T - 0.090\,T^2$.

As Table A-1 suggests, just as was the case in dealing with rural roads, travel time per mile and hence travel-time costs vary substantially with the volume/capacity ratio on city streets.

TABLE A-1. Travel Time and Travel-Time Costs per Vehicle Mile on One-Way and Two-Way Urban Arterial Streets

| Volume/Capacity Ratio | One-Way Streets | | | Two-Way Streets | | |
| | Travel Time/Mile (minutes) | Travel-Time Cost at Travel-Time Value of | | Travel Time/Mile (minutes) | Travel-Time Cost at Travel-Time Value of | |
		$1.55/hr.	$2.80/hr.		$1.55/hr.	$2.80/hr.
0.1	2.5	6.5¢	11.7¢	2.5	6.5¢	11.7¢
0.2	2.7	6.9	12.5	2.7	7.0	12.6
0.3	2.9	7.4	13.4	2.9	7.5	13.5
0.4	3.1	7.9	14.3	3.1	8.0	14.5
0.5	3.3	8.5	15.3	3.3	8.6	15.5
0.6	3.5	9.1	16.4	3.6	9.2	16.6
0.7	3.8	9.8	17.7	3.8	9.9	17.9
0.8	4.1	10.6	19.2	4.2	10.8	19.5
0.9	4.5	11.7	21.2	4.6	11.9	21.5

TABLE A-2. Chicago Area Transportation Study Speed-Related Cost Parameters
(cents per vehicle mile)

Miles per Hour	Operating Costs	Accident Costs	Total Costs
5	4.80¢	6.75¢	11.55¢
10	3.69	4.25	7.94
15	3.10	2.70	5.80
20	2.78	1.80	4.58
25.	2.57	1.15	3.72
30	2.41	0.80	3.21
35	2.36	0.55	2.91

two-way streets respectively at zero traffic volumes. The latter value clearly seems too high. For this reason the two-way street travel-time values implied by the equations listed above are 1.08 minutes per mile greater than those indicated by the computed relationships.

In the Chicago study, as noted earlier, both vehicle-operating and accident costs were found to be negatively related to the average speed for that segment.[40]

Table A-2 summarizes the operating and accident cost factors determined in this study. By combining the travel-time cost estimates contained in Table A-1 and the data summarized in Table A-2, total travel costs at various traffic levels can be estimated. Table 3 (page 248) contains such estimates for various volume/capacity ratios on two-lane streets.

Traffic Densities and Optimum Tolls

Table A-3 contains data on the average traffic densities estimated to have prevailed in an area of 890 square miles surrounding Minneapolis and St. Paul during the period covered by the Twin Cities Area Transportation Study in 1958.[41] The daily average density in the area was approximately 32,000 "passenger vehicle equivalents." Average half hour densities ranged from lows of less than 1,000 vehicles (3 percent of the daily average) at 3 to 4 A.M. to highs of over 70,000 vehicles (more than 225 percent of the daily average) in the afternoon rush hour.

Actually, these averages underestimate the magnitude of the peak-load problem, and probably by a substantial amount. During most of the day, traffic on most roads in the area has an approximately even directional split. That is, even for quite short intervals of time, the number of vehicles traveling in one direction that pass any given point on any given road is very close to the number traveling in the other direction. This is not true of peak hours, however. In the morning rush hour, the preponderance of traffic heads toward employment concentrations, particularly the central business districts; in the afternoon, the preponderant flow is in the reverse direction. At distances of 5 to 8 miles from the CBD's, as

[40] Haikelis and Joseph, *op. cit.* (see footnote 27 above).

[41] I am deeply indebted to Dean Wenger and Donald Carroll, who are respectively Traffic Analysis Engineer and Economic Studies Supervisor of the Minnesota Highway Department, for many helpful suggestions, and for providing access to a great deal of unpublished data from the Twin Cities Area Transportation Study. Data quoted throughout this study without reference to a source are from these unpublished tabulations.

TABLE A-3. Average Weekday Traffic Densities on Twin Cities Metropolitan Area Streets, July 8–December 1, 1958

Half Hour Beginning	Passenger Vehicles[a]	Trucks and Taxis[b]	Buses[c]	Total (Passenger Vehicle Equivalents)[d]	Half-Hourly Density ÷ Daily Average
12:00 a.m.	6,266	479	25	7,362	23%
12:30	4,098	401	25	5,038	16
1:00	2,989	371	25	3,869	12
1:30	1,762	279	25	2,458	6
2:00	918	175	25	1,406	4
2:30	671	143	25	1,095	3
3:00	566	127	25	958	3
3:30	522	137	25	934	3
4:00	778	250	25	1,416	4
4:30	909	375	25	1,797	6
5:00	1,892	483	25	2,996	9
5:30	3,966	915	25	5,934	18
6:00	9,581	2,153	625	17,325	54
6:30	22,676	3,693	625	33,500	104
7:00	33,268	5,811	625	48,328	150
7:30	49,103	8,413	625	69,367	215
8:00	31,356	9,651	225	51,896	161
8:30	21,332	8,884	225	40,266	125
9:00	20,962	11,717	225	45,634	142
9:30	16,764	12,094	225	42,190	131
10:00	22,922	14,202	225	52,564	163
10:30	19,235	11,837	225	44,147	137
11:00	23,970	12,378	225	49,964	171
11:30	24,177	11,343	225	48,101	149

[a] Average of densities prevailing during individual six-minute periods.
[b] Estimates derived from a tabulation of truck and taxi trip midpoints occurring during six-minute intervals by weighting each trip by the approximate average travel time for such a trip, 17.8 minutes.
[c] From information supplied by Lynn Gerber, Director of Schedules, Twin City Lines.
[d] In arriving at passenger vehicle equivalents, a truck-taxi and a bus were respectively treated as 2 and 5.5 private passenger equivalents. The former value is one commonly employed by Highway Department analysts. The latter was based on estimates by Walker and Flynt that a bus displaced an average of 3.3 private passenger vehicles at intersections during morning and evening rush hours and that it takes buses 1.65 to 1.75 as much time to cover a mile as a passenger vehicle during these periods. See William P. Walker and Roy A. Flynt, "The Efficiency of Public Transit Operation in the Utilization of City Streets," *Public Roads*, October 1957, pp. 239–243.

TABLE A-3 (continued)

Half Hour Beginning	Passenger Vehicles[a]	Trucks and Taxis[b]	Buses[c]	Total (Passenger Vehicle Equivalents)[d]	Half-Hourly Density ÷ Daily Average
12:00 p.m.	24,631	8,382	225	42,633	132%
12:30	20,624	9,759	225	41,380	128
1:00	28,677	11,055	225	52,025	161
1:30	23,256	10,633	225	45,760	142
2:00	26,882	10,609	225	49,338	153
2:30	25,398	10,000	225	46,636	145
3:00	31,596	9,842	225	52,518	163
3:30	36,972	8,386	225	54,982	171
4:00	40,911	7,608	625	59,565	185
4:30	61,663	5,919	625	76,939	239
5:00	59,597	5,081	625	73,197	227
5:30	42,485	3,371	625	52,665	163
6:00	38,883	2,737	150	45,182	140
6:30	31,044	1,696	150	35,261	109
7:00	35,253	1,948	150	39,974	124
7:30	30,741	1,609	150	34,784	108
8:00	30,160	1,193	150	33,371	103
8:30	26,140	964	150	28,893	90
9:00	24,680	1,191	150	27,887	86
9:30	18,665	1,163	150	21,816	68
10:00	14,052	716	150	16,309	51
10:30	10,628	612	150	12,677	39
11:00	9,585	597	150	11,604	36
11:30	6,732	534	150	8,625	27

much as 75 percent of total traffic moves in the direction of the heavier flow. That is, at these distances, volume/capacity ratios on one side of a road are as much as three times those on the other.

Average weekday volumes on seventeen of the most heavily used arteries in the Twin Cities metropolitan area totaled 362,500 vehicles per day. The estimated aggregate "practical capacity" of these arteries is 336,000 vehicles per day, where daily practical capacity is defined as hourly practical capacity times the ratio (approximately 9.9) of total daily traffic to traffic during the 4 to 5 P.M. rush hour.[42] "Practical capacity," in turn, is approximately 80 percent of "possible capacity"—the relevant denominator for the volume/capacity ratios used in this study. The *possible* capacity of these arteries would therefore appear to be about 420,000 vehicles per day and their volume/capacity ratios between 4 and 5 P.M. to average approximately 86 percent.

To repeat, these seventeen streets are among the most heavily used in the Twin Cities area. The average volume/capacity ratio for the area as a whole is probably on the order of 75 percent between 4 and 5 P.M. The pre-freeway travel cost estimates were generated by using as cornerstones this ratio, the traffic densities listed in Table A-3, and the travel-time and cost-volume/capacity relationships summarized in Tables A-1, A-2, and 5.

More specifically, for a volume/capacity ratio of 75 percent, equation (6) implies a travel time of 4.04 minutes per mile, or an average speed of 14.9 mph. At this speed, the CATS data summarized in Table A-2 imply a vehicle operating cost of 3.1 cents per vehicle mile, while a travel-time value of $2.80 per vehicle hour implies a time cost of 18.8 cents per mile. The average number of vehicles on the area's highway network during the 4 to 5 P.M. period was 68,300. Since traffic density divided by space mean travel time equals traffic volume, during this period a volume of approximately 16,900 vehicle miles per minute and a possible capacity of 22,600 vehicle miles per minute are implied.

As for determination of vehicle operating and time costs during the remainder of the day, again noting that volume equals density

[42] State of Minnesota, Department of Highways, *Twin Cities Area Transportation Study*, p. 15.

divided by travel time, equation (6) gives:

$$- 1.90 + 1.02\ T - 0.090\ T^2 = \alpha = N/C = D/TC.$$

Substituting 22,600 for C and the individual half-hour values for D in these equations enables determination of the travel time per mile associated with each level of traffic density.

Comments

Allen R. Ferguson, *Office of International Aviation, Bureau of Economic Affairs, U.S. Department of State*

Herbert Mohring's paper is an extremely valuable one. I assert this at the outset of my comments, because some of what I shall say later may seem to indicate that I do not perceive this value—which is found especially in the first part of the paper where Mohring rigorously establishes that, under two major limiting assumptions, marginal cost pricing can be expected both to induce optimal utilization of existing highways and to recoup the full resource cost of an optimal facility.

My major comments deal with the two basic assumptions underlying the analysis and conclusions: (1) that there are constant returns and no discontinuities (which I lump together); (2) that the only time variations in demand that need to be coped with are fluctuations about a stable mean. Apart from being unrealistic, these assumptions may lead to faulty conclusions.

The constant-returns assumption is a highly special case. Mohring, himself, gives some evidence of economies of scale in the production of individual highway segments. In addition, the evidence seems to suggest large discontinuities in producing highway facilities. For the bulk of the analysis the lack of realism in assuming constant returns and no discontinuities in providing of individual highway routes or segments is, in principle at least, very damaging. However, before his conclusion as to the optimality of marginal cost pricing could be applied to highways as a whole, one would have to examine whether there are economies in providing highway nets, as well as individual highways. An effort should be made to extend the analysis of scale economies to networks. It could turn out that the economies and discontinuities observed segment-by-segment disappear when segments are combined into nets, with alternative routings between major points.

Further, rapid secular change seems to characterize the demand for transportation facilities. Mohring's treatment of "variable demand," important as it is in dealing with oscillations in demand, does not justify the conclusion that marginal cost pricing will in fact meet his criteria where demand is rapidly growing or declining, while the life of facilities is long. Yet such characteristics are typical of much transport investment. The greater the rate of growth of demand, the larger the discontinuities or the greater the economies of scale, and the lower the discount rate, the greater will be the disparity early in the life of an asset between the revenues from tolls which meet Mohring's short-run rule and the average annual revenue required to recoup the cost of the facility. Late in the life of the asset the opposite will, of course, be true.

Until it is possible to cope with these two problems, we have no basis for a clear-cut policy statement regarding the benefits of highways. In particular, Mohring's conclusion that "greater benefits could be realized by using some part of highway user taxes to subsidize mass transit . . ." does not follow. It is based first on the economies of scale in rapid transit—ignoring those in highway construction and operation—and on an exceedingly weak second-best argument deriving from the practical difficulties of imposing optimal highway tolls.

Some doubt should also be expressed about the method of estimating the value of one's time from realized speeds in an unrestricted traffic flow. I suspect that these speeds are technologically limited and especially, within any given technical environment, by each individual's highly imprecise and subjective estimates of the risk and consequences of accident. The value of time as estimated by Mohring may be inconsistent with Gary Fromm's estimate of the value of life.[43] In any case, Mohring should not overlook such a relationship in his estimate.

Finally, I would like to challenge the notion that the study can be applied to arterial urban streets, although I am not clear just how directly the author expects it to apply. First, although the

[43] See pp. 193 ff. above.

assumption that maintenance costs as a function of use can be neglected may be applicable to freeways, those costs surely cannot be neglected in the case of arterial streets. Second, any conclusion about the appropriateness of highway investment as a whole—for example, Mohring's assertion that the Twin Cities have under-invested in highways—depends on an inevitably hazardous esti-mate of the value of time.

PAUL SITTON, *Office of International Aviation, Bureau of Economic Affairs, U.S. Department of State*

Herbert Mohring's consideration of public policy criteria for highway investments is a prelude to more intensive discussions of these issues as the present Interstate program nears completion. Even under current assumptions, annual levels of highway capital investments are estimated to rise from a national total of $7,791 million in 1963 to $9,115 million by 1970. Of particular concern will be questions on desirable levels of future investment ac-companied by reappraisal of highway-user charges.

Conditions in the 1970's may be dissimilar to those of 1956 when the present program was accelerated. In 1956, the general consensus was the existence of an absolute shortage of highway facilities requiring major capital expansion. In 1970, if highway congestion remains the critical factor, need for expansion of physi-cal facilities may be replaced by need for more selective rationing of use.

An interesting speculation is whether indefinite growth in high-way investments under present user-charge concepts could result in overexpansion of the highway network in terms of broad social consequences. Conditions implying this situation could arise out of political support for sustained or increased highway investments; or a "self-compelling" system of government capital accumulation with outgo related to a revenue-expenditure balance, which could work counter to optimum allocation of resources in a neutral framework; or traffic generating influences at work in the economy arising from physical restructuring of the society which results in supply creating new demand. The net result could be an inex-haustible supply of capital for highway investment unrelated to desired levels that might otherwise be formulated from policy planning considerations at the local and national level. Further growth in federal investment levels and the consequent expansion of federal responsibility could also continue to erode local government responsibility for establishing a desirable mix of regional or urban transportation services as between alternative modes.

Major program expansion in 1956 was made possible by the federal system of earmarked user taxes. This user-charge schedule seems to have only a coincidental relationship to optimum capital investment levels since charges against an individual user cannot be clearly related to specific benefits provided to him at specific cost and ignore the broader impact of highways upon other development decisions. A system of toll financing as an alternative to the present program might well have provided a more selective expansion of the highway network geared to meeting most critical volume/capacity requirements.

The concept of federal user charges has increased the economist's task of establishing a rational benefit-cost relationship for the two levels of independently determined user charges (exemplified by the difficulties which faced the U.S. Bureau of Public Roads in its cost allocation study). In addition, a system of neutrality (exclusive of noneconomic consideration) for government investments in highway transportation as it affects competitive transportation modes becomes impossible.

Mohring has presented an objective and well-reasoned outline of the long- and short-term guidelines which policy-makers can use to maximize the benefits of highway investments. The most significant short-run problem is the highway congestion phenomenon which he would solve through a more rigorous pricing system of "efficiency tolls" as contrasted to the presently followed concept of "free roads." The long-range problem concerns the measure of benefits as a basis for investments in new or improved facilities. The author has framed a theoretical marginal cost-pricing solution for long-range investments but has properly identified its restricted use in establishing highway investments. The analytical techniques are then applied to actual benefit measurement of a real world situation.

The analysis of measurable highway costs is restricted to capital costs of freeways, relationship between travel time and operating costs as related to capacity utilization, and the value of travel time. In measuring benefits of urban freeway investments in the Twin Cities area, the author concludes that the benefit-cost ratio supports further expansion of freeway investments and increased user taxes, assuming constant returns to scale and short-run marginal cost pricing. "Noncompetitive" elements are identified which qualify accuracy of the measurements. The conclusion is that they do not weaken the desirability of need for expanded highway investments. The author then proceeds to develop an estimate of incremental dollar benefits based upon costs of operation, accident, and time.

Additionally, Mohring briefly considers other significant problems of highway investment policy. Among these are consideration of circuity problems in urban traffic flow; impact of the federal-aid program; identification of alternatives for alleviating traffic congestion other than facility expansion; and impact of economies of scale on subsidy investment decisions for highways or transit.

My comments on this paper are limited primarily to a discussion of assumptions which permit one to conclude that highway transportation investment benefits can be measured as if they are either directly or indirectly reflected in market transactions. I have separated the two distinguishable problems considered by Mohring that are of particular concern to policy decisions. These are (1) optimization of existing highway network utility and (2) what optimization to consider in evaluating new facility investments. It is with respect to the latter that my major concern arises.

Utility Optimization

Mohring, among others, recognizes that the congestion phenomenon of highway transportation creates the presumption for more rigorous pricing systems to maximize utility benefits. The pricing method suggested for equating individual to social costs would result in a system of "efficiency tolls." Social costs are defined not only as costs to individual users, but also as costs imposed by the individual on others simultaneously using the facility. The technological and political problems involved in establishing such a pricing system are not easy to solve in a society which assumes "freedom of the highways." Yet such economic constraint could temper the unrestrained concept of facility expansion as providing the only effective solution to the problems of urban congestion. An alternative approach for optimizing traffic flow would be engineering and electronic traffic control systems. At this time, however, technological problems prevent satisfactory application. This latter approach would meet with less political resistance.

Nevertheless, Mohring's proposition of formulating a pricing system of efficiency tolls seems clearly convincing. But successful application would seem practicable only where demand can be shifted by the availability of an alternative mode with similar service efficiency. Without this alternative, would demand be significantly responsive to the suggested pricing system, since congestion is a peak-hour problem of home-to-work movement to be satisfied by services available within certain time limitations?

Contrary to Mohring's observation, I can see the possible desirability of toll charges upon the urban highway user which exceed his total marginal social costs if the results increase demand for public transportation. Further analysis would be necessary to determine whether the shifts to public transportation induced by this action and resultant investment expansion in public transportation would create economies of scale with greater welfare implications, which would offset nonoptimum costs of decreased highway utility. This difference from the author's conclusion also assumes that other "nonmarket" considerations are of significance in such public investment decisions.

Optimizing New Facility Investments

The basic fact of governmental intervention in resource allocation implies a strong presumption that the private market is unable to achieve an efficient allocation.

Mohring concludes that these "nonmarket" factors have limited significance as they relate to transportation investments decisions. If his assumptions are accepted, he can perhaps justify his theoretical approach for optimizing resource allocation to highway transportation. Unfortunately, these assumptions raise questions of their application outside the theoretical world in which they are framed. The principal questions are whether most transportation investment benefits are directly or indirectly reflected in market transactions and whether optimization of highway services to users results in optimization of other community objectives.

The political formulation of the community's collective aspirations for capital expenditure programs in the public sector of the economy must assume existence of a more profound impact upon the social and economic order of things than is normally imposed by investments in the private sector. This concept of the community and its welfare as governing the problems of public resource allocation, however, does create a conflict between rationalization of individual consumer preferences operating under the constraints of private market situations and public interest preferences operating through political channels.

Those who advocate computing highway investment values on an individualistic market economy basis must choose between (1) continuing the use of analytical tools for measuring benefits to the individual and assuming that nonmarket effects are incidental, or (2) broadening their techniques to account in economic terms for total community effects. The latter may be unattainable because of inadequate economic information. However, existing cost-benefit techniques contain enough arbitrarily assigned values and variable considerations to raise questions of whether policy decisions derived on the basis of value judgments may not have equal validity. Even if existing techniques were more accurate, I am not persuaded that assumptions on well-ordered individual preferences as a measure of the general welfare are likely to answer the policy-maker's concern for community consequences.

The author has, however, chosen to measure benefit-cost relationships of highway investments as private market transactions limited by certain "noncompetitive" factors. Assuming away nonmarket influences allows questions to arise as to whether optimization of benefits to the highway user simultaneously optimizes the general welfare.

I think that several questions will illustrate my concern:

1. Can income redistributional effects be ignored as having a minimal applicability to public investment expenditures?

Mohring concludes that "income redistribution effects of transportation programs does not in itself appear to be a major end of governmental transportation expenditures." Although income redistribution is not a major objective of highway development, public highway investments have distinct redistributional effects that significantly influence decisions on allocations and seem to assume an objective of wealth redistribution. Federal highway grants are not apportioned to states on a proportionate basis of revenues collected through user charges, and state distribution of highway funds often reflects the disproportionate effect of rurally dominated legislatures. The President's Appalachian Regional Commission has stated, as one major element of its regional program studies, the conclusion that improvement in the economic position of depressed areas requires major investment in key road development. In this last instance, expanded investments for highways would thus become justified because of the redistributional impact of new industries growing out of improved transportation services. It is questionable that access roads create adequate revenues to justify their pricing under competitive market conditions, and they must probably, therefore, be subsidized because of other developmental objectives. A major objective of such investment allocation is the redistributional effect—the geographic redistribution of jobs and production in the total economy. The implication is that policy judgments on highway investment allocation in this area elude definition of market transactions.

2. Is it realistic to conclude that subsidies have minimal applicability to transportation expenditures justified on the basis of esthetic, humanitarian, national defense, and other "nonmarket benefits" arguments? Would, for example, the interstate highway program be at its present investment level unless defense justification (whether real or not) had been used?

More specifically, in 1959 the Secretary of Commerce decided, strictly upon the basis of asserted defense requirements, to raise the minimum height of bridges on the interstate system at a substantial increase in system costs. For esthetic and humanitarian reasons, future interstate highway projects in urban areas may be extensively modified, or rerouted with significant cost

ramifications, to avoid adverse effects upon community environment. These convenience and psychological values may have a
growing impact on guiding urban highway investments, unrelated to highway utility, particularly as they concern nonhighway users upon whom the facility and its use may work a tangible or intangible discomfort. The conclusion seems questionable that policy-makers consider these contributions small and
that few clearly defined groups are seriously harmed or benefited
by these expenditures. These secondary objectives are often significantly at variance with the engineer's and economist's desire
for developing optimum benefit-cost ratios to the user in terms
of reduced travel time and designs at lowest unit costs.

Another difficulty is how to assure that a highway user, paying
for facilities under a market price mechanism, reflects in his decisions the impact upon competitive transport modes. Probably
the user cannot do this, but the government must consider these
implications from the standpoint of, for example, the desirable
railroad service levels for national defense purposes. The policy-
maker would face the choice of either subsidizing rail services
which have been reduced to an uneconomic operating position
or controlling future investments in highways to avoid major
erosion of this alternative mode.

3. Is the conclusion accurate that with few exceptions 100
percent user responsibility for highway investment is firmly
established?

The conclusion may be correct in principle, but can hardly be
said to exist in practice. In 1963 total highway expenditure dis-
bursements for construction, maintenance, and administration
were estimated to be $12,973 million, of which $3,341 million will
be financed from nonhighway user sources. At the federal level,
an estimate made in 1961 indicated that the Highway Trust
Fund annually received about $60 million of nonhighway user
taxes and the federal government annually expended from general funds about $140 million for road construction of benefit
to highway users. A recent analysis of total highway expenditures
for the Washington (D.C.) metropolitan area found that, for
the period 1958-62, total investments (construction and main-

tenance) and services provided to highway users were approximately $550 million, of which about $170 million were not derived from highway user revenues. It would be reasonable to assume that similar conditions exist in other parts of the country. Although contributions by abutting property owners because of access value derived from the facility can perhaps be justified, unless these values can be isolated from highway-use values, we have no basis for measuring what part is a market transaction for the highway user.

The above factors, when combined with admitted qualifications to market pricing techniques assumed by Mohring (indivisibilities, economies of scale, stage construction diseconomies), make it nearly impossible to apply market pricing techniques to establish appropriate cost-benefit relationships for evaluation of investments in new highways. To the extent that these considerations are valid, exclusive reliance on the tools employed by economists in placing value on investments becomes impracticable unless broadened to account for those factors.

Broader Implications of Highway Investment

Assuming, as does Mohring, that the above "nonmarket" factors can be ignored, does his method of measuring transportation benefits and cost through market transactions adequately measure all social costs or benefits of such investments? He concludes that a pricing system which equates private user costs with social costs provides an optimum method of resource allocation, and he has accepted that optimization results only if the individual user can be made to feel not only those costs which he incurs directly, but, in addition, all costs that his trip imposes on others. Would his conclusion extend to general community economic consequences, unrelated to highway use that may be identified but which cannot be measured objectively—such as captive markets, inconvenience to pedestrians, noise, air pollution, and displacement hardships?

Another economic impact area of significance to policy formulation are the costs of other governmental services considered on an overall regional basis resulting from highway development. For example, expansion of urban highway facilities results in land-use changes from which springs new demand for community services such as sewage, schools, and utilities. What is the total net unit cost effect of new community services in suburban developments when compared with unit costs of services that must continue to be provided in areas adversely affected by redistribution effects of highway development? Effects of redistribution in land use are of rather large proportions and should be integrated into the analytical process for obtaining a true measure of transportation investment benefits. In this same area, can the individual highway user who operates on a tax-exempt facility rationalize his decision so that a market price effect would reflect the impact upon the tax base value of the railroads which supply significant levels of revenues by which state and local governments finance many services?

While models may have been developed to measure these additional socioeconomic consequences, no integrated package of economic facts is available which permits accurate answers as a basis for planning decisions. In a sense, the frustrations which face both the planner and the economist represent a failure to integrate objectives and arrive at mutually agreeable standards to measure the effects of community development and resulting reconfiguration.

Another question related to the measurement of highway user benefits that Mohring employs in his computation of consumer surplus benefits is the extent to which derived savings of time and efficiency resulting from new highway investments over the long range are accurately reflected in the calculated net benefits. On this point, a recent observation was made that calculation of rates of return on investments for urban expressways may be overstated because of the failure to consider over the long run the congestion-shifting effect and the impact of induced traffic upon imputed time savings growing from highway improvements. I do not wish to imply that decisions on highway investment cannot be geared to economic analysis. On the contrary, government officials charged with responsibility for decision-making are grateful for any rationale which justifies or rejects recommendations affecting the public interest.

Application of Benefit-Cost Framework to Twin Cities Area

Mohring concludes that the assumptions made in his theoretical model for an optimum transportation system do not exist in urban transportation. He recognizes (1) indivisibilities and scale economies in the provision of highway facilities, (2) stage construction diseconomies, and (3) seemingly insuperable legal, political, and cost obstacles involved in applying marginal cost pricing.

However, by applying the competitive benefit estimation framework to the Twin Cities metropolitan highway network, he concludes that expansion of the highway investment program is desirable since the effects of the "noncompetitive" factors discussed above probably indicate that benefits are of greater magnitude than would be derived from his analysis.

Here arises my concern over the limitation in the analytical methods used in reaching this conclusion. Costs imposed upon nonusers by development are often presumed to be small; however, this presumption may arise from lack of understanding of such effects. Increased study of highway impact on social configuration and community living is evidence of growing concern and desire to examine more carefully these broader consequences. Models now used to describe optimum benefit to highway users can perhaps be replaced by some broader model which describes total social optimization as a better reflection of the public interest.

Summary

The "optimum" situation sought by Mohring's analysis is found by market pricing measurement. Shortcomings are presented as "technical" difficulties which do not destroy the general theoretical conclusions. If policy aims in highway investments are broader than optimizing individual user choice, the measurement approach used here is not sufficiently broad.

The question to be answered is this: do real world situations require that highway investment problems be examined in the context of policy objectives which recognize plurality of aims, rather than segmental optimization of individual choice, as the starting point?

The rigorous application to highway investment decisions of economic factors—such as time, costs, and convenience—that are reflected in individual user choices stands in danger of assuming away some of the significant problems with which planners and policy-makers struggle in deciding the kind of urban community the American people desire to construct. The provision of an urban transportation system must be geared to these aims. Individual choices cannot be segmented from the surrounding social environment and other government objectives without a sure understanding that such choices not only optimize the product selected by the individual but also make an optimum contribution to the general welfare. The techniques of valuation used by Mohring tell us very little of total welfare by their denial that highway investment has significant repercussions beyond the satisfaction of individual user choices which must be considered in analysis.

Even if assumptions in the paper have limited impact on the measurement of highway investment benefits there still remains a most serious flaw in this approach to highway financing. As was pointed out in a recent article on environment as a new focus for public policy,

. . . our national tendency is to deal with environmental problems segmentally, as specialists whose frequently conflicting judgments require compromise or arbitration. . . . And so in the administration of environmental problems he [the public decision-maker] has been compelled to seek some calculus of objectives that would pass as defensible rationality and simultaneously afford room to maneuver among the fixed or conflicting political forces—the "pressures" of public life.[44]

The problems discussed here do, I believe, point to the need for some method by which comprehensive planning policies may be formulated which take into account the environmental consequences which they produce.

[44] Lynton K. Caldwell, "Environment: A New Focus for Public Housing?" *Public Administration Review*, Vol. 23 (September 1963), p. 134.

Planning provides the comprehensive pattern for considering public policies in totality rather than by focus upon component parts. Component analysis is still required, but it should be guided by the constraints of a broader discipline. A multidisciplinary synthesis must incorporate the talents of engineers, economists, sociologists, and other professions. Such a synthesis could then perhaps bring dollar measurement into a practical relationship with nonmeasurable welfare objectives.

Concluding Statement

HERBERT MOHRING

The question, "Do the freeway rates of return estimated in this paper overstate the desirability of investments in them?" seems to have dominated the discussion of my paper. Perhaps understandably, the consensus appeared to be that a "yes" answer was in order. The arguments given in support of this answer are too numerous to permit dealing adequately with all of them. I will therefore restrict my attention to two broad groups of arguments that arose with particular frequency: those dealing with highway externalities and those which appeared in one way or another to reflect the fact that highways are primarily providers of intermediate rather than final goods.

Automobile traffic, regardless of whether it takes place on freeways or city streets, involves technological external diseconomies. Travelers render disservices in the form of noise, smells, and occasional physical injury to those adjacent to the routes they follow —disservices for which no compensation is paid. A welfare loss is clearly involved in our failure to have these costs reflected in user charges.

It does not necessarily follow, however, that additional freeway construction would add to the aggregate of these nonuser costs, although it would affect their distribution. True, increased freeway capacity can naturally be expected to produce increased vehicle mileage. Trips presently made would tend to follow more circuitous routes. In addition, the freeway-associated reduction in travel costs would likely generate new trips in the short run and, in the long run, a more transportation-intensive organization of urban activity—especially lower residential densities and broader marketing areas for commercial and industrial centers. However, the nonuser cost of a vehicle mile is much higher on a city street than on a freeway isolated from its neighbors, as it typically is, by a more or less substantial buffer of land. Thus, even though freeways would generate more vehicle miles, it is by no means clear that their construction would add to the aggregate of nonuser costs.

Several discussants quite rightly noted that, in addition to the externalities of vehicular traffic, poorly planned freeways can do, and likely have done, serious damage by fragmenting communities, disrupting existing communication patterns, and the like. The failure of the courts to recognize the external diseconomies of freeways in awarding damages to those harmed by them has unquestionably made this sort of damage more likely. At present, compensation appears only to be made for property incorporated in a right-of-way or property to which access has been seriously impaired. Less tangible damages, real though they may be, are rarely redressed.

Currently, about the only inducements highway planners have to look upon parks as expensive rights-of-way or to regard the cost of a tree as something more than the expense of cutting it down are the exhortations of sociologists and community planners that greater heed be paid to community values. How successful these exhortations are is debatable. A pricing system that more nearly reflected the true cost of freeway rights-of-way would, I am sure, provide a much more effective goad to improved highway planning.

In any case, it would seem reasonable to increase the denominator of benefit/cost ratios to reflect the windfall losses in property values and other presently uncompensated costs of freeways. At the same time, however, well-designed freeways can promote (and likely have promoted) community values. Well-chosen rights-of-way can serve as buffers between conflicting land uses. Freeways do reduce traffic levels and hence traffic externalities on surface streets. It would therefore also seem reasonable to increase stated freeways benefits to reflect the windfall gains of, e.g., property owners along surface streets who are benefited by reduced traffic externalities. Here, too, then, it is by no means clear that taking externalities into account would yield lower rate of return estimates.

Urban transportation is primarily an intermediate rather than a final good in that most trips are taken, not for the utility they themselves provide, but rather as an inevitable cost of the utility to be derived at their destination. That this is the case may be the source of some of the curious anti-highway arguments presented during the floor discussion. For example, several speakers commented on what one of them referred to as "a sort of Parkinson's Law, that congestion rises to meet capacity." We typically draw demand curves sloping from northwest to southeast. We naturally expect them to shift to the east through time in a growing economy. The intermediate good aspect of urban transportation may explain why these common characteristics of demand relationships are regarded as a source of disapprobation in dealing with highways, but not with parks, concert halls, or, for that matter, apples and oranges.

That urban transportation *is* an intermediate good, however, does pose at least two serious problems: first, the nature of the final good with which trips are associated is not always apparent; second, the transportation network provided has a direct and substantial effect on the composition and prices of the bundle of final goods available. Highway planners have, to a considerable degree, failed to face up to either of these facts.

To cite an example of the former: in the late 1930's, a bridge was built across the lake separating Seattle from the land to its east. The resulting substantial decline in access time to the central business district induced considerable residential development. Indeed, by the late 1950's, traffic had become so heavy that an expansion in bridge capacity was deemed in order. The highway engineers proposed paralleling the existing structure. Quite rightly, I think, the city planners maintained, in effect, that the relevant commodity was *not* trips made over the existing bridge but rather land suitable for low-density residential development at a small travel-time distance from the CBD. They therefore argued in favor of building the new bridge at a substantial distance to the north of the existing structure.

As for the second problem, construction of rail rapid transit systems would almost inevitably lead to higher density residential development than would a continuation of existing freeway programs. Similarly, freeways would likely increase the number of integrated shopping centers and reduce the importance of existing strip commercial zones along major arterials. Unfortunately, little consensus appears to exist about the effects of alternative transportation systems on the bundle of final goods—let alone about which of the alternative bundles is preferable. Many of the criticisms of my rates of return estimates seemed to reflect an implicit judgment that the bundle associated with freeways is inferior to its possible alternatives.

I doubt that this is the case, if only because of the great and apparently growing demand for low-density housing. Such residential activity is simply incompatible with high-quality mass transit as we presently know it. True, invention of a mass transit system that preserved not only the speed but also the flexibility of the private passenger vehicle would end this incompatibility. Regrettably, however, such a system seems unlikely to appear in the near future.

JEROME ROTHENBERG*

Urban Renewal Programs[1]

THE FEDERAL URBAN RENEWAL PROGRAM was born in 1949. By the end of 1961, 587 local projects had major contracts under the Urban Renewal Administration. For 574 of them the total gross project cost was $2.9 billion, of which $1.3 billion was federal money, and $0.7 billion local government money. The private development resources which were, or will eventually be, associated with these projects are probably three or four times these amounts. And the many resources involved are used in ways that make diverse and far-reaching changes in the lives of millions of people. It is important to be able to evaluate the program and its components, making use of the experience of about fourteen years.

This paper attempts to set forth a conceptual framework for making benefit-cost analyses of portions of this program. It will be concerned with what was originally the pre-eminent component of the program—urban redevelopment—but which has recently been somewhat de-emphasized, although it has accounted for the lion's share of resources since the program's inception. My attention here is largely devoted to trying to specify the na-

* Department of Economics, Northwestern University.
[1] An extensive treatment of this whole subject is contained in my study, "Economic Evaluation of Urban Renewal," in preparation for the Brookings Institution.

ture of the benefits from a single redevelopment project, and how they may be measured. The problem of specifying the real resource costs involved is not examined in any depth, although the final section presents a numerical illustration of the application of the procedure, and crude money cost figures are given. These are intended for illustration only, and do not pretend to be derived from a serious analysis.

To begin with, I shall briefly sketch the operation of a typical redevelopment project under the federal Urban Renewal Program. A "local public authority" (LPA)—sometimes organized specifically to formulate projects under the program, sometimes already in existence with other operations—declares a certain area blighted and legally condemns it. Then, by the right of eminent domain, the LPA buys up all sites within the area, paying the owners "fair value." Land assembly is followed by demolition of structures in the site (clearance) and preparation of the land for redevelopment. This includes the local government's adjustment of streets, sewers, schools, and other utilities. In addition, any families displaced as a result of clearance are relocated elsewhere by the LPA, under an explicit relocation program. Public processing of the area under the auspices of the LPA is now complete. The next step is the sale or lease of the area to a private redeveloper; but sometimes part of the area is given to a government agency for construction of public housing or other public purpose. Under the private redeveloper, the area is typically subjected to significant change in land-use—often high-rise or large-scale housing projects for people in the upper-middle-income bracket, and, more recently, modern commercial centers.

The role of the federal government in this process is to approve the advance planning for the overall project, to lend working capital funds for detailed planning, surveys, and acquisition, and to finance up to two thirds of the LPA's "net project cost," which is defined as the LPA's total expenditure on the project less the receipts realized by selling prepared land to redevelopers. Under the 1949 Act the federal government was authorized to spend $1 billion in loans and $500 million in grants over a five-year period. Amendments in 1955, 1957, 1959, and 1961 steadily increased the amounts of the grants; as of the 1961 amendments, grants could be as large as $4 billion.

In the version of benefit-cost analysis used in this paper, the policy alternatives are specified, and one of them—the status quo policy—is treated as the point of departure alternative. The consequences of the other alternatives are considered to be the positive and negative changes in well-being of everyone affected. Where possible, these changes are calculated in money terms. Where this is not possible, changes are specified in terms of types of events which are presumed to affect well-being. Complete specification of consequences may therefore involve a multidimensional vector of effects, only one dimension of which is monetary.

Many benefit-cost studies distinguish between total income (production) and distribution of income effects, typically including the first and excluding the second in the effects to be considered. The presumption is that undesirable distributional effects stemming from some policy can be offset through auxiliary income transfer policies unrelated to the specific policy area at issue. This presumption may be warranted for policy areas at the federal level, since the federal government has powerful means available, such as the income tax and other transfer programs, to achieve many redistributional aims. Most local governments do not have such means; distributional consequences of major local programs must typically stand uncorrected. In the case of urban renewal, which involves local projects, we shall therefore specify both total income *and* distributional effects of alternative policies.

The alternative policies that must be compared in this area are three:

1.   The status quo policy of allowing land-use to be determined by private market decisions, subject to the various present governmental regulations and enforcement, including tax and credit policy.

2.   Redevelopment projects, designed to supersede market decisions in the way described above.

3.   A composite of ameliorative public actions other than re-

development, to "improve" on private performance in those respects where private performance is suboptimal. (This third alternative is not unique, and we do not attempt to specify any one such composite for evaluation in this paper. But the nature of some possible components for such a composite are strongly suggested by the analysis of the kinds of benefits to be derived from redevelopment. And once a composite has been specified, the methods presented in this paper can be straightforwardly applied to evaluate its benefits.)

Every benefit-cost analysis requires aggregating effects for groups of individuals to facilitate computation, and every such method of aggregation is an arbitrary compromise away from the "true" value of the social welfare change. We propose to aggregate together, as a single group, all individuals who are similarly affected by each type of consequence of a given policy. Moreover, for each of these groups a subgrouping may be performed in terms of a few stereotype characteristics most perceived in local political decision-making: for example, rich vs. poor, central city dwellers vs. suburbanites, minority groups vs. majority groups. A rationale for this procedure is that the lumping-together of benefits and costs in the analysis will parallel the kind of lumping-together that actually informs local political choice.

The selection of this type of benefit-cost analysis automatically decides an important question about evaluative criterion. The criterion is not maximization of property values within a certain area but maximization of welfare among a relevant population. Choice of the relevant population is not easy. In formal terms, since the federal level is involved in at least part of the financing, the well-being of the entire United States population should be consulted. But the truly distinctive benefits and costs of urban renewal are almost entirely contained within the population of the pertinent metropolitan area. It is unwise to narrow the relevant population further—to the central city, or even project neighborhood—because specific alterable features of projects can have significant effects on individuals residing outside such narrower boundaries. (Projects designed to attract middle- and upper-income groups back to the city from the suburbs are of this type.)

Our decision about a relevant population may seem to introduce an anomaly. Renewal projects are typically creatures of central city planning and execution (with federal cooperation). Noncentral city residents of the metropolitan area have no close representation in the projects. This introduces a methodological slippage into our procedure. It is tacit in benefit-cost analysis that policy-makers will be attracted to benefits and repelled by costs. But where these policy-makers have no legal responsibility for— or feel no political pressure from—a numerous portion of the affected population, costs and benefits to the latter will fail to carry the appropriate incentive force. Indeed, they may bear perverse force—policy-makers attempting to impose costs on and deny benefits to the outlying populations. However, many projects do in fact offer benefits to suburbanites, in the form of an enlarged set of city housing alternatives. Costs are typically imposed on them, not by specific structural characteristics of renewal projects, but via the general redistribution of federal taxation to support its financing share. Nonetheless, the divergence between policy responsibility and policy consequence very likely leads to suboptimal choice. By including the wider population in our criterion we point up the need to change jurisdictional boundaries—whether formally or informally—or otherwise directly modify planners' incentives in projects of this sort.

The Nature of Redevelopment Benefits

The consequences of redevelopment are varied, extensive, and complicated. It is not obvious *a priori* which environmental indications one would have to observe to measure any relevant individual's well-being. As a short cut, we look for the directions of significant influence by examining the purposes of the redevelopment program. Such an examination, in addition to suggesting the distinctive types of benefits to be found from redevelopment, can suggest alternatives to redevelopment by clarifying the status quo deficiencies that redevelopment is designed to correct—and, therefore, what other measures can also be so designed.

Redevelopment has had a variety of explicit and implicit aims. The more significant ones seem to be:

1. The elimination of blight and slums.
2. The mitigation of poverty.
3. Provision of decent, safe, and sanitary housing in a suitable environment for all.
4. Revival of downtown areas of the central city.
5. Attraction of middle-income families from suburbs back to the central city.
6. Attraction of additional "clean" industry into the central city.
7. Enhancement of the budget balance of the central city government.

Of these purposes the elimination of blight and slums has been officially considered by far the most important, but there are indications that local policy-makers have substantial incentives for some of the others, especially in more recent projects. We shall argue that only the first of the aims, and to a more limited extent the fourth and fifth, provide benefits of the sort that an economic benefit-cost evaluation can consider. The others largely involve public subsidies to achieve "public goals," the treatment of which our approach does not handle. This does not mean that such goal achievement does not count as a benefit in an overall evaluation of urban redevelopment. It means only that valuations must be given by decision-makers themselves, since they cannot be calculated within the traditional economic benefit-cost focus.

Elimination of Blight and Slums

Neither this goal nor the very definition of "blight" and "slums" is self-evident. But both a functional, judgmental definition and a more operational, structural one employ the rubric "substandard." Under the judgmental definition, "blight" and "slum" refer to dwellings which are not "decent, safe, and sanitary."[2] They are likely to be dilapidated, overcrowded, filthy, vermin-infested fire traps. Under the structural definition, architectural characteristics such as absence of bathtub or toilet, or of adequate wiring, comprise the components of "substandard." Despite its greater judgmental latitude, the former definition is more appropriate for our purposes, since the latter often operates as a criterion of architectural obsolescence rather than of housing quality.

The benefits inhering in the goal of eliminating blight and slums are not self-evident, because it is not at once obvious that what is gained exceeds what is lost. Slum dwellings represent low-quality housing. The usual overcrowding of such dwellings implies that the typical dwelling unit per household is also of very low quantity. Furthermore, these units are typically inhabited by poor families. The nub of the ambiguity is simply that, given the technical characteristics of the commodity which supplies housing services, so long as there exist poor families the existence of a significant supply of low-quality, low-quantity housing units might represent an optimal use of resources in this market. If so, their elimination would not render even *positive gross* benefits, but would introduce suboptimal resource utilization —i.e., it would render *negative gross* benefits.[3]

On the demand side, the existence of poverty makes the demand for low-quality, low-quantity housing eminently reasonable. Housing, being such a considerable part of the total budget, is a good place to economize with a highly limited budget, especially for families newly arrived in cities, who had experienced rural or foreign housing conditions that were considerably worse. On the supply side, the durability of housing makes it most efficient to meet the demand for such housing, not through new construction, but by means of a "filtering" downward of the existing housing stock through aging, structural conversion to permit occupancy of smaller units, and depreciation of maintenance.[4]

---

[2] "Blight" refers to a process, and can refer to one or more structures, whether residential or not. "Slum" refers to a cluster of structures, usually residential, in an advanced stage of "blight."

[3] "Gross" in the sense that it is exclusive of the resource costs of elimination through redevelopment.

[4] This treatment of "filtering" is heavily dependent upon William G. Grigsby,

"Filtering" is not restricted to the lowest portion of the housing stock; the great majority of the stock goes through a life cycle downward through lower and lower quality uses. Most new construction occurs in the upper half of housing use, and most retirements in the lowest use categories; thus, most units pass through a wide range of use levels during their lifetime. Since housing mobility is greater at the higher-income levels than at the lower, filtering is largely initiated by middle- and higher-income groups releasing their present accommodations in favor of newly constructed dwellings. Conversion occurs via changes in vacancy rates and relative price changes. Thus, a stock of old, worn housing, cut up into small units in the process of downward conversion, is the market's efficient way of providing housing services to the poor. If the overall pattern is unsatisfactory, the problem is poverty, not inefficiency in the housing market.

The above argument applies to low-quality housing. A slum is low-quality housing, but it is something more as well. And it is in the respects in which slums are something more that we may find the source of benefits from slum elimination. I shall argue that there are at least three respects in which slums may represent suboptimal resource use:

1.   There exist important neighborhood effect externalities in land use; these are likely to be especially significant in slum areas.

2.   The profitability incentives to produce the particular *type* of low-quality housing that characterizes slums rest upon market biases which are either ethically disapproved or the inadvertent result of public policies.

3.   The functioning of slums entails the creation of important social costs which are externalities to the actors involved.

NEIGHBORHOOD EXTERNALITIES. There are important externalities in the nature of housing services and hence in the value of the property which provides them. The housing consumed by a household consists not only of occupancy of a specific dwelling but also of the location of the dwelling and its neighborhood.

Housing Markets and Public Policy (Institute for Urban Studies, University of Pennsylvania, 1962).

The neighborhood consists of other residential dwellings, of commercial and industrial establishments, of public services such as schools, street lighting, and police protection, of recreational and cultural amenities, and, most of all, of people. In the simple model presented by Davis and Whinston, the quality of the housing services associated with a particular dwelling depends on the character of the dwelling, the amount of maintenance and repair devoted to it, and also on the character of dwellings in the neighborhood together with their state of maintenance-repair.[5] For each of the n pieces of property comprised by the neighborhood, the owner obtains the highest return if his property is undermaintained, while all or most others are well maintained. He obtains a smaller return if his, as well as all or most others, are well maintained, less if his and all or most others are poorly maintained, and least if his is well maintained while everyone else's is undermaintained. This is the payoff matrix of the "Prisoner's Dilemma" type of strategic game. Each owner has an incentive to let his property be undermaintained while others maintain their property well. But the very generality of this incentive means that it cannot be realized: all property will tend to be undermaintained. Yet this outcome is less satisfactory to all owners than the only other attainable outcome—namely, high maintenance for all. This latter could not be attained by atomistic behavior, since each owner singly would shy away from high maintenance, but it is an outcome which all could bring about simultaneously: i.e., each could agree to it contingent on everyone else's agreeing to it.

Thus, the important externality of neighborhood means that the outcome arrived at by atomistic choice will typically be suboptimal, in the sense that nonatomistic coordinated choice could make all owners better off than they would be when acting atomistically.

This type of externality is very general. It is by no means restricted to slum areas, and may be supposed to affect property in all areas. Moreover, neighborhood effects may not everywhere be such that low maintenance is suboptimal; in some situations,

[5] Otto A. Davis and Andrew B. Whinston, "Economics of Urban Renewal," *Law and Contemporary Problems*, Vol. 26 (Winter 1961), pp. 105-117.

high maintenance may well be suboptimal. All of it depends on the specific payoff matrix appropriate to the particular market. Is the market any better able to adjust land uses to the complicated pattern of such externalities in upper-income areas than in lower-income areas? At first blush it might seem that suboptimal land use can occur anywhere, at any level, in any direction, and to any extent, depending only on accident. This conclusion may seem even more persuasive when one considers the many forms of land misuse: undermaintenance; or the vying character of housing services, such as single vs. multiple-occupancy dwellings, high-rise vs. walkup apartment buildings; or even the residential-commercial-industrial mix.

Yet there is a special link between this type of market suboptimality and slum areas. Neighborhood effects are generally recognized. Indeed, special social mechanisms have been devised to minimize their most adverse effects. Building codes and zoning are attempts to moderate some of the worst effects of land use externalities; they typically stipulate permissible lower limits in quality and quantity—never upper limits. The slum-blight linkage stems from the fact that zoning and building codes are likely to be less effective for low-quality areas than for high-quality areas. Low capital availability, high population density, and householders' ignorance and, often, lack of urban disciplines are among the reasons for this.

PROFITABILITY BIAS IN THE PRODUCTION OF SLUMS. Slums do not simply happen. They represent a pattern of resource use which is made by man; they are produced. They are produced because they are profitable. It is the contention of the present section that this profitability stems partly from market circumstances which are not "normally approved" but rather are generally desired by the electorate to be publicly rectified. It also stems partly from certain existing public policies, which also are capable of rectification. The profitability of slums is not inevitable, nor does it rest on market forces that are ethically neutral.

Slums are produced both intensively and extensively. Intensive production means conversion of property to lower and lower use —and then, for lowest uses, to lower and lower quality levels of

service. Extensive production means extending the spatial boundaries of slum concentrations. The two aspects often go hand in
hand. The most important types of intensive production are converting dwellings to increasingly overcrowded occupancy, and allowing the state of the property to deteriorate progressively.

To say that slums are profitable to produce is not to say that
all property owners obtain excess returns. It means rather that
high rewards are available to "innovators" and, initiative having
already been taken, it pays others to follow suit, sometimes
against their personal preferences. Profitable opportunities exist,
and the dynamics of contagion through neighborhood effects
magnifies the drift. Profitable opportunities exist because of characteristics of the demand for inferior housing and because of legal and financial policies.

On the demand side, slum dwellers have historically been not
only poor but disadvantaged: large clusterings of recent immi-
grants and underprivileged minority groups, radically uninformed about the market and their legal rights, and too poor to
invest in information. Moreover, the rest of the community has
often superimposed on these disabilities the artificial disability
of discrimination. This has become especially onerous in the
present generation, when it is the Negro who is preponderantly
the slum dweller. The problem of slums is not at present solely
a problem of the Negro, but that it *is* the Negro who disproportionately inhabits the slums aggravates the problem. Discrimination against the Negro in employment and housing has been
especially severe. In housing it takes the form of segregation;
large areas of cities are effectively closed to Negro occupancy,
despite the willingness to pay stipulated rents.

The impact of all this is a large, uninformed, highly inelastic
demand for low-quality housing in concentrated areas. Expansion
of quantity comes most profitably from conversion to overcrowding. Moreover, the higher cost of accelerated depreciation is not
made up in rentals, but in the less noticeable way of neglecting
to keep the property up to legally required standards. Thus,
conversion to slum use often increases revenues without increasing costs and even sometimes decreases actual expenses. It is not
at all atypical for property to be run down profitably to a state

where the cost of bringing the structure up to the legally required minimum level exceeds the value of the property. The
property in this case has *negative* social value.

The profitability of slum production based on these characteristics is akin to the adulteration of commodities based on consumer ignorance. The market response in such situations is not
considered to be due to the interaction of "normally approved"
market forces. The consumers are considered to be "unfairly"
disadvantaged. Political consensus in the United States has typically called for protection to offset such disadvantage. Thus, one
can argue that slum profitability is to this extent "socially inadvertent."

A number of "artificial" factors on the supply side enhance
the profitability of slum production—"artificial," in the sense
that they are inadvertent consequences of public policies rather
than inherent characteristics of the market. The federal income
tax law is one such factor: a landlord can advantageously report
accelerated depreciation on his slum property, without attempting to offset the depreciation by maintenance outlays. Property
completely depreciated for tax purposes will, despite the lack of
maintenance, still retain most of its competitive market position
relative to the rest of the neighborhood, because of neighborhood
externalities and disabilities already mentioned on the demand
side. The property can be profitably sold, since the buyer can
take depreciation on the property anew while failing to maintain
it, and resell it profitably in turn. Thus, the same property, many
times dilapidated, is kept in lower and lower occupancy use while
continuing to record depreciation for tax purposes. Slum property tends to be kept in use far longer, and to a much lower level
of quality, than it would in the absence of these tax advantages.

The profitability of slum property attracts speculators to buy
up and convert many nonslum structures to slum use. The properties are refashioned to make overcrowding possible, thus setting the stage for faster depreciation, and then the speculators
typically sell them at higher prices, the aim being capital gains
rather than slum incomes.[6]

[6] David Laidler, "The Effects of Federal Income Taxation on Owner-Occupied
Dwellings" (doctoral dissertation in process, University of Chicago, 1964).

The property tax is another factor that encourages slum use and discourages unslumming. Rarely reflecting the profitability of slum use, it *is* raised when investment is made for upward renovation and maintenance. Such asymmetric assessment response acts as a drag on capital expenditure to improve the quality of slum and near-slum property. The tax biases resource use into less capital-intensive uses of land. Other things being equal, there is a tendency to choose the lower of two capital intensities, and with it, lower maintenance and lower quality use.

Capital rationing has an even stronger effect. Credit is essentially unavailable for remodeling and repair in slum and near-slum areas.[7] This aborts attempts at unslumming or even at maintaining existing housing quality. Credit sources apparently turn down each application because they judge it separately on the marginal basis that, alone, it will fail to offset downward neighborhood pressure and will thus be itself overwhelmed by adverse neighborhood effects. This neglects the potentially significant impact that could be created if the whole set of otherwise creditworthy applicants were approved.

Besides discouraging the maintenance and upgrading of property, the credit squeeze discourages homeownership in the area, encourages emigration, and thus tends to increase the supply of dwellings for cheap, quick sale—further encouraging speculative accumulations for slum creation. The dynamics of such creation is aided by asymmetry in the working of neighborhood externalities: low-income minorities are undesirables to majority groups, but not vice versa. The minorities can drive out the majority group just by their presence, but the opposite movement is much more difficult to produce: it requires providing attractions great enough to offset often considerable antipathies.

A final "superfluous" factor stimulating slum production is ethnic prejudice itself. First, it enhances the *internal* production of slums by aggravating the poverty, immobility, and weak market condition of slum dwellers, as noted above. Second, it enhances the *external* spread of slums because the resulting concentration pressure within the segregated area brings about differen-

[7] See Jane Jacobs, *The Death and Life of Great American Cities* (Random House, 1961), Chaps. 15-16.

tially higher housing prices (for given quality) and lower quality than elsewhere, creates a strong pent-up demand for better quality and for lower-priced housing, and thus results in spill-outs into majority areas, when the prospect of differential profits from minority-group members is enough to offset the prejudice of the seller or renter. Even a single incursion into a hitherto "poor" area is often enough to precipitate panic flight by majority-group members, followed by large-scale replacement by the minority group. Once such a breach has occurred, the pent-up demand pressure of the minority is often great enough to make conversion of the area to slum use profitable, whatever the area's initial socioeconomic character. Panic flight makes for property bargains attractive to speculators—bringing about an ownership pattern highly conducive to slum creation. Moreover, rapid extensive turnover of property itself favors slum creation, since it destroys stable neighborhood expectations.

Internal and external production of slums is favored by the same factors, since the strength of the incentives toward both depends on the extent of price differentials (for given quality) between "ghetto" and outside areas. On the other hand, given the strength of these enhancing factors, successful external spread acts as a safety valve, tending to decrease internal production. Thus, the greater the barrier to spread, the greater the internal production.

Our discussion has implicitly assumed a "tight" housing market. If the market is "loose," the above factors may be more than offset by supply ease, and thereby slums may remain unexacerbated, or unslumming may actually take place. In the latter case a spatial spread of minority groups, for example, would not be equivalent to a spread of slums. Even so, the presence of the factors listed here makes for a greater quantity and intensity of slums than would otherwise be true for the same overall supply and demand conditions. There is some indication that the market was tight in the 1940's through the late 1950's, and that this, together with substantial emigration, favored slum creation. Since the late 1950's a loosening has been in evidence, either stabilizing or even diminishing the extent of slums.

Insofar as exploitation is absent, quality level and money price are substitute forms of higher real prices. We argue that the former tends to be favored by gratuitous forces where "natural" forces dictate the level of real prices. Even if we "accept" the desirability of the real price level as part of an optimal resource allocation, this acceptance does *not* lead to social indifference about the form it takes. That it takes the form of slums may well have great social significance, since the mere physical existence of slums and the living patterns thereby favored have been strongly asserted to entail heavy social costs through important externalities. Thus, slums are neither natural nor inevitable, and they may not be desirable.

THE SOCIAL COSTS OF SLUMS. It was indicated above that the existence of slums and hence their eradication could well have far-reaching effects on living patterns in the city. For many years slums have been alleged to generate important social evils: physical, psychological, and health hazards to inhabitants and passers-by; heavy resource costs to inhabitants of the rest of the city. These social costs presumably arise out of important externalities and are therefore "inadvertent." The interdictions in much of this literature are something as follows:

1. Slum dwellings are likely to be fire traps, significantly increasing the probability of general conflagration.
2. Given overcrowding, filth, and inadequate sanitary facilities, slum areas are likely to be a health menace, increasing the frequency and severity of illness both to inhabitants and outsiders (through contagion).
3. Slums breed crime.
4. Slums create personality and social adjustment difficulties.

It is beyond the scope of this paper to expatiate on these contentions. But the allegations are clearly serious and deserve careful evaluation. Substantiation would furnish strong grounds for holding that slums per se represent suboptimal resource use. The net elimination of slums would then qualify as rendering social benefits, independent of any other function it performed.

SUMMARY. Old, low-quality housing is not in itself undesirable, especially if poverty exists. Its elimination carries no *a priori* benefits. Slums are, however, distinguishable from an optimal natural market response of this sort. They represent especially—unnecessarily—low-quality clusterings (if only because of overcrowding), and are largely brought about by important externalities and the side-effects of alterable social policies. Their mere existence may result in social costs which stem from externalities in the functioning of slums. Thus, elimination of slums could rectify a suboptimal market response. Redevelopment is one method of elimination. Our analysis indicates, however, that it is not the only one, since it shows that the slum-enhancing factors are amenable to social control. Thus, policies alternative to redevelopment can be derived from the same analysis that establishes the potential desirability of redevelopment.

Other Goals of the Urban Renewal Program

The elimination of blight and slums concerns only what is to be removed or destroyed. The other goals of the urban renewal program concern largely what is to be substituted. Do these goals carry additional sources of net benefits? The limited scope of this paper precludes any detailed examination, but a few conclusions can be sketched out.

On the whole, slum removal differs appreciably from the other goals. It can in principle involve almost exclusively the real incomes of the present slum landlords and their tenants—for example, by policies directed toward internalizing the externalities in the housing market. This is true of both redevelopment *and* nonredevelopment approaches to slum removal. Whatever income transfers (as opposed to aggregate effects) are involved concern primarily slum landlord/slum tenant relationships. But these other goals involve income transfers far more intrinsically, and the transfers often extend considerably beyond the parties to slum transactions. In general, slum clearance aims to correct market distortions; the other goals aim to subsidize particular land and/or consumption uses of particular groups.

The distinction is not hard and fast, however. There is reason to believe that slum clearance sometimes involves large income redistribution effects relative to its aggregate effects. Moreover, subsidization can in a dynamic setting have aggregate effects, and these several goals differ in their ratio of aggregate to transfer effects. Finally, I am chary of laying too great a methodological distinction between the level of income and its distribution. In any case, the present form of benefit-cost analysis enables the distinction to be made without underscoring.

MITIGATION OF POVERTY. Intrinsically, redevelopment is neutral toward the mitigation of poverty. In practice, many projects may indeed have exacerbated poverty, by eliminating numerous units at one end of the housing stock while adding a smaller number much nearer the other end. In any case, the impact of a project on poverty depends on the particular features of the project, not on the fact of redevelopment. Features designed specifically to mitigate poverty are likely to involve a separable marginal cost. Thus proper evaluation of the program from this point of view must ask whether comparable resources applied outside the redevelopment format could have as much impact on poverty.

DECENT, SAFE, AND SANITARY HOUSING FOR ALL. The housing portion at least, of this goal, can be met if the level of new construction becomes and stays high enough to accelerate filtering and hasten retirements from the housing stock. Since incentives for overuse and overextended structure lifetimes exist, new construction must be great enough to cause a substantial "oversupply" of housing in order to meet the goal. This necessitates shifting large amounts of resources to housing that would otherwise remain elsewhere—i.e., it involves subsidizing a particular consumption area—presumably at the expense of other areas. Such subsidization may at special times seem to carry an aggregative weight— as during a period of drastic housing shortage (relative to the availability of other goods). Such a situation does not seem to apply today. This goal is therefore treated as generating predominantly transfer effects, and an additional benefit category is not introduced for it.

REVIVAL OF DOWNTOWN AREAS OF CENTRAL CITY. This represents an attempt to enhance the economic attractiveness of the central city's downtown area relative to suburban areas, the former having been substantially damaged attendant on the process of suburbanization. To a large extent this goal envisages simply a subsidized transfer effect. A kind of aggregative effect, however, may also result. A major city represents a scale of population concentration large enough to make possible forms of specialized services—opera companies, museums, zoos, specialty shops—which would not be viable with lesser concentration. To the extent that the decentralizing impact of suburbanization decreases group concentrations below the relevant critical masses, subsidized enhancement of downtown can help reattain these endangered scale economies.

It must be noted that implementation of this goal competes with that of mitigation of poverty, since it aims to displace low-income residential and commercial uses by much higher income uses. In general, the list of goals does not form a consistent set—the specific fashioning of a project to enhance one goal frequently results in meeting the others less well.

ATTRACTION OF MIDDLE-INCOME FAMILIES FROM SUBURBS BACK TO THE CENTRAL CITY. Like the preceding goal, much of this one's effect is simply that of income transfer. But there is an aggregate effect here too. The goal is advanced as a way of rectifying a progressive narrowing (homogenizing) of the population base in the central city toward lower income and social minority groups. The achievement of a more balanced population distribution can improve commercial diversification, the use of public services, and cultural and political vigor. Again like the preceding goal, implementation of this is likely to come at the expense of mitigation of poverty.

ATTRACTION OF ADDITIONAL "CLEAN" INDUSTRY INTO THE CITY. Since we operate on the assumption of full employment (or, at any rate, if this be dropped, on the assumption that the urban renewal program is too clumsy to be used countercyclically), attraction of clean industry can be treated as a pure transfer activity, subsidizing one land use on a specific site at the expense of other uses and other sites.

Implementation of this goal typically comes at the expense of most of the others we have mentioned.

ENHANCEMENT OF THE BUDGET BALANCE OF THE CENTRAL CITY
GOVERNMENT. There are two distinct perspectives within which
changes in the fiscal balance of the planning government are interpreted as generating net benefits. The first postulates that any marginal budgetary surplus enhances the ability of the local government to carry out valuable social coordinating functions and therefore renders net social benefits to the community. The second asserts
that the health and functioning of the local government are not
at issue at all, but rather a marginal budgetary surplus indicates
that the redevelopment program instituted by the government
created more market value than it used up, and created therefore
net benefits. The difference between the two interpretations is
fundamental. The second measures the values already created,
the first measures the ability to create new values in the future.

The first stems from a model of local government finance
somewhat as follows. The expenditure and revenue systems of
the local government are strongly constrained externally. There
exist important public expenditure needs; the tax base available
to the government to finance these needs is limited. Therefore,
when events affect the public fisc adversely this hampers the ability of the government to do what the electorate wants it to do.
One might call this a model of imperfect responsive government:
it adverts a crucial imperfection in the governmental process.

The application of this model to the present case involves the
same trend of population movement that we have noted above.
The growth of slums and suburbanization have tended to erode
the city's property tax base, to give a lower rate of growth of
sales tax revenue, and to bring in a population disproportionately
constituting the poor, who require a larger volume of public services (for example, welfare and health services) than the groups
whom they replaced. In addition, the central city has continued
to render important services to suburbanites who work and shop
in the central city and are currently outside the reach of *quid pro
quo* taxation. Indeed, traffic, road, and direct transportation
(e.g., city-operated transit systems) services and facilities have
increased substantially in the past decade. Thus, the expenditure
needs of the central city have risen and its traditional revenue

sources have lagged behind. In this context it is argued that governmental programs which produce a marginal budgetary surplus (raising tax yields more than they incur outlays) are producing net benefits for the community, because they make possible additional desired public actions which were otherwise rationed by financial stringency.

This argument must be rejected. To support it would imply that the government ought to undertake programs, or reformulate existing uses, for the purpose of making "profits." This would seriously distort the role of government, since in a representative democracy that role is to do things which are explicitly undesirable to leave to profit-seeking actions of the private market. The governmental role would be significantly abdicated.

Moreover, the supposed dilemma begs the question. If existing modes of taxation prove inadequate under changing trends, why is there not resort to other forms? If present expenditure programs are crushing, relative to the willingness to pay, cannot some of these programs be reconsidered? Admittedly, some of the problems stem from the complex interrelationships of a metropolitan area with important externalities, yet without metropolitan government. But these call for a solution within the political process itself—to make that process more responsive to the balance of wants and resources. To corrupt the process further by changing the criterion for public action would seem to be folly. Thus, while I recognize the present financial difficulties of some city governments, I shall not count the impact of redevelopment on the city's treasury as either net benefit or loss.

The case for the second perspective is quite different. In this, the production of a marginal budgetary surplus has nothing to do with benefits to the government. It concerns the size of the costs incurred by the redevelopment project—public outlays—on the one hand, and the size of the value created by the project, as allegedly exemplified by changes in assessed valuation and thence by public revenues, on the other. The value is enjoyed by redevelopers of the prepared site, but the government is assumed to exact a payment in the form of higher taxes equivalent to this created value. Thus, what is involved here is not a kind of benefit different from those which we have already discussed, but only a special way by which benefits of the sort already considered can be counted.

SUMMARY OF OTHER GOALS. Of the six goals other than slum elimination, all have important elements of subsidy, involving real income redistribution; but only two—revival of downtown areas and attraction of middle-income families—were judged to be capable of generating general benefits. Mitigation of poverty is very important. Its achievement is by no means intrinsic with redevelopment; it requires additional separable subsidy components. The general benefits which allegedly may stem from downtown revival and attraction of middle-income families are broad and amorphous, so that measurement will be extremely difficult, if not even impossible in principle.

In addition to these characteristics, analysis of the six goals suggests to us here, just as it did for slum elimination, the kinds of suboptimality which redevelopment can help solve. It therefore also suggests that alternatives to redevelopment can also solve some of these same problems. Under mitigation of poverty, for example, a simple alternative is to make general grants to the poor with funds equal to the amount of the subsidy for the poor involved in any particular redevelopment project. And regarding provision for adequate housing to all, an alternative approach is more rigidly to enforce health and building codes. This would lead to spot rehabilitation and selective rent increases in the slum area. It would also tend to displace some of the very poor, as under the typical redevelopment project, and would improve housing quality in the relevant areas. But since the quality improvement would be far less radical than under typical redevelopment, relative housing prices would not change so adversely in the low end of the housing stock, and no price declines would occur in the upper end. The resource cost of such a package would be considerably less than under redevelopment.

The Structure of Benefit-Cost Comparisons

It has been argued above that the proper benefit-cost procedure is to compare three alternatives: redevelopment (R), the status quo (S), and a particular package of policy measures (M) designed to perform many of the same kinds of functions as R. Only if R performs favorably with respect to *both* S and M is it a desirable policy. It was noted that M is not unique; there may be many possible substitute packages. I do not attempt the selection of an appropriate M here, but the previous analysis makes it possible now to give examples of elements that might appear in M.

To combat slum formation we may consider reform of property tax assessment procedures; provision of mortgage credit for dwellings in blighted and nearly blighted areas; more rigid enforcement of health and building codes; spot rehabilitation or demolition of individual dwellings not brought up to code regulations. To combat poverty and internal familial problems we may consider general-purpose monetary grants; informational services; programs for special education (like recent dropout educational programs in the city of Chicago and elsewhere); additional social work services. Other inclusions might be more adequate garbage disposal services and police protection, and open-occupancy programs to combat the segregation pressure that adds to the profitability of slum creation.

In computing benefits among S, R, and M, we must specify differences, not simply in their immediate impact, but over their expected future course. This is of real importance. We have indicated that slum creation is closely associated with a normal adaptation of durable capital stock in the housing market—that there is a "life cycle" through which items in the housing stock pass. Housing on present slum sites will be gradually retired on private initiative, the rate of retirement and nature of the replacement depending upon supply and demand conditions in the market. Moreover, unslumming (significant uncrowding and structure rehabilitation) sometimes voluntarily occurs, and at least one observer alleges this to be more than a rarity.[8] Thus, the status quo policy with respect to a particular site does not imply a permanent commitment to slum use.

In the same vein, redevelopment does not imply a permanent commitment to nonslum use on the site in question. Indeed,

[8] Jacobs, *op. cit.* (see footnote 7, above), Chap. 15.

there are reports that certain types of redevelopment—with or without public housing—may accelerate the process of slum formation, not only by pushing slums elsewhere in response to dislocatees' demand, but by decreasing the nonslum lifetime of new structures.[9] These types of redevelopment encourage high crime rates and behavior that aggravates depreciation of property in neighborhoods where property is still new. Thus, some of the behavioral phenomena associated with slums may occur in environments dissociated from old, obsolescent structures.

One of the dimensions in the evaluation of benefits and costs is, therefore, the forces determining the useful lifetime of items in the housing stock. Other things being equal, running down the useful lifetime is a cost, and elongating it is a benefit. The relevant variable is the rate of depreciation of the stock. This is not congruent with either obsolescence or the rate of filtering, both of which more likely represent enhanced usefulness in the stock.

The Measurement of Benefits

We classify the housing stock into H_1 (low-quality units) and H_2 (nonlow-quality units).[10] H_1 consists of standard (H_{12}) and substandard (H_{11}) units. All of H_2 is assumed standard. A redevelopment project destroys h_{11} units of H_{11}—the redevelopment site—and substitutes h_{12} units of H_{12} and/or h_2 units of H_2. Commercial and industrial units may be involved, in addition to residential units, in either the demolition or the replacement.

The Impact of Redevelopment

Three types of impact are postulated:[11]

[9] See Harrison Salisbury, "The Shook-Up Generation," series of articles (March 24-30, 1958) in the *New York Times;* and Jacobs, *op. cit.*, throughout.

[10] The official definition of "housing unit" is "accommodation designed for the residence of a single household." More than one family may thus inhabit a single unit; moreover, the size of the unit is not fixed. Thus, the "unit" is not really a standard *quantity* of housing.

[11] I omit highly diffuse impacts, such as architectural aesthetics or population balance, as not subject to measurement. They should not be excluded from consideration, but should be brought to the attention of the electorate. It is for them to call out a valuation for these impacts.

1. Improved resource efficiency through internalizing market externalities.

2. Differential real-income effects according to location, income level, owner-tenant status, and functional classification of property, as a result of changes in numbers and location of housing units and commercial and industrial property.

3. Changes in slum-generated social costs due to destruction of slum property.

The first and third impacts are aggregate income effects, the second is a distributional effect. The first arises from the existence of important neighborhood externalities. Land use under the market is suboptimal because, where many landowners are involved, a lack of coordination leads them to act singly in a way that results in an inferior overall situation. Coordination would correct this. But coordination is costly, since it involves production of informational and organizational services, and undependable, since it must cope with the individual profitability from cheating. This problem of enforcement under voluntary agreement is especially difficult. Common ownership is the only complete answer.

This conclusion holds even more strongly when we broaden the type of externality involved to include not only maintenance decisions but also the more fundamental decisions about what type of structures to erect. Such decisions require, not simply coordinated decisions, but a single, integrated decision. For example, the choice may be between many small structures on individual plots or a single high-rise structure for the entire site. Only common—i.e., single—ownership of the entire site makes it possible to act on such a choice.

Thus, land assembly is required. Land assembly is costly and time-consuming, all the more so the more individual parcels that must be purchased. It is beset by private obstructionism and lesser (but also costly) forms of bargaining pressures. These costs can easily be great enough to dissuade private entrepreneurs from undertaking projects designed to internalize externalities.

Government redevelopment substantially cuts the cost of assembly by invoking the right of eminent domain. This bypasses the obstacle course of bargaining sequentially with individual parcel owners. Eminent domain enables land to be inexpensively assembled and employed in units large enough to internalize important neighborhood externalities. Thus, government site preparation—assembly, demolition, clearance—is in effect the creation of a new type of land input: land in neighborhood-sized lumps. It is as if a technical innovation were made, transforming the units of a certain input to new units with higher productivity. This transformation is the source of the externality benefits. It represents net social gain.

Where externalities impose poor land use, redevelopment should increase the productivity of land in the redevelopment site, and therefore its value. But redevelopment also involves a replacement of h_{12} and/or h_2 for the loss of h_{11}. How does this affect the enhanced value in the site? Analysis of this takes us through the second, or redistributional impact, of redevelopment.

I argue, first, that as a good first approximation redevelopment does not affect either the total low-quality or total nonlow-quality demand for housing in the metropolitan area, since I assume that aggregate money income for each group of consumers is essentially unaffected, and that the only significant income redistribution occurs through income effects, which are fully reflected in a given demand function.

In the typical redevelopment project, most demolition occurs in H_{11}, a small amount of replacement occurs in H_{12}, and most replacement occurs in H_2, at a considerably higher quality level. The total number of all types of units constructed is less than the number destroyed.[12] Thus, typically, the supply falls in H_1 and rises in H_2, again with no change in demands. Both

---

[12] The Urban Renewal Administration (Housing and Home Finance) places the figures (as of 1964) at 0.8 unit of planned replacement (mostly in H_2) for every 1.0 unit demolished (mostly in H_1). More lopsided figures of .25 to 1.0 given in Martin Anderson, *The Federal Bulldozer* (MIT Press, 1964), reflect only actual replacement as of the date tabulated, and fail to allow for the substantial lag between demolition and final replacement. Actually, the long period of gestation means that both short-run and long-run analyses of redistribution impacts should be undertaken. The two would show quite different results. Planned public housing (h_{12}) is approximately 8 percent of planned replacements. In our simplified model we pretend all demolitions occur in H_{11}. Actually, some occur in H_{12} as well—redevelopment tracts can have less than 100 percent blight.

supply functions have an elasticity greater than zero because a high price encourages either new construction or conversion of existing units, the first being more important for H_2 housing, the second more important for H_1 housing. Under H_1, much of the downward conversion is not from H_2 to H_1, but within H_1 toward slum use. The elasticity is not infinitely great for H_2 because of the usual reasons, or for H_1 because of imperfect substitutability among different units and the costs of conversion. Under these assumptions, H_2 prices decline and H_1 prices rise. This induces some conversion, both from H_2 into H_1 and within H_1 itself, down toward slum use (through increased crowding). But no change in the demand for land should occur, since the conversion is accomplished by filtering of existing units. Thus, the change in relative unit prices should leave land prices unchanged.[13]

One type of effect on land prices is possible. If the above shifts affect the location of significant commercial, industrial, and public service properties in broad sections of the metropolitan area, they can affect the relative locational advantages of different areas, and thereby cause relative shifts in land prices. For simplicity, it is assumed that the sum of any such price changes over the whole area is zero.

Thus redevelopment increases the productivity of land in the redevelopment site (an aggregate effect); it differentially affects the size of, and therefore the price of, different portions of the housing stock (a redistribution effect), but without necessarily affecting land prices; and it may, where significant commercial and public service facilities are moved, affect relative locational advantages (a redistribution effect), thereby affecting land prices between the redevelopment site and elsewhere. Our purpose is to measure the first effect. But land values in the redevelopment site may show the effect on locational advantages as well. To isolate the first, we make use of our assumption that the sum of locational effects is zero, and that these effects are the only source of land price changes elsewhere. Then:

[13] An aggregate effect on land prices is conceivable, but likely to be unimportant. As a result of the net decrease in housing units, a substitute for filtering is new construction on low density land, especially in suburbs. This might raise land prices very slightly. I shall neglect this possibility in what follows.

(1) $$\Delta P^l_s = \Delta P^l_{sE} + \Delta P^l_{sL} ; [14]$$

(2) $$\Delta P^l_{sL} + \Delta P^l_{\bar{s}L} = 0;$$

(3) $$\Delta P^l_{\bar{s}} = \Delta P^l_{\bar{s}L} ; \text{ so}$$

(4) $$\Delta P^l_s = \Delta P^l_{sE} - \Delta P^l_{\bar{s}} ; \text{ or}$$

(5) $$\Delta P^l_{sE} = \Delta P^l_s + \Delta P^l_{\bar{s}} ;$$

where $\Delta P^l_s \equiv$ land price changes in the redevelopment site; ΔP^l_{sE} $\equiv$ land price changes in the redevelopment site due to internalization of externalities; $\Delta P^l_{sL} \equiv$ land price changes in the redevelopment site due to changes in locational advantages; $\Delta P^l_{\bar{s}} \equiv$ land price changes elsewhere than in redevelopment site (but within metropolitan area); $\Delta P^l_{\bar{s}L} \equiv$ land price changes elsewhere than the redevelopment site (but within the metropolitan area) due to changes in locational advantages. ΔP^l_s and $\Delta P^l_{\bar{s}}$ are observables, but ΔP^l_{sE}, ΔP^l_{sL}, and $\Delta P^l_{\bar{s}L}$ are not. Equation (5) derives the desired nonobservable as a function of observables only.

The third type of impact, relating to social costs of slums, depends on what happens to slums. From our analysis, there are three influences. First, slum property is physically eliminated. Second, some households dislocated from these destroyed units move into other already overcrowded slum units, increasing the degree of overcrowding (the rise in housing prices leads to their purchase of less housing than before). Third, there is conversion of property down to slum level for the first time. Thus, the net effect depends on the relative magnitudes of elimination, worsening, and spread of slums. On balance, a smaller physical area, and fewer total units, are likely to be at slum level. This effect will be smaller than the area and number of units demolished by redevelopment, and some exacerbation will have occurred in existing slums.

The influence of this on slum-generated costs depends on the particular composite of effects. Fire hazards depend on geographic extent and degree of overcrowding; health hazards depend more on overcrowding than on geographic area; personality problems depend more on the size of the real income effect and on

---

[14] Assuming no property taxes. These taxes will be treated later.

household turnover. Thus, the overall impact depends on specifics. One complex that has general relevance, however, is neighborhood stability and diversity of opportunity. It has been argued that social rootlessness is more to blame for slum-generated behavior than is physical dilapidation or overcrowding.[15] Some types of redevelopment, by moving many people en masse, badly disrupt supportive social relationships. In addition, where projects emphasize extreme homogeneity they lay the groundwork for future destabilizing migration waves. Such redevelopment emphases can accelerate depreciation of property. They entail social costs.

Thus, the third type of benefit (and cost) from redevelopment is highly complicated. Some benefits are likely to be produced; but there are possibilities for substituting one kind of social disutility for another. The fact that one or both kinds may prove difficult to measure does not excuse planners and outside observers from giving them attention. We shall return to consider them more fully below.

Internalization of Externalities

Under this heading we seek to measure the increased productivity of land in the redevelopment site. A first approximation involves observing the difference between the highest price for which the local public authority (LPA) can sell the prepared site at competitive bidding and the price it paid for it, and then making the adjustment in equation (5) to remove any effect of locational advantage. Now, we must adjust for property taxation as well. The market price will reflect the diminution in value resulting from capitalized property tax. Since increased productivity will be reflected in higher taxes, the capitalization factor in the LPA's sale will exceed that factor in its purchase. To measure the full productivity change we must add back a tax adjustment factor to the price change. The following equations, all referring to land in the redevelopment site,[16] make this explicit:

[15] For example, see Jacobs, *op. cit.*, Part I and Chap. 15.

[16] Therefore, for simplicity we temporarily suppress superscript referring to "land" and subscript referring to site.

Let V_E be the productivity value of site land attributed to non-locational factors (factors which are subject to the externalities which the LPA project internalizes); and let P_E be the market value corresponding to these factors. Then P_E will be less than V_E by the capitalized value of anticipated taxes on this property (associated with nonlocational factors):[17]

$$(6) \qquad P_E = V_E - t(P_E),$$

where $t(P_E)$, is the capitalized value of anticipated taxes, assuming that these taxes depend chiefly on market value.

Writing the change between before-and-after redevelopment values as ΔP_E, ΔV_E, we have:

$$(7) \qquad \Delta V_E = \Delta P_E + t(\Delta P_E).$$

Substituting equation (5) in equation (7) gives:

$$(8) \qquad \Delta V_E = (\Delta P_s + \Delta P_{\bar{s}}) + t(\Delta P_s + \Delta P_{\bar{s}}),$$

where

$$\Delta P_s \equiv \Delta P_s^l$$

and

$$\Delta P_{\bar{s}} \equiv \Delta P_{\bar{s}}^l.$$

Thus, to calculate the relevant change in value we must know the change in land prices for the site and for the rest of the metropolitan area, and the change in capitalized value of taxes for both. This is subject to a number of pitfalls. The price at which the LPA sells the prepared site may diverge substantially from the competitive market price. Absence of competitive bidding, disposition by convention to a single redeveloper, and specification of necessary redevelopment characteristics enormously complicate the problem of discovering the true market value of the site.

We implicitly assume that when the LPA buys property under eminent domain it shifts the same capitalized tax liability backward to the seller that a private purchaser would—thus, in equa-

---

[17] This is derived from the algebraic partitioning of total productivity and market values of land into a locational and nonlocational component. Thus, where

$$P = P_E + P_L \text{ and } V = V_E + V_L,$$
$$(6a) \qquad P = V - t(P), \text{ and hence}$$
$$(6b) \qquad P_E + P_L = [V_E - t(P_E)] + [V_L - t(P_L)].$$

tion (8) $V_{E0} = P_{E0} + t(P_{E0})$ and $V_{E1} = P_{E1} + t(P_{E1})$. The rationale is that the LPA expects to lose through backward shifting when *it* disposes of the land: the expected value of the land is less by the capitalized value of newly expected tax. But actual practice may diverge from this.

It will be noted that our measure of increased value in the redevelopment site refers only to the value of land, not to the value of improvements on the land. This diverges from other treatments.[18] One main rationale for exclusion of changes in structure values is that we assume full employment: that is, that any investment in improvements on the site displaces other useful employment for the resources so used. Thus, we reject the explicit use of redevelopment as a countercyclical instrument of fiscal policy; it is too clumsy for that.

The full-employment assumption alone does not suffice to determine the benefits from site investment. Resources might be used here more advantageously than elsewhere, especially where lumpy ventures are involved. However, we largely eschew such differences, assuming competitive capital and construction markets—effectively treating site investment as representing only a marginal resource shift. Its productivity is thus assumed equal to that in other uses. This assumption, whenever incorrect, biases measured redevelopment benefits downward.

While changes in the value of site structures are excluded by our treatment, there are changes in the value of some structures that are not excluded. Redevelopment improves the neighborhood. Hence it improves the real value of housing services furnished by given bundles of land and improvements near enough to the redevelopment site to be affected. The value of the land sites and their structures thereby becomes enhanced. This is a real externality, not a pecuniary one, and should be counted as a benefit.

Unfortunately, actual measurement of this enhancement in value is enormously complicated, because the relative price repercussion resulting from redevelopment (due to dislocation and relative stock changes) and the relative location effects—both of which are distributional effects—are likely to impinge especially on just the units most affected by this externality; and the effects are likely to en-

---

[18] See, for example, Davis and Whinston, *op. cit.* (see footnote 5, above).

hance the value of the land and structures as well. Thus, the observed increased value of adjacent land and structures will be a composite of three forces, only one of which is an aggregate productivity effect. In principle, one can disentangle the separate influences by specifying the characteristics of redevelopment projects which differentially influence the strength of relative stock and dislocation, location, and neighborhood externality, forces. But in practice such specification will exceed the subtlety of available data.

A simple procedure that might be used is to determine whether the project is locationally neutral, or small enough to have only trivial relocation and relative housing stock effects. If *both* conditions are met, then the entire change in adjoining structure value can be treated as the result of externality enhancement. Otherwise, the value changes can be treated as entirely redistributional. *Erroneous* choice of the first option overstates benefits; *any* choice of the second understates benefits, mostly when it represents the worse side of the dichotomy.

Even so simple a procedure is faced with a serious problem— that of determining the boundary of significant neighborhood effect. If one decided in a specific case to include neighborhood enhancement, which structures should be included? Since many factors in addition to redevelopment are constantly at work changing relative prices of land sites and housing stock, the setting of boundaries by observing which prices rise is dangerous. Yet no simple but dependable method for setting boundaries suggests itself. Perhaps one would have to adopt the conservative procedure of including only a few blocks in each direction of the projects, and even here excluding directions in which structural or functional barriers to interplay with the project area exist.

In sum, we measure productivity changes by measuring changes in the redevelopment site land, adjusting for locational advantage changes and for tax capitalization. Changes in the value of improvements on the site are excluded, largely through the assumption of full employment, but spillover changes in the value of neighborhood land and structures are included to the extent that they represent externality enhancement.

A final point should be noted. We have discussed only the effects of redevelopment, not of the composite alternative to redevelopment and the status quo (policy M). Clearly, before redevelopment is practically decided on, some such concrete composite should be formulated and its consequences evaluated as with redevelopment.

Relative Changes in Housing Stock

Under our model of redevelopment, the replacement structures are basically different from those they supersede. One population moves out, essentially another moves in. The relative changes in the housing stock (subtractions from H_{11}, smaller additions to H_2) create the following groups of affected individuals:

1. Individuals with specific types of assets or goods and services previously being consumed but now removed as objects of choice.
2. Individuals with the same range of commodities open to choice but facing increased prices for some of them.
3. Individuals with specific types of assets or goods and services, hitherto not available, but now made available, and chosen.
4. Individuals with the same range of commodities open to choice, but facing lowered prices for some of them.

Group 1 applies largely to residential, commercial, and industrial inhabitants of property which is to be demolished as part of the project. Group 2 applies largely to inhabitants of structures which are close substitutes for those which are to be demolished. Group 3 applies to individuals who are potential customers for the new residential dwellings or commercial or public-service facilities produced as part of the project. Group 4 is similar to Group 3, except that its members need not actually move to the project site to be affected and, where they do make such a move, it is to structures and services which are similar to those already available—only the total *quantity* of such structures is affected by the project. Group 3, on the other hand, is involved only where the project has made available novel types of residential, commercial, or public-service facilities and contains as members only individuals who have actually moved into them. Since location is sometimes an important aspect of such facilities, a combination of a type of such facilities already available elsewhere but in a distinctively different location might qualify as a novel facility.

An important asymmetry must be noted among the groups. Whereas in Groups 2, 3, and 4 any change in behavior is a voluntary response to changes in market signals—lower or higher prices for goods hitherto available, or availability of new goods—in Group 1 the change is coerced. Inhabitants of demolished structures are forced to move out. They alone are *precluded* from unchanged consumption in housing; others are simply induced to change by the desire to make a better adjustment, but need not if they do not wish it.

The welfare impact on these groups is as follows. If we assume first that all individuals had been in equilibrium with full information about alternatives prior to the project, then Groups 1 and 2 suffer welfare losses and Groups 3 and 4 experience welfare gains.

Group 1 suffers three types of loss:

1. Elimination of the chosen alternatives and others in the same neighborhood from the opportunity set of residential inhabitants—the necessity to choose less preferred alternatives.
2. Loss of accumulated capital in specific neighborhood adjustment by both residents and commercial enterprises.
3. Rise in the absolute price of the type of housing typically purchased by these inhabitants.

The size of the first factor depends on the heterogeneity of the housing stock. We must allow that differences believed by residents to be important do exist even in small sectors of the housing market. Thus significant discrepancies may be involved between first and lower choices.

The second factor concerns the adjustment of each household's whole pattern of nonhousing consumption and social interaction (for example, the making of friendships) to the opportunities presented by residing in a specific housing unit in a specific neighborhood. These investments in knowledge and decision-making are largely lost when a family moves out of a neighborhood. Not all such loss represents a welfare loss. Some such changes are voluntarily sought to obtain novelty or a widening of experience. But the involuntary mass eviction attendant on redevelopment is likely to represent an unwanted change for most dislocatees. An implication of this factor is that a family's preference for its chosen—and lived-in—housing over its previous second choice is greater after the family has dwelt there some time than at the time of its choice.

The first and second factors together constitute "moving costs" over and above the physical cost of transporting property and person, but are independent of one another. The first is a flow cost, the second a stock or capital cost. The first can be *directly* offset by the attractiveness of the new destination, the second much less so. We assume that physical moving costs are reimbursed as an integral part of redevelopment. This is a fair approximation to actual practice.[19] "Psychological" moving costs are not reimbursed.

Group 1 includes business establishments which are forced to terminate or relocate. Assuming that physical moving expenses are reimbursed, components of good will not bound up with specific location are transferable; locational advantages of the original site are, however, lost. For enterprises which terminate, the whole of good will is lost. These are of course distributional, not aggregate, effects.

The last factor, price rises, has a conventional effect. Those affected suffer a welfare loss measurable in terms of consumer surplus. One of the components of this effect, however, relates to the first factor. Families dislocated from their first choices in the precluded site do not in fact select what their second choices might have been if no price rises had occurred. Instead, higher housing costs induce them to choose less housing than previously —or less probably, lower-quality housing.[20]

The above consequences assume that the members of Group 1 were in full equilibrium before dislocation occurred. But there

[19] Martin Millspaugh, "Problems and Opportunities of Relocation," *Law and Contemporary Problems,* Vol. 26 (Winter 1961), pp. 8-11.

[20] Less probably, because of the emphasis on higher quality in the relocation function of redevelopment projects.

are real reasons, associated with significant lack of information on their part, for believing that their prior situation may have substantially diverged from equilibrium. If so, this will make a difference in the size of the first factor, but not of the other two. Eviction now need not necessarily lead to an inferior choice by each family. The explicit search, perhaps aided or entirely assumed by relocation authorities, could frequently turn up improvements over the original unit, in which case some families may actually benefit (perhaps some of those admitted to public housing). But the larger the number of families who are simultaneously dislocated and the tighter the housing market, the larger will be the proportion of dislocated who must settle for inferior choices. Moreover, the displaced will still lose accumulated neighborhood adjustment capital, and will still be faced with generally higher housing prices.

Measurement of these losses is difficult. The third factor could be measured by the conventional method of estimating consumer surplus via the demand function for housing, if the first factor were not important. If it *were* important, this would complicate the interpretation of movements along such a demand function. In view of our discussion about prior disequilibrium of residents, we may simplify without too much error by assuming that the first factor is negligible. Thus we treat housing transfers within a broad quality level as taking place between essentially homogeneous units. This assumption can be tested by sampling actual dislocatee experience.

Estimates of social capital adjustment costs will be of the money value of the minimum housing package improvement necessary to induce families to move from one neighborhood to another. The sample used should exclude families that like novelty.

Group 2 suffers only a price-rise effect. We measure this by the same means employed for Group 1. Any moves made by these families are voluntary—so no coerced adjustment cost is involved.

Group 3 experiences a quality-change effect (i.e., new housing opportunities), a price effect, and an adjustment effect. Since members of this group move voluntarily, they gain by the new first choice/second choice discontinuity. Since they are subject to neighborhood adjustment costs, the net gain is less by this amount, but the gain must exceed the cost. This is a weak constraint, since those who become members of Group 3 may do so because they like novelty. For them adjustment costs will be negligible or even negative. We shall treat such costs for Group 3 as a whole as zero. This suggests that we may measure the first gain by assuming that the differential price of the new housing is just high enough to equal the marginal purchasers' evaluation of the quality appreciation. Actual computation of consumer surpluses is of course very complicated and requires drastic simplifications.

Group 3 is subject to the effect of price declines—a gain for them—even though they do not remain in dwellings whose prices have fallen. Since they could have remained, their move indicates that the quality-price package represented by the new housing is preferred to the old housing at lower prices. Thus, the measure of "quality appreciation" should include this price effect, and will be in effect a composite measure of quality *and* price change.

Group 4 is nearly symmetrical with Group 2, Group 4 being gainers while Group 2 are losers. For this group quality differences resulting from redevelopment are not important; we can deal largely in price and quantity changes. The biggest effect here, then, is the increase in supply (H_2) with its resulting decrease in prices; members of this group benefit whether or not they move into the redevelopment site. Any adjustment costs resulting from moving are likely to be trivial, as with Group 3— since members will voluntarily move within the same neighborhood or to a "better one." The price effect can be measured by the method used for Group 2.

We have largely discussed effects on residents. Effects on businesses are much the same, requiring translation of effects into profits. If we include property owners in this class, we can classify them as follows: owners of property on the site; owners of improvement properties patronized by Groups 1 and 2; and owners of nonsite improvement property patronized by Group 4. The first class of owners gains a producer's surplus in anything but a perfectly discriminating arrangement with the LPA. The second gains a quasi-rent as a result of increased demand; the third loses symmetrically as a result of decreased demand.[21] Methods of calculation are essentially the same as for consumers.

The discussion of Groups 2, 3, and 4 has assumed prior optimal adjustment. Unlike the treatment of Group 1, dropping this assumption here makes no difference. Since all moves by these groups are voluntary, they represent anticipated improvements. It makes no difference that the groups could have been better off than they were before the project. Insofar as their subsequent move is induced by the project, it is *this* improvement only that we seek to measure.

Effects on Social Costs of Slum Living

This is an extremely difficult area for appropriate measurement—in fact, pessimism has often gone so far as to suggest that measurement attempts be abandoned. But something can be known, and important distinctions about order of magnitude can be made.

The problem of measurement here is, briefly, that the causal process between slum living and these dimensions of social cost is complicated and interrelated with other causal factors; it has only a probabilistic influence; outcomes are difficult to read; changes in the arguments of the functional relationship typically have only minor short-run effects, the important ones being long-run; lastly, even where proper effects are isolable, they are likely to be observable only qualitatively. Exhaustive relevant dollar measurement is far from obviously feasible.

Our four categories of social cost are (1) fire hazards, (2) health menace, (3) crime, (4) personality and social adjustment difficulties. For each of these, our task is to measure, not the total costs generated by slums, but the change in these total costs, throughout the metropolitan area, brought about by the redevelopment complex of demolition, dislocation, and replacement. The analytic

---

[21] Owner-occupiers are composites of property owners and consumers. Their gains as the one are offset by their losses as the other. On balance, their owner role predominates since their opportunities are capitalized (frozen) there but not for their consumer role.

building block of such measurement is to find, where possible, the money value of resources which the affected individuals would be willing to incur to avoid the costs imposed on them. Where this is not possible, we attempt to find the magnitude of the cost consequences in terms of the original dimensions—such as number of serious illness days, or of deaths.

FIRE HAZARDS. Costs here involve two elements: the slum-connected differential loss and damage to property and human life; the differential fire-protection costs to prevent life and property losses from being greater. For property loss, fire insurance records are appropriate. Measurement of the value of human life is a highly controversial question, but a number of relevant procedures can be suggested. (This particular measurement problem is common to most areas in the public sector, and extended discussion is beyond our present scope.) Human disability is not well measured by medical expenses alone since, like fire-protection services, these limit rather than prevent damages. A component that would measure lost productivity and the personal valuation of suffering should be added.

Differential fire-protection costs can be approximated roughly by actual fire department expenditures, although this may entail a bias. It is claimed that actual service levels are inadequate in slums, relative to nonslum areas.[22] If we measured the change in fire-protection needs due to redevelopment by a before-after service comparison we would understate the difference, since the "before" (and presumably higher) figure would have been below actual needs. But this bias is offset to some extent by the fact that inadequate protection leads to higher damages. One might even crassly argue that the political determination of service levels marginally equates the extra cost of service with the "social value" of additional expected fire damage—so that the bias is exactly offset. Less crassly, the absence of the necessary data makes the assumption attractive for purposes of simplification.

A basic complication of our procedure is that we must separate out the fire hazard effect of the slum itself from that of its special

[22] For example, see Max R. Bloom, "Fiscal Productivity and the Pure Theory of Urban Renewal," *Land Economics*, Vol. 38 (May 1962), p. 140.

population. Redevelopment demolishes structures, but redistributes the population elsewhere. Insofar as kind of use influences fire hazard and is influenced by type of population, a gross association between fire and slum will overstate the net relationship. Isolating the effect of slums proper should be amenable to multiple regression, in which fire hazard per $1,000 assessed valuation is made a function of family income level, population density, and percentage of substandard dwellings, the last reflecting slums.

HEALTH HAZARDS. As noted above, medical services do not prevent illnesses perfectly, or even cure them perfectly, but they limit the ravages more or less. Thus, as with fire hazards, our measure contains two components: differential protective, therapeutic services; and differential morbidity and fatality.

Measuring these components is considerably more difficult than the corresponding ones for fire. The first should be measured by the value of medical services, but this value is not easy to obtain for the specific geographic population. Moreover, sliding scales of compensation for such services, extending down to free care, complicate valuation. On the other hand, it is not easy to point out a bias in differential adequacy of *public* health and medical services; if anything, a disproportionate amount of these goes to slum inhabitants.

Morbidity and fatality figures are even more difficult to obtain for the slum population, since such households are likely to seek medical care in a smaller percentage of illnesses than nonslum households do. Specific sampling studies might be required. The problem of evaluating the social cost of human illness and death is the same as with fire.

Also as with fire, we must isolate the effect of slums proper from that of the selective characteristics of the slum population, since the latter effect remains after redevelopment. The population influence is especially important here, and especially difficult to separate out. As with fire, multiple regression should be used, but observations would be expressed in per capita terms rather than, as with fire, in assessed valuation terms.

CRIME. Public protective services here (unlike the preceding categories) do have substantial deterrent effect. The level of police protection authorized by the political process does bear a closer relation to the amount the public is willing to pay to *avoid* losses due to crime than do services under the preceding categories. The relation is far from perfect, since it is probably criminal apprehension as much as deterrence that is increased by enlarging the level of police services, and not even apprehension is dependably related to budget size. But if we posit that the police budget, say, is a fair first-approximation measure, then we are able to bypass some very difficult problems concerning the human cost of crime—assault, robbery, rape, murder, etc.—where such cost probably far exceeds the value of the property involved.

Our method here, with a very healthy dose of error (especially because of population selectivity), compares slum with nonslum per capita police costs. The measuring procedure is much like that for our earlier categories, with per capita units preferable to area or value of property units.

PERSONALITY AND SOCIAL ADJUSTMENT DIFFICULTIES. For this category we possess especially poor measures. The more personal, psychiatric types of difficulties are rarely deterred and—for this socioeconomic level of the population—rarely treated or cured. Only when they reach the level of severe psychosis is even a third level of action resorted to: custodial or mildly adjustive care. For what may well be a wide ocean of unhappiness, despair, misery, frustration, anger, and fright, we have no operational measuring rods. Broad qualitative judgments about both psychic states and the processes which might bring them into being are all that we possess. We cannot either stringently establish the truth of slum-generated costs here or measure them. Since I personally believe the qualitative argument that slums do generate costs in this category, I am prepared to suggest that our exclusion of any measure here imparts a downward bias—possibly an important one—to our specified total of external social costs due to slums. But we are unable to specify the direction of the bias in estimating how these costs are affected by the redevelopment project as a whole, when account is taken of the pattern of relocation.

For the more obviously family and neighborhood group types of difficulty the problem is somewhat easier. These are difficulties which *are* attacked by explicit actions—in this case, family social work, settlement house activities, neighborhood clubhouse activities, and so on. Some deterrence is present and some amelioration. On the basis of slight information, one is inclined to believe that the combined deterrent-ameliorative efficacy of these services—even in prospect—is less than for the other categories. Moreover, these services are the only ones which seem to give us an operational entree to measurement in this overall category. If we use total social work budgets as we have other protective services, we shall avowedly be omitting the submerged portion of the iceberg.

To summarize the discussion of changes in slum-connected social costs, I believe these alleged costs probably exist, but even a rough measurement is extremely difficult. It may be even more difficult to find out how they are affected by redevelopment, when relocation is taken into account. Some components are easier to measure than others, but all results should be taken with real reservation.

Reservation about even the direction of change stems from the possibility that the effects of massive relocation in spreading slums could significantly offset the effects of redevelopment in eliminating them. If in particular projects, or even in general, findings should indicate that the former effect is slight, our procedures become more useful—since most remaining biases then run in the same direction of understating benefits from redevelopment.

A final point should be made. Some observers (probably including myself) believe that this social cost category is the most important of all: urban renewal programs will stand or fall ultimately on how significant are these benefits. Yet we have been deeply pessimistic about being able to obtain relevant dollar figures to use as offsets to project costs. To base important policy decisions on only the dollar amounts that *can* be computed, just because these *are* the only dollar amounts, would be most dangerous; the underestimation of the most truly distinctive benefits of the program might be crucial. Wherever possible, therefore, money magnitudes should be supplemented by the vector of nonmonetary consequences, e.g., serious illness days, murders, incidence of psychosis. While the investigator may not be able to discover unique consensual tradeoffs in the community among these different kinds of consequences, society has the option of discovering them by a form of simulation—the governmental decision-making process.

An Illustrative Numerical Example[23]

The numerical illustration briefly presented in conclusion is intended, not as a serious empirical application, but only as an exercise designed to highlight some of the statistical problems that may be encountered. The scope of this study was too small to enable us to collect important bodies of data; in what follows below, therefore, we indicate some of the places where a larger study could reasonably hope to improve data quality and quantity appreciably.

The example relates to the three projects in Chicago which had been terminated early enough to make data available: Michael Reese, Hyde Park "B," and Blue Island. Of the three, only Michael Reese is a large project; Hyde Park "B" is a small strip in the much larger overall Hyde Park-Kenwood Project. The first two began in 1956, the third in 1959; they all terminated in 1961 or 1962. For each project we measure all three types of consequence, but to facilitate estimation we omit from internalization benefits the spillover effects on the value of neighborhood land and structures (these are extremely difficult to obtain within the limited scope of our empirical study).

Internalization Benefits

For simplicity, we use the simple model of equation (5) above[24] (omitting superscripts):

(5) $$\Delta P_{sE} = \Delta P_s + \Delta P_{\bar{s}}$$

---

[23] Data accumulation and calculations were performed by Robert Puth.

[24] Thereby omitting the tax capitalization adjustment of equation (8) above. This imparts a downward bias to our estimates. It can be rectified by estimating a tax capitalization factor.

where ΔP_s is the land price change in the redevelopment site; ΔP_{sE} is the land price change in the redevelopment site due to internalization; $\Delta P_{\bar{s}}$ is the land price change elsewhere in the metropolitan area (location advantage adjustment). Further, only Michael Reese is large enough and strategically enough located to warrant any consideration of a locational advantage shift. This is not likely to be important enough to be picked up by our statistical technique.[25] But the present scope precludes any attempt at calculating the correction. A larger empirical study should invest resources here to attempt the correction, since this is a marginal case. Our present procedure, then, will simply assume that $\Delta P_{\bar{s}} = 0$.

Our purpose is to calculate site land-value changes due to internalization, measured as the difference between 1955 and 1962 (Michael Reese and Hyde Park "B"), or between 1958 and 1962 (Blue Island). These periods are long enough, however, so that general forces affecting land values in the city as a whole could have had noticeable effect. We must adjust site-value changes for these general influences, to isolate the effects of redevelopment. Our method is to "explain" part of land value movements in terms of per capita income and population, and to subtract this explained portion of site-value changes from the gross change in order to obtain the site-value change due to redevelopment. Thus, the explained portion

$$(9) \qquad\qquad _{Y,N}P = a + b_1 Y + b_2 N,$$

where $_{Y,N}P$ is "Y-N explained" land prices, Y is per capita income, and N is population.[26]

Direct estimation of equation (9) requires data on site values, population, and income which are not available. A major empirical study could construct series for site population and land prices, since elaborate and laborious operations are involved. But this is certainly beyond the scope of even moderate applications. We therefore substitute indirect estimation, depending heavily on interpolation from Cook County data. Our model is:

$$(10) \qquad\qquad _R\Delta P_s = \Delta P_s - {}_{Y,N}\Delta P_s,$$

[25] The method suggested in my "Economic Evaluation of Urban Renewal."

[26] Our estimation actually used Y as aggregate income, because of data problems.

where $_R\Delta P$ is redevelopment induced land changes; and post-subscript s refers to the redevelopment site.

Our indirect estimation from Cook County data is obtained from the assumption that:

$$(11) \qquad \frac{Y,N\Delta P_s}{Y,N\Delta P_c} = \frac{\text{mean}(P_s)_{0,1}}{\text{mean}(P_c)_{0,1}},$$

where postsubscript c refers to Cook County data and subscripts 0,1 refer to preproject and postproject dates. We estimate (9) for Cook County and then calculate $Y,N\Delta P_c$. We can obtain mean $(P_c)_{0,1}$. The one type of data we possess about the site is $\Delta P_s/\text{mean } (P_s)_{0,1}$. Algebraic manipulation of (10) and (11) shows that from these we can determine $_R\Delta P_s$.

Our data to estimate the coefficients in (9) are for 1947-1961: income and population figures from *Sales Management,* land values from the Cook County Tax Assessor's Office. The latter are assessed valuations inflated to a market value basis. Following official assessment practice, we assume assessments are 40 percent

TABLE 1. Land Prices, Aggregate Income, and Population, Cook County, 1947–1961

Years	Land Prices (P_c)	Aggregate Income (Y_c)	Population (N_c)
1947	$18,399,558	$ 7,416,982	4,225.7
1948	19,666,323	8,176,791	4,305.8
1949	19,574,018	8,425,620	4,522.7
1950	19,984,798	8,137,803	4,490.7
1951	22,045,478	8,967,124	4,548.8
1952	22,052,150	9,173,481	4,601.8
1953	22,448,690	9,583,494[a]	4,607.1
1954	22,839,805	9,993,506	4,667.5[a]
1955	22,947,800	10,769,380	4,727.9
1956	25,716,518	11,476,197	4,866.1
1957	27,345,985	11,684,475	4,881.8
1958	28,316,530	11,757,304	4,944.8
1959	31,196,125	12,814,366	5,049.1
1960	31,543,940	13,428,844	5,119.8
1961	32,199,125	13,352,979	5,165.7

Source: Column 1 from Cook County Tax Assessor's Office; columns 2 and 3 from *Sales Management.*
[a] Value interpolated.

TABLE 2. Changes in Standard-Lot Land Values During Redevelopment Project, 1955, 1958, 1962

(land values per standard lot)[a]

Changes	Hyde Park "B"	Michael Reese	Blue Island
Preproject Land Value[b]	$12,500	$5,750	$5,000
Postproject Land Value[c]	20,000	10,500	7,250
Change, Absolute .	7,500	4,750	2,250
Percent Change (mean land value as base)	46%	58%	37%

Source: Olcott's *Land Values Blue Book of Chicago,* volumes for 1955, 1958, 1962.
[a] 100×125 feet.
[b] For Hyde Park and Michael Reese this is 1955; for Blue Island, 1958.
[c] 1962 for all projects.

of true market values. Our land value series is clearly inadequate. Somewhat better results can be obtained from direct approximations to market valuations in *Olcott's Land Values Blue Book of Chicago,* but these are available only on a lot-by-lot basis, and aggregation is terribly cumbersome. A larger study than the present one should certainly attempt to get better land figures. The data appear in Table 1.

TABLE 3. Changes in Redevelopment-Site Land Values Due to Redevelopment $(_R\Delta P_S)$, 1955, 1958, 1962

Changes	Hyde Park "B"	Michael Reese	Blue Island
1. Actual Percentage Change ΔP_s/mean (P_s)[a]	46%	58%	37%
2. Percentage Change Explained by Y and N $_{Y,N}\Delta P_s$/mean (P_s)[b]	23%	23%	13%
3. Percentage Change Due to Redevelopment $_R\Delta P_s$/mean (P_s)(line 1 − line 2)	23%	35%	24%
4. Absolute Change Due to Redevelopment, per Standard Lot $(_R\Delta P_s)$	$ 3,738	$ 2,844	$ 1,470
5. Initial Total Market Value of Acquired Land (L_0)	$49,506	$1,596,433	$45,559
6. Terminal Total Market Value of Acquired Land (L_1)[c] Adjusted for Y and N	$79,080	$3,315,618	$74,333
7. Total Change Due to Redevelopment (L) (line 6 − line 5)	$29,574	$1,719,235	$28,774

Source: Line 1 from Table 2 above. Line 2 from Table 1 and equation (9) above. Line 4 from Olcott's *Land Values Blue Book of Chicago,* 1955, 1958, 1962. Lines 5–7, *Ibid.,* 1955, 1958, 1961, 1962; from Chicago Land Clearance Commission, *Michael Reese-Prairie Shores Redevelopment Project: Final Project Report,* 1962 (for column 2); and line 3.
[a] Actual site mean: for Hyde Park, $16,250; Michael Reese, $8,125; Blue Island, $6,125.
[b] Explained Cook County mean: Hyde Park, $28,434,954; Blue Island, $29,704,660.
[c] For columns 1 and 3, L_0 is the figure adjusted for Y, N; for column 2, L_1 is the adjusted figure.

Equation (9) was estimated as:

(12) $\qquad Y,_N P_c = -\ 10{,}857{,}084 + 1.79Y_c + 3554.76N_c.$

Our direct estimation of gross site-value changes is shown in Table 2, and is based on market values for a standard lot (100 by 125 feet). Given these, we compute land value changes attributable to redevelopment in Table 3.

Redistribution Effects

We deal here with real income redistribution through the (largely) price effects of relative changes in the housing stock. We have suggested that dislocatees lose most, and those replacing them in the redevelopment site gain most. The former, but not the latter, are likely to be poor and members of racial minorities. Thus, the redistribution is politically potent.

But dislocatee worsening is not inevitable. Significant prior suboptimality, along with active relocation efforts by the LPA, could reverse this. Besides, real worsening will occur only if the number of simultaneous dislocatees is large relative to the size of the market. A small project will have little or no effect on housing prices elsewhere.

In addition to the question of the actuality of real influence is that of circumstances where such influences can be detected. Redistribution effects are most noticeable if the market is tight. If the market is loose, dislocation effect may appear solely in terms of decreased vacancy rates, or, at most, in prices falling somewhat more slowly than they would have in the absence of redevelopment. Such effects either do not much affect well-being, or are too subtle to be detected by most statistical procedures.

In the projects studied here, all three factors are present: we suspect prior disequilibrium, the number of dislocatees is small relative to the market and, most important of all, the overall market was significantly loosening during the period. It is doubtful whether, in this context, the small, if any, tightening effect of redevelopment could be detected.

For actual measurement, two procedures are suggested by the constraints of our small scope, first, to obtain differences in actual rentals paid by relocatees from relocation authority records, second, to compare preproject site rentals with postproject rentals in areas heavily affected by dislocatees. Both prove inconclusive. Under the first, relocatees from Chicago projects somewhat more recent than those we are studying paid slightly higher rents after dislocation (median rents rose by $6),[27] but most moved from substandard to standard housing in the process; thus no estimate of even the direction of real rental changes can be given. This is an area, however, where a larger study might well improve the quality of data. The relocation experience of specific families might be followed.

The second procedure is no more promising, even for larger studies, since it rests on finding areas comparable except for relocation impact. The combination of substantial market loosening, and locational and population differences among areas superficially similar in terms of housing, will probably overwhelm systematic redevelopment impact by statistical "noise." We have been unable to produce figures that adjust for these complications.

The data are not persuasive, let alone decisive. But for all their paucity, they may be consistent with the tenor of our introductory remarks: dislocatees in these projects may not have been much hurt, if at all. We shall therefore proceed on the assumption that at least *negative* redistribution effects were not important. This assumption is being made, *not* on the basis of lack of data, but on the strength of the moderating circumstances mentioned above which we are pretty sure were operative in the present case. Lack of data is not a general warrant for assumption of zero effects, since the attendant circumstances might well favor strong impacts. The lack of direct measurement of them would not justify their exclusion, especially if their direction were known.

Effects in Slum-Generated Social Costs

We have probed some of the voluminous literature on the twin problems of slum-generated costs and of the effect of redevelop-

[27] Relocation Office, Urban Renewal Administration, Chicago, Illinois. Figures from our projects are not available.

ment and public housing on these costs. We found suggestive treatments of parts of these problems, but no definitive studies, no sources from which quantification, even in dimensions other than money, can reliably be made. To attempt such a study ourselves is beyond our scope—and would probably be even beyond the scope of an expanded study on these lines. Still needed are additional studies out of the allied social sciences.

But we may sketch some gross tendencies abstracted from the literature. Slum concentration very likely increases fire hazards appreciably; physical configuration and use characteristics strongly suggest this. Since the sheer spatial area of concentration is important, redevelopment, on balance, probably decreases this hazard.

Overcrowding and filth demonstrably increase morbidity. Relocation into standard dwellings would decrease this hazard, but health hazards due to poverty and group practices will be untouched. The sketchy data of the last section indicating appreciable dislocatee movement from substandard to standard housing suggests that redevelopment does decrease social costs relating to health.

Crime is more related to poverty and specific population than to physical surroundings. But spatial configuration may have relevance, by increasing personality difficulties, or by affecting productivity (through illness, deterioration of aspirations, etc.), or through homogeneity of deprived population. Redevelopment might therefore have some latitude. If redevelopment, including relocation, improves neighborhoods on balance, then it might well decrease crime. On the other hand, the dislocation process may itself generate an offsetting impact: uprooting people from their accustomed locale may have traumatic personality effects, encouraging crime. Such an effect may be only transitional, however, lasting only until new roots are fixed.

On balance, then, redevelopment may slightly decrease crime; but this effect will become apparent only after some time as offsetting transitional effects disappear and as the long-run impact of slums on behavior is gradually moderated (especially in new generations).

Much the same can also be said about the last category: poverty and population will account for most of the difficulty. But spatial situational factors will count too, since depressed, unhealthy, overcrowded surroundings can warp individual and family interactions. Moreover, under these circumstances, unhappiness and mental illness may be easier to evoke than crime—but may also be harder to measure. Redevelopment should have an ameliorative effect over time, depending, as with crime, on relocation's achieving some situational improvement. The eventual impact on human happiness may be very important. But numbers are not available.

In sum, if we conclude that relocation did not much spread or aggravate slums elsewhere, then we may tentatively infer that redevelopment probably decreased social costs in all five categories, most probably in fire and health, least in crime. But no quantification can be offered here, not even in the dimension natural to each kind of social cost.

These benefits may be very important. When we bring together our *quantified* benefits with a rough estimate of the order of magnitude of project costs in the next section, the latter far outweigh the former. Redevelopment is not justified in terms of land productivity alone. Justification must rest most strongly on the importance of this present category, the most distinctive type of benefit under renewal. We thus cannot use lack of data as an excuse for omitting this category: to do so would give redevelopment a seriously biased hearing. This warning holds especially when we compare redevelopment with the status quo. But it also holds, though in lesser degree, for comparison with the modified public-policy package as well, because the configuration of effects on slums differs under the two alternatives.

Summary of Benefit-Cost Analysis

We are now in a position to summarize these numerical estimates. They are presented in Table 4, which is structured to give the reader an idea of the magnitude between overall resource costs and benefits.[28] Spillover benefits and benefits associated

[28] The basic organizing relationships are:
(1) $$GPC = AC + R$$
(2) $$AC = L_0 + I_0$$

TABLE 4. Summary Table of Benefit-Cost Analysis

(*in thousands of dollars*)

Project Costs and Benefits	Blue Island	Hyde Park "B"	Michael Reese
I. Resource Costs of Project			
1. Gross Project Costs (GPC)	$396	$638	$6,235
2. Less Initial Value of Land (L_0)	46	49	1,596
3. Equals Total Resource Costs (TC)	350	589	4,639
II. Benefits Produced by Project			
1. Increased Productivity of Site Land ($L_1 - L_0$)	$29	$30	$1,719
2. Increased Productivity of Neighboring Land and Improvements (Spillover)	+	+	+
3. Decreased Social Costs Associated with Slums (ΔSC)	+	+	+
Total Costs Not Offset by Site Land Benefit (I.3 minus II.1)	$321	$559	$2,920

Source: Line I_1 from *Urban Renewal Characteristics* 1962, Urban Renewal Administration, 1962. Line I_2 (columns 1 and 2) *Olcott's Land Values Blue Book of Chicago*, 1956, 1962 issues; column 3 from Chicago Land Clearance Commission, *Michael Reese-Prairie Shores Project: Final Project Report*, 1962. (Assessed valuation figures converted to market value on assumption that former is 40 percent of latter.) Line II.1 from Table 3, line 7.

with decreasing the social costs of slums are marked with pluses.

To simulate a decision-making context, site-land benefits are subtracted from total resource costs on the bottom line, so as to indicate how much spillover and social cost benefits would have to be worth to decision-makers in order that the total exceed total costs from the projects listed. Benefits associated with subsidization to achieve "public goals" (like population heterogeneity, university expansion, architectural beauty) would appropriately be added at this point to determine the grand balance.

---

where $GPC \equiv$ Gross Project Cost
$AC \equiv$ Cost of Acquired Real Estate in Site (Acquisition Cost)
$L_0 \equiv$ Market Value of Land Acquired in Site
$I_0 \equiv$ Market Value of Improvements Acquired in Site
$R \equiv$ Resource Expenditures in Project Other than AC
(3) $\qquad TC = I_0 + R \qquad$ where $TC \equiv$ Total Resource Costs of Project
$\qquad$ (since I_0, but not L_0, is lost to society through Project)
so
(4) $\qquad\qquad TC = (AC - L_0) + R = GPC - L_0$
(5) $\qquad\qquad TB = (L_1 - L_0) + (\text{spillover}) + (\Delta SC)$

Comments

ANTHONY DOWNS, *Real Estate Research Corporation*

In presenting his cost-benefit approach to urban renewal, Jerome Rothenberg stated that a relevant analysis would have the following basic steps:

1. Specification of alternative policies to be examined, which assumes that these alternatives are aimed at goals which are known. (A series of goals for urban renewal was included in the paper.)
2. Designation of one of the alternatives as the *base* alternative. This is usually the status quo.
3. Analysis of the amount of deviation from the base alternative involved in each of the other alternatives—the deviation being estimated by using static equilibrium theory. A series of critical variables is established and divided into those which are and are not beneficial.
4. Measurement of all variables that are measurable and some statement about the net outcome of the various alternatives in relation to the base alternative.

My comments will be devoted to a discussion of several classes of problems regarding an analysis of the sort exemplified by this paper.

Problems of Omission

The first type of problem involves *the omission of important considerations*. In setting forth a list of alternatives, there is always danger of leaving out some that are just as important as those which are examined. Rothenberg mentions that he necessarily omitted some important alternatives, because his paper could not be of infinite length. I think there is almost always good reason to omit some plausible alternatives, but there is a definite cost in doing so. In particular, many of the benefits of urban renewal might be obtainable without any renewal program at all if we used alternative approaches—for example, more rigid zoning enforcement or changes in the property tax structure and assessment procedures. Such alternatives are typically left out of analyses of urban renewal, which generally focus instead on a comparison between a specific urban renewal project and the status quo.

Another kind of omission which has appeared frequently in the papers presented at this conference is a failure to include or examine a crucial variable. For example, one of the most important purposes of urban renewal is the protection of cultural institutions now existing within our larger cities. Among these institutions, universities are probably the most significant; they involve a very large sunk capital investment, often in older neighborhoods experiencing both population transition to low-income groups and deterioration of physical property. Universities faced by these conditions have encouraged urban renewal in the hope of maintaining their ability to attract good-quality students and faculty members and thereby to function efficiently. They have attempted to gain the locational advantages of side spillover effects from urban renewal—effects which Rothenberg regarded as unmeasurable and therefore did not take into account in his analysis. He apparently believes that protecting downtown areas is the principal purpose of urban renewal, whereas I believe that protecting all types of areas within the city is the principal function, and that protecting downtown areas is merely a subfunction. Thus, omitting important variables sometimes leads to misleading results, as does omitting important alternatives.

Problems Regarding the Beneficiaries

Who benefits from a policy being considered? What particular group of people should we use as our "basic citizenry" in judging costs and benefits? Rothenberg discussed this issue extensively in his paper: should we analyze the net impact upon citizens of the city, or the suburbs, or the metropolitan area, or the state, or the nation as a whole? Exactly whose costs and whose benefits are we discussing?

Most of us who have dealt with this problem follow Rothenberg's proclivity to accept the metropolitan area as the most logical unit for measuring costs and benefits, although a large number of analysts prefer the nation as a whole. However, as Rothenberg points out, the actual decision-making involved in urban renewal is carried out on the one hand by the federal government and on the other hand by central-city governments. Moreover, the federal government has followed a policy of substantial decentralization of decisions by leaving considerable authority in the hands of central-city governments. It may be intellectually satisfying to point out that the central-city government is not the appropriate organization for measuring costs and benefits in the metropolitan area, especially since this assumption renders irrational a great deal of urban renewal activity actually being carried out, thereby enabling us as critics to satisfy our urge to find something quite wrong with the world.

In many cities where we have conducted studies of urban renewal, the efforts of urban renewal authorities are directed at maintaining resources within the limits of the central-city rather than allowing them to shift to the suburbs. From the point of view of the metropolitan area as a whole, investing resources in this type of allocation-influencing activity is meaningless. However, from the point of view of the central-city government which is making the decisions, retaining these resources within its boundaries may be of critical importance to its very fiscal survival. The fact that central-city boundaries make *theoretically* poor limits for testing costs and benefits does not help the people who are actually making decisions. They must accept the facts (1) that they cannot alter existing city boundaries and shift to a metropolitan government, at least in the short run, and (2) that they face immediate fiscal problems caused by their boundaries which they must somehow solve.

It is certainly true that economists can and should perform the function of pointing out that maintenance of political boundaries and fiscal programs along present fractionalized lines is irrational from the point of view of society as a whole. However, once we have pointed that out, I believe we should not abandon the decision-makers in our central cities who are faced with the difficulty of living with this suboptimal situation. They certainly cannot institute metropolitan governments themselves. Therefore they must make decisions based upon their interest in maintaining the political units they live in. We should give *these* decision-makers advice on a cost-benefit basis, as well as advising the theoretical overlords of metropolitan government who really **do not exist.**

For us simply to call for metropolitan government, or state that we should resort to other tax forms, as Rothenberg does in this paper, is not enough. Most local governments cannot resort to other tax forms such as the income tax, not only because of legal restrictions (which are often important factors), but also because of the mobility of the resources they are attempting to retain in the city. If any individual city adopted a heavy income tax, it would have a competitive disadvantage with other cities in attracting these resources or keeping them. As a result, many mobile resources would soon flee the area and others would fail to enter it, thereby making the city's problem worse.

Because of the very real and frustrating difficulties of reforming the basic structure of political boundaries in metropolitan areas, I believe we must spend at least as much time giving advice to the people who actually must make decisions as we do to fictitious people who would make them in the best of all possible worlds. We should do this even if we do not believe the present situation is optimal in the long run.

My second point concerning "who benefits and who pays" is that the recipients of benefits from urban renewal are not necessarily the same people who pay the costs. It may be that the people who move into an urban renewal project are the major beneficiaries, as Rothenberg contends, but federal taxpayers in rural Wyoming also bear some of the cost. It is true that metropolitan areas as a whole bear most of the costs of urban renewal, because the people and firms in them pay most of our federal income taxes. Nevertheless, even within metropolitan areas, urban renewal causes a tremendous redistribution of income because people who get the benefits are not paying the costs.

I think there is a constant implicit assumption in many of our cost-benefit analyses that we can add up utilities by means of interpersonal comparisons which count A's gain against B's loss. We are either assuming (1) interpersonal comparability of util-

ity, or (2) a random redistribution of benefits and costs over
time with everybody balancing out in the long run, or (3) the
existence of some kind of overall social-welfare function by which
we maximize social utilities. I do not raise this issue as either a
trivial or a strictly theoretical point. Rather, I believe that the
necessity of making interpersonal comparisons shifts the nature
of the decisions involved from essentially economic to essentially
political. Once we grant that the decisions are essentially political, the concept of maximizing social welfare becomes a rather
naive way to analyze the decision-making process. Neither the
politicians nor the bureaucrats involved in city, local, and federal
governments make decisions by means of the criteria associated
with theories of maximizing social welfare. By the very nature of
their activity, they must take into account political forces such as
those I have dealt with in *An Economic Theory of Democracy*.[29]
Thus, assuming that the decision-makers maximize votes is probably much more accurate than assuming they maximize social
welfare when they are dealing with urban renewal.

If it is true that the decision-makers involved are using political rather than economic criteria, then I believe we should be
examining costs and benefits in the framework of political criteria. This would have a definite impact on the way we set up
the cost-benefit analysis. For example, we should perhaps follow
suggestions made by Julius Margolis and identify, in the total
population, those specific subgroups which are most likely to be
affected by a certain program. If our identification properly iso-
lates groups which are relatively homogeneous regarding their
relationship to urban renewal programs, we can perhaps specify
approximately how *much* each group will benefit and how *much*
it will pay regarding a specific program. We can then leave it
up to the politicians to decide how to compare the net gains or
net losses of these groups in selecting among alternatives, since
that is essentially an ethical—and therefore political—decision.

My third point concerning "who benefits" is that urban renewal programs tend to impose so-called benefits upon certain
citizens—raising the question whether a person who is forced to
accept better housing standards is really receiving a benefit from

---

[29] Published by Harper in 1957.

his point of view. Most of the urban renewal projects undertaken in the United States have resulted in the destruction of more housing units than were subsequently created. Therefore, residential density on the land concerned almost always declines, and density elsewhere rises, since there is a drop in the number of housing units in society, *ceteris paribus.* However, qualitative changes initiated by urban renewal are just as significant as quantitative changes. Generally, the residences demolished in an urban renewal project are of very poor quality, and the people who leave them can only afford to live in relatively poor housing. Conversely, the new housing units which are created are of quite good quality, and the people who move into them generally move from existing units of reasonably good quality. Thus, a whole "filtering-down" process is set in motion. Existing units outside the urban renewal project must be made somewhat worse in quality to accommodate the low-income people displaced by the project. These people move from formerly very bad units now demolished into not-so-bad units "filtering downward" in the existing supply. As a result, almost every individual family who was badly housed before gets better housed. But the supply of existing units (excluding those demolished) also becomes somewhat more deteriorated, although this is partly offset by the construction of the new units.

The statistics Rothenberg cites show that 98 percent of the people who moved out of demolished units had been in substandard units, and that 82 percent of those people relocated in standard units. Thus, there was a distinct upgrading of quality in the housing of the people originally living in the urban renewal area. This is clearly one of the principal benefits of urban renewal.

However, some of the people whose situations are thus improved may really not want better housing. The substandard housing such people were occupying may have seemed quite desirable. Because a good many of them were recent arrivals from various southern states, where they lived in shacks, what would appear to us as slums, would appear to them as very decent places to live. Further, according to this argument, they wish to spend their income on things other than housing, since their relatively low standards re housing are satisfactorily met by the substandard units which urban renewal destroys. In this view, then, slums are merely housing for poor people—and, so long as we have poor people, we *ought* to have slums.

From a purely economic point of view this is a correct analysis if we assume that (1) consumer sovereignty ought to be allowed full weight, (2) there are no imperfections in the housing market, and (3) the external costs generated by such housing do not exceed the external benefits thereof. Rothenberg has pointed out that there are indeed some imperfections in the housing market, principally because the people who are displaced lacked information about the alternatives to the housing they previously had lived in. However, I find it hard to believe that these people were able to make a transition from 98 percent occupancy of substandard housing to 82 percent occupancy of standard units if they were really all that ignorant. True, urban renewal did nudge many of them out of their normal patterns of living and therefore widened their horizons. Yet that so many would be simultaneously enlightened by such a process seems incredible. I am inclined to believe we are essentially forcing higher housing standards on these people than they desire. We are destroying the old slums and forcing the rebuilding of standard, up-to-date, modern units on the same sites. Moreover, we are not allowing existing units elsewhere to be downgraded sufficiently so that they become slums; therefore, these displaced people have no choice but to live in better housing than they did before. Society is imposing its own decisions about standards upon slum dwellers, whether they like these standards or not.

Is it really a benefit to a man if we force him to move into a better housing unit, even if he does not want to, and thereby make him pay more of his income on housing and less on other things that he prefers? Does this make him better off? I am not sure that I know. Perhaps we can justify imposing our standards on his behavior because his preferred line of action (living in slums) imposes net external costs on the rest of society. Nevertheless, I believe this aspect of the matter must be carefully looked at when we are appraising programs like urban renewal. We should not assume that the flows of costs and benefits in such programs are entirely voluntaristic. Instead, we must look at some of the activities that society considers benefits, even though it imposes its standards on other people who do not necessarily share those standards.

In analyzing the changes in land values caused by urban renewal, Rothenberg assumes that there is no change in the values as a whole as a result of urban renewal: all the people displaced from bad housing units move into newly worsened existing units, and a complete reshuffling occurs with all changes canceling out. But in my opinion, there is a double change in land values. On the one hand, the value of land in the project itself declines because the possibility of exploiting poor people in very high-densities is eliminated through the imposition of higher standards of quality—and lower standards of density—on utilization of the urban renewal site. However, this decline in values is swamped by an increase in value resulting from the assemblage of many small parcels of land into one large bundle. The use of eminent domain for such assemblage usually creates an increment of value much larger than the decline caused by requiring lower densities. But the enforcement of standards on the urban renewal site that differ from those in surrounding properties causes a relative decline in land values on that site and a rise in other land values, *ceteris paribus*. Thus, the enforcement of social standards through nonmarket devices, by restricting the alternatives open to developers, has important repercussions upon urban renewal that should be considered in our analysis of it.

Problems of Dynamics and Scale

In the kind of analysis such as Rothenberg has provided here, one is often forced to use a static equilibrium approach because of the great difficulties of dynamic models. Nevertheless, a static approach in an essentially dynamic world tends to make us give weight to the wrong variables. In the past few years, for example, the housing supply in the United States has been rising much faster than the number of families; hence, the market has been shifting toward a condition of relative surplus. Therefore, any impact of urban renewal programs which tends to decrease the supply of housing and raise rents and values elsewhere is usually swamped by the dynamics of the housing market as a whole, thanks to supply changes not directly related to urban renewal. This is especially important because most urban renewal projects are small compared to the total inventory in any given area. Normally, their impact on the total supply in a metropolitan area is utterly trivial, but the impact regarding certain kinds of housing or certain neighborhoods may not be trivial. Even so, any decisions we make about the effect of urban renewal on housing markets as a whole should take into account the fact that these projects are currently being conducted on a minute scale compared to other changes in the overall market.

Another disadvantage of using static equilibrium analysis is that one of the alternatives we are analyzing is the status quo. In reality the status quo is rapidly changing and does not provide a meaningful alternative to compare with proposed urban renewal programs, since the many changes will alter it regardless of whether urban renewal occurs. I have in mind such things as the spatial extension of the housing market into the suburbs, rising population, transportation changes, and changes in mortgage markets. These elements of dynamism should be very important parts of our cost-benefit analysis of urban renewal. We should therefore, rather than comparing the status quo with and without urban renewal, try to compare various *future* states into which these dynamic elements have been incorporated. Unfortunately, this will tend to make the analysis even more difficult than it is already.

Problems of Differential Measurability

The final problem that concerns me is the use of variables that are simply not measurable and their incorporation into the analysis. This does not apply to Rothenberg's paper alone; many of the studies presented at this conference fall into a category of analysis which I call a horse-and-rabbit approach. There is an old joke which says that a fifty-fifty proposition offered by a shyster resembles a horse and rabbit stew: each partner puts in one unit —you contribute the horse and he contributes the rabbit. In the analyses we have been hearing about here, this joke takes on a new slant. The tiny rabbit is minutely examined and exhaustively analyzed with sophisticated techniques before it is placed in the stew—but then we go out and get any old selected-at-random horse and throw him in. Thus it often seems that we are doing an exhaustive analysis of certain variables which are capable of being so analyzed, then making wild guesses about relatively incommensurable variables which are in fact much more important in determining the outcome. We are making a stew with a scientifically-prepared rabbit and a randomly-chosen horse. The quality of such a stew is bound to be rather indeterminant.

Consequently, it seems to me that this type of analysis cannot be directly applied to decision-making. In many cost-benefit analyses containing relatively immeasurable variables, the size of probable errors in estimating the variables is larger than the variation necessary to change our policy from one alternative to another. Hence we are not in a good position to make decisions on this type of analysis alone. About all we can do is to (1) specify all the particular effects the decision-maker should look at, (2) indicate which effects are measurable and which are not, and (3) try to measure the effects which are measurable. This will at least tell the decision-maker what things he ought to consider and give him a rough measurement of those which can be measured. Since the decision-maker is generally a politician, it will then be up to him to make up his mind. He will have to estimate the immeasurables and balance the gains of some groups against the losses of others. I think that cost-benefit analysis, if approached from this point of view, can perform an extremely useful service indeed.

Frederick O'R Hayes, *U. S. Office of Economic Opportunity*

We have now had nearly fourteen years of experience with federal assistance for local urban renewal projects. With the federal government at present committing about $700 million annually for grants and with projects under way in nearly 700 communities, it is time that we developed some standard calculus for evaluating costs and benefits of the programs. The need has been underlined by the widespread emergence of renewal as a local political issue and by the increasing number of critiques of renewal now appearing from various sources. Rothenberg's paper represents welcome and worthwhile progress toward this objective.

The Limits to Measurement

We should, however, be clear on the nature of Rothenberg's contribution. I think he would agree that it is less a solution than a refined statement of the problem. This is best indicated by the summary table at the end of the paper, showing the results of the application of his technique to three Chicago projects. The net benefits to justify the expenditure of public funds on renewal must arise almost entirely from the overspill impact on areas outside the renewal site and from the social impact on the people involved, neither of which was Rothenberg able to quantify.

Most of the tangible and identifiable costs and benefits of renewal are covered here, but, in my opinion, Rothenberg devotes too much attention to worrying about the question of changes in land value, which proves to be of negligible importance as soon as he applies it to a concrete example. The obvious but important point is that public investment can seldom be justified solely on the basis of market considerations.

Some part of the support must be found in other benefits. In renewal, a major part of support rests on an assumption of net social benefits, some immeasurable and others not, some clearly identifiable and others representing the satisfaction of a kind of itch on our collective super-egos. The balance sheet will be a statement in a number of different currencies without specification of exchange rates. The net of all of the costs and benefits measurable in dollars can only be compared or judged against a residual listing which cannot be so measured.

The most careful effort to analyze the costs and benefits of renewal is, therefore, unlikely to produce a definite answer. It should, however, result in an identification of the issues which provide a far more adequate basis for political judgment and for rational governmental behavior. I doubt that I have any disagreement with Rothenberg on this score.

The Diversity of Urban Renewal

The major weakness of the paper is its concentration on the classic redevelopment project—where residential slums are cleared and the cleared site is used for the construction of new housing. Such projects have been a minority of those initiated in recent years. Currently, over one third of the new projects being approved involve the clearance of nonresidential areas for non-residential reuses.

There are, in addition, a substantial number of projects in which residential slums will be cleared for nonresidential reuses. Lastly, there are rehabilitation and conservation projects where housing and neighborhoods are upgraded without substantial clearance. Further, some mention should be made of special problems such as the sale of land for low-rent public housing (where there is no local sharing in the writedown) and for middle-income housing (where the price is set below market value). Hospital and university expansion projects constitute another important group of projects.

This great diversity of the urban renewal enterprise vastly complicates the problem of determining costs and benefits and introduces questions and issues not fully covered in Rothenberg's analysis. Moreover, an overview of the whole spectrum of urban renewal would probably support a somewhat broader statement of its purposes than the one Rothenberg uses.

This broad range of renewal activity makes more apparent the limitations of two of Rothenberg's assumptions: (1) that no net benefits result from any redistribution of income produced by renewal projects; (2) that all costs and benefits must be determined in a context of full employment of resources. But renewal projects are being undertaken in an economy characterized by significant noncyclical unemployment, and, more important, by a concentration of a significant proportion of this unemployment in areas characterized by long-term economic and employment problems. The economic development of these areas has become, under the Area Redevelopment legislation, a national objective. The favorable impact of renewal upon this objective must, in my opinion, be regarded as a benefit even if there is no net income creation on a national basis. However, there is a persuasive argument that the movement of jobs to people can create, not merely net benefits in the comprehensive sense, but even net additional income through reduction in the total cost involved in its production or creation.

By the same token, I am not quite as chary as Rothenberg of accepting as net benefits some of the increment in taxes resulting from redevelopment in the urban renewal area. Any increase in taxes must, of course, be reduced by the cost of servicing the new redevelopment and also by the loss in taxable investment in some other part of the metropolitan area. But if the redevelopment does improve the financial situation of a municipal government, it also improves the government's ability to deal more effectively with the needs of the people who live in the city. Given the great difficulty of effecting a more rational solution through changes in government structural or tax policy, any step that can increase the tax flow to the central city is likely to be advantageous. The ultimate net benefit involved, after all discounts and allowances, may not even be quantifiable but it is nonetheless real.

The Spillover Benefits

The nature of the spillover impact of renewal activity upon areas outside the site also deserves some elucidation. Rothenberg tends to emphasize neighborhood impact and to underplay community impact of spillover effects. Any renewal project is tied to the entire structure of the community by a complex set of linkages. Generally, but not always, the significance of the project to any other activity follows the gravity hypothesis and declines sharply with distance from the site. An example of an exception might be the removal from certain locations of transportation-congesting activities which will have very important effects upon accessibility factors throughout a wide part of a metropolitan area and with consequent strong effects on the activities which can be conducted in those locations.

By the same token, the renewal project can establish a core of linked activities, particularly those of areawide importance— which will support and modify related economic activities and behavior throughout the metropolitan area. By way of illustra-

tion, the provision of a site for a public university in a sector
with strong transportation services will ultimately have a strong
impact upon the character of residential and other locational
preferences over a wide part of the area, probably with some
definite net income effects. The basic point is that urban renewal
projects are often of a scale or incorporate activities of such wide
strategic effect that they can have an important impact upon
the functioning of an entire community. These impacts, however,
are not easy to get at analytically.

Let me illustrate specifically some of the points made in the
paragraphs above. Say that a number of urban renewal projects
are displacing the produce wholesaling center in the heart of the
city. This center has tended to remain in the very oldest part of
the city because the linkages between different merchants are
so strong that the parts of the center must be moved (if at all)
very nearly en masse. Hence, they have not moved, despite an
increasing tendency for existing locations to be suboptimal, especially for servicing and distribution. The displacement, with
appropriate public guidance and assistance, has made it possible
to establish new produce and food centers in sites more adequate
and efficient in both internal design and location within the metropolitan area. The benefits to the merchants are beyond question. This has, in turn, resulted in a sharp reduction in transportation congestion in the streets of the old produce area and the
adjacent neighborhoods, with widespread small savings in the
cost of movement. At the same time, sites have been made available for high-quality housing in areas with pronounced locational
advantages. This, in turn, produces a close-in population with
relatively high incomes and also with tastes conducive to the support of some of the basic cultural and unique commercial facilities of the entire region. The same people if located thirty miles
away in exurbia would produce demonstrably less support for
these institutions. These benefits represent opportunities that are
available to everyone in the region, but the value that should be
placed upon the whole result as a benefit is most difficult to determine. This is particularly so since the renewal project is not necessarily a unique solution to the basic problem.

Another example is the renewal project related to a university. In a number of cities, urban renewal is providing sites for the expansion of urban universities. This is clearly a service to higher education, but it is also an important aid, particularly in our older urban areas, in the adaptation of the economic structure of the urban community to changing conditions. At the same time, land adjacent to the university will be provided for research businesses and other related activities, which will, hopefully, establish a complex of linked activities that will attract a larger proportion of the national total of business and research in such fields. There will be effects (certainly not wholly attributable to urban renewal) on the economic base of the community, on the level of education of its population, and on the internal efficiency of the university and activities related to it. Again, cost-benefit determination is not easy.

Renewal and the Metropolitan Housing Market

I come back now to Rothenberg's key observations on the classic residential renewal project. One important point should be underlined: the costs and benefits of a renewal project which clears a residential area will, in very large part, be a function of conditions in the metropolitan housing market. Or we can say, alternatively, that urban renewal is not a complete program in itself but is dependent upon housing resources already available in the market or produced under other program auspices. The general concern—entirely proper on theoretical grounds—that family displacement may result in the creation of new slums and an increase in the supply price of the residual low-quality housing inventory has not been borne out in practice, simply because the housing resources of the community supplemented by other public projects were adequate to meet the need of displacement. Rothenberg notes in his conclusions on the Chicago projects that the combination of small size relative to the market, prior suboptimality, and a substantially loosening market suggests that dislocatees were not much damaged, if at all, by relocation (except for their attachment to their original dwellings.) The general case is, in fact, one of general improvement in housing conditions for relocatees at marginally higher rents and a reduction rather than an expansion of slums.

Social Values and Problems

One of the most difficult problems on residential renewal projects is the determination of the cost which should be attributed to the destruction of a neighborhood with a strong sense of identification and social cohesion. The strong anti-environmentalist turn in sociology has produced a number of papers which go to the very brink of saying that the physical environment is so unimportant and the social values of the place so great that urban renewal can never be justified. We need some means of placing social neighborhood values in proper perspective and also of recognizing the sharp differences among neighborhoods in this respect. There is some hope that the degree of neighborhood cohesion can be roughly measured or estimated through simple survey techniques, but the weight which should be assigned these values in the total calculus is extraordinarily difficult to determine.

The impact of renewal on the problem of racial discrimination should also be mentioned. Given the goals of the democratic society, the expansion of housing opportunities for groups with restricted choices is a valid and important objective. (We regard this as one of the major benefits from the Lake Meadows and Michael Reese projects in Chicago, where opportunities were provided for good housing available to middle-class Negro families in an integrated neighborhood.) The problem of discrimination and restriction of opportunity can even be important in terms of the urban areas designated for clearance. The strengthening of existing neighborhoods through rehabilitation and conservation may be reinforcing the results of housing discrimination, and the disruption of any segregated neighborhood may contribute ultimately to more open occupancy in housing. This is, to say the least, a subtle issue.

Renewal Expenditures

A word should be added on the character of urban renewal expenditures and the extent to which all of these should be attributed to the core project operation. The gross project costs of an urban renewal operation include costs for schools, parks, and other supporting facilities to serve the population of the new area. Many of these are facilities which, in the absence of the urban renewal project, would have had to be provided somewhere in the city. They may be regarded, in some senses, as subject to separate cost-benefit evaluation rather than as an integral aspect of the renewal project. There is, on the other hand, a strong argument for considering the whole complex of activities as contributing to one basic result.

Program vs. Project

Finally, the cost-benefit analysis may prove to be more applicable to city-wide long-term improvement and renewal strategy than to individual projects. Nearly 100 cities are now preparing community renewal programs aimed at the development of long-term programs of this kind. With the time phasing of a recommended program established and the interrelationship between renewal projects and other developmental programs defined, there is an opportunity to examine the costs and benefits in a context which better emphasizes the long-term objectives and problems of renewal. The problems tend to emerge in sharply different form. For example, displacement from the typical project is ordinarily so small in terms of its impact upon the rest of the city that it can be, as Rothenberg found in the Chicago projects, ignored in cost-benefit calculations. On the other hand, a succession of proposed projects emphasizes a long-term problem of rehousing displaced families at a very substantial level compared to the existing supply of housing and points up the potential long-term need for stronger direct support from the public sector.

In concluding, I want to commend Professor Rothenberg for a solid contribution to the state of the art. His paper has gone a long way on the road toward establishment of a cost-benefit structure for urban renewal. I only hope he is encouraged to travel the remainder of the distance.

Concluding Statement

JEROME ROTHENBERG

The remarks of the two commentators and general discussion by the group raised a number of issues. I shall consider a few of them in the limited space available.

1. Some of the remarks concerned delineation and measurement of the real costs involved in redevelopment. However, my paper did not attempt a systematic consideration of the cost side of the picture. Only for the purpose of giving an idea of the orders of magnitude involved in the final statistical summary did I draw on cost data, and for this I used simply the dollar value of "net project costs." This was for convenience only—it made no pretensions to profundity.

2. I had argued in my paper that the population for which it was most proper to calculate benefits and costs—particularly benefits—was that of the metropolitan area. Mr. Downs and, to a lesser extent, Mr. Hayes argued that it should rather be the population pertinent to the decision-makers involved, and, indeed, that benefits and costs to this population should count only insofar as and how they affect the gcals of the decision-makers. Specifically, only the central-city population counts, and the financial "well-being" (in terms of tax inflow) of the government of the central city counts as a benefit too. The argument was fortified by the claim that the advice of economists is useless unless it is relevant to the needs of the decision-makers.

Certainly our analyses must be useful to decision-makers in order to be useful. But there are two senses in which usefulness may be interpreted in the present context, and in both of them I believe that the practice of my paper is the preferable one. Who are the relevant decision-makers in a model of this kind? I think there is both a long-run and a short-run answer. The long-run answer is the electorate of the nation. Given a particular type of problem calling for public policy, this electorate makes some attempt to delegate it to that level of government whose decision-

making focus can best encompass the distinctive pattern of benefits and costs engendered by the problem. An existing delegation
may not be optimal at some particular time. One of the important functions of an applied welfare economic analysis such as
benefit-cost is to diagnose any suboptimality of such an existing
pattern of governmental delegation, by showing that significant
externalities are produced by present jurisdictional boundaries.
A function directly connected to this is to indicate how boundaries may be improved. In the present context the allegation that
a city-wide focus is less desirable than a metropolitan focus fulfills these functions.

But Downs and Hayes refer largely to the short-run situation,
where decision-making delegation is given and is relatively in-
flexible. In this context I argue that the city government is not
in fact the chief decision-maker; the chief decision-maker is on
the national level. The program is a national program, with national aims. At least two thirds of the financing comes from the
national level.[30] The program is meant to subsidize people,
not particular political jurisdictions. Interpersonal comparisons
of utility are, of course, involved: geographic, urban-rural income
redistribution is intended by the program, so gains to some
groups are explicitly intended to be able to offset losses to others;
this is what allows benefit-cost analysis, with its aggregation of
different individuals' gains and losses, to be at all pertinent. But
the national focus means that benefits and costs to *all* groups significantly affected must be considered. Since we are treating only
the benefit side, we may limit the affected population to the metropolitan area. But it *is* the metropolitan area, not the central
city.

The central city enters because the federal authorities in effect
are delegating to it some important administrative responsibilities. Yet important as these are, they are within the mandate of

---

[30] The real marginal federal contribution is likely to be considerably more than
two thirds, since cities are permitted to meet their legal "at least one-third" share
by contributions in kind—public utilities addressed essentially to the use of the
project site. These contributions often turn out to be a provision of facilities that
would have been provided anyway, and whose services go significantly to benefit
a population outside the project site—whether the benefit is direct or indirect.

the goals of the national program: the city is to act as the agent of the federal government in formulating and evaluating alter native projects—for federal approval. This means that the city government is to act in these matters as if it were guardian of the total affected population.

Admittedly, this is far from actual practice. The city authorities in fact think of redevelopment as their own program, for their own ends. But this seems in itself a palpable distortion of the national program. To present these authorities with estimates based on the total significantly affected population, and based on the assumption that the tax intake of the city government is not a benefit per se, is not therefore to contravene the legally constituted focus of decision-making, but indeed to contribute toward bringing that focus back into relevance.

3. Downs complained that my procedure was restricted to a comparison between redevelopment and the status quo, whereas he believed it important to compare redevelopment with other possible public policies that meet similar goals, such as better code enforcement, rehabilitation, etc. I agree entirely with him on the importance of such comparisons. Indeed, in my paper I made this the explicit cornerstone of the benefit-cost analysis that should be performed. That my own empirical illustration did not show this is due only to the avowedly simplistic character of that illustration, forced on me by limitation of scope. The desirable task is very difficult, but it *should* be performed when a concrete question of public policy formation is at stake. The resources for such a study will of course have to be considerably greater than those that were available to me.

4. Downs emphasized that protection to special institutions is an important proximate goal of redevelopment; Hayes stressed the existence of many different types of redevelopment, each type with its own set of goals. I certainly agree with both on the existence, significance, and diversity of goals other than those I explicitly stipulated. However, the apparent omission of other goals was not inadvertent. The "aggregate" real-income effects of some of them—e.g., protection of institutions—are logically similar to comparable effects for protection of downtown areas

(i.e., a maintenance of unique service and facility mixes). So to this extent they are included.

Their distribution effects, on the other hand, are particular to the specific types of redevelopment involved and can scarcely be generalized other than to say that they involve some group or constellation of services or function or location which is designed to be worth subsidizing at the expense of other groups or constellations of functions or locations. Society (i.e., you and I) sometimes lays great value on such types of subsidy, which then must be considered to constitute a social benefit that offsets real costs. The problem of their inclusion in our analysis, however, is not their inadmissibility but that they would have to be enumerated singly with their complete context expressed and their social value—a value judgment entirely—specified. Whenever a concrete question of public policy is encountered, and alternatives of this sort are to be considered, then of course these social goals must be explicated. But even in such a context, the social decision-makers, not the economists, are the ones to stipulate the value of the goals involved. Thus, these goals are "excluded" just as the various other goals I do explicitly mention in my list are excluded—not to be expunged from social choice, but simply to be separated out from a generalizable economic analysis.

5. Downs criticized what he felt to be an undesirable staticness in my "equilibrium" analysis, arguing that the analysis ought to take into account that the housing market is always changing, is never standing still. In particular, he decried my analysis that, in principle, redevelopment plus dislocation can decrease the housing stock, raise the rent level for low-income dwellers—especially the dislocatees themselves—and accelerate downward filtering. He pointed out that in the Chicago projects the dislocatees came predominantly from substandard dwellings and relocated predominantly in standard dwellings, arguing that this upgrading is an effect to be ascribed to redevelopment.

This sort of criticism tends to confuse a dynamic force *generated* by the policy in question and a set of exogenous "dynamic" forces onto which the effects of the policy are juxtaposed. Downs' example seems largely of the latter sort. Yet he used it as though it were of the former. Certainly the price-raising, quality-lowering forces unleashed by redevelopment can be more than offset by outside forces tending to loosen the housing market. But one can—must—still try to measure the contractive effects of redevelopment by seeking the difference between what the housing stock is and what it would have been, what rentals are and what they would have been, in the absence of redevelopment. The "equilibrium" mode of analysis does not itself assert that external forces are absent. It simply provides a way of isolating what the effects would be *if* they were absent, so that actual occurrences can be understood as a composite of these tendencies and the tendencies stemming from external changes.

Even if redevelopment with its associated relocation policy did really *generate* the relocation from substandard to standard dwelling units, this would not call for the kind of calculation of benefits that Downs seems to want. For one thing, the accelerated downward filtering predicted in a tight market would take time to occur and would therefore not show up in Downs' figures. However, let us suppose that such filtering does not take place. Because of relocation policy, the dislocatees have been *induced* to move to higher-quality dwellings. This has two types of benefits: the improvement in quality, and the decrease of social diseconomies associated with slum living. As to the first, there is nothing in the analysis to suggest that the real cost of housing (i.e., per unit of housing at given quality) will have gone down for these dislocatees *as a result of redevelopment*. So higher quality is being financed by the combination of a tendency toward higher rentals (the figures show that dislocatees on the average did pay a higher rental) for higher quality—a quid pro quo—and a tendency toward lower rentals as the housing stock increases for reasons not associated with redevelopment. Thus, there is no consumer surplus on this account which stems specifically from redevelopment. As to the second type of benefit, this is in fact included in my analysis as the decrease in social costs of slums.

One might object that coercing erstwhile slum-dwellers to increase their budgetary allocation for housing represents a socially desirable form of education for these people: they are ignorant. However, insofar as their ignorance was of slum-generated externalities, this again falls under the second head and is already included. A different factor is introduced, however, if the "ignorance" involves a presumption that outsiders know objectively that certain families ought to devote more of their income to housing (and thus less to other items of consumption) than they want to. Outsiders often argue that this is so, and on some items of consumption (e.g., narcotics, illicit sexual relations) the composite outsider consensus known as Society argues it strongly. But these are really value judgments that, while admissible for social decision-makers, are not amenable to economic analysis.

A different characterization of "ignorance" *is* amenable to economic analysis, but appears not to be very useful. One could argue—and I have some sympathy with this—that the housing market is full of imperfections of information and immobility. Families know too little about alternative opportunities and are too timid or lazy to try to improve their situation. Redevelopment projects then serve periodically to force or dramatically encourage a game of musical chairs, to get people into the housing market where, happily, they then realize how easy it would have been all along to improve their situation—especially in a market where conditions are loosening markedly. In this situation everyone can in principle be made better off. This argument may be sound, but redevelopment is an extremely expensive way for society to transmit information about market conditions. It is like Lamb's elegant suggestion that one can burn down one's house to roast a pig. Considerably cheaper alternatives are available.

Thus, the novel categories of benefit generated by a supposed redevelopment encouragement of quality improvement are highly controversial. It is not clear that Downs would wish to claim that their exclusion is very damaging to the analysis.

6. My assumption of full employment has been questioned. Certainly resources and investment projects are heterogeneous, certainly resources are unemployed. Thus, redevelopment can affect overall employment of resources (or aggregate investment) in any given locality or even the nation as a whole. But any government or private expenditure project will have similar kinds of effects, depending on the specifics of the project and of resource heterogeneity and unemployment pattern. It is the spe-

cifics one assumes, therefore, that determine whether a particular project will have net employment effects relative to alternative uses of resources. But what specifics should one assume? It seems safe not to consider a particular type of project in the context of relative employment generator unless it possesses characteristics which are uniquely advantageous for that use.

Redevelopment appears to be an especially clumsy tool for such a purpose, because of its long and uncertain gestation period and its notable inflexibilities. It is for this reason that I assume full employment, meaning that any resources (capital or labor) brought into this sector would have been equally well used elsewhere. Thus, for example, any aggregate increase in housing construction that might arise from a demonstration effect of attractive new construction on redevelopment sites is not counted as a benefit.

7. Lastly, but perhaps most importantly, some questions were raised about measurable and immeasurable benefits, and their importance to this kind of evaluation. In my empirical illustration of three Chicago projects, almost all of the work of estimation went into measuring land-value changes, and these turned out to be quite small. There was little or nothing to show about spillover effects on neighboring property, about changes in social costs of slums, about distribution effects due to changes in the composition of the housing stock, and about achievement of the diversity of specific goals other than slum removal. But this does not mean that the benefits actually "measured" were the only ones that either can or need be measured.

In particular, I believe that the benefits falling under the rubric "social costs of slums" are the truly distinctive ones arising out of redevelopment. Moreover, I believe that they, as well as spillover and distribution effects, are approximately measurable in principle, and in my paper I attempted to suggest some avenues by which to approach their measurement. (Indeed, my treatment of distribution effects was not to consider them immeasurable, but to argue that in the special circumstances of the cases treated, adverse effects could probably be approximated by the number zero. This did leave out the positive gains of individuals whose housing options were enhanced by redevelopment, gains which are difficult to measure but approximable in principle.) But such measurement is extremely difficult, and while I strongly believe we must do the best we can, that best will for some time be quite far from what we should like. One advantage I see here is that many of the evaluations we must make—such as the value of a human life, or the human cost of illness—are not uniquely related to redevelopment. They are categories that appear and reappear as effects of many different kinds of public policy. A cooperative venture pooling resources and insights from many quarters would seem to be called for. The rewards for success are large.

In the interim, before we have anything approaching satisfactory figures for these in-principle measurable categories, we can at least formulate our analyses to make clear that the categories must be considered when making a concrete policy choice. Here we may use a procedure which is useful with regard to the treatment of achievement of special goals whose social valuation is a direct responsibility of the ultimate decision-makers. We simply specify the extent to which the several goals are achieved under the project being studied, these "extents" being definable in terms of their own natural dimensions. For some of the components of social costs of slums, as I noted in the paper, we would indicate number of crimes of different types affected, number of patient days of medical care, number of lives "saved" in terms of age-specific death rates, and so on. Thus the impact of redevelopment would be multidimensional, not solely because distributional benefits would be broken down for different groups, but also because the terms in which types of consequences were listed had no obvious common denominator. Alternative policies, giving rise to different outcome vectors, would achieve the single-dimensionality of a preference ordering only through the evaluative mechanism of the ultimate decision-makers.

HERBERT E. KLARMAN*

Syphilis Control Programs

THE AIM OF THIS PAPER is to measure the economic benefits of syphilis control programs. There were several reasons for selecting syphilis control as an example of a health program:

1. It was supposed that there would be ample and complete data to provide a basis for measurement, since syphilis is a reportable disease in every state of the union.

2. Expert advice would be forthcoming on the treatment of the disease and on the administration of the control programs, both of which have long been objects of study at the Johns Hopkins University Medical Institutions.

3. The incidence of syphilis is reported to be rising in this country, as elsewhere, after a dramatic postwar decline; the problem is, therefore, of immediate importance.

The benefits of a particular disease program are the costs of the disease that are averted by it. Costs comprise three elements: loss of production; expenditures for medical care; and the pain, discomfort, etc., that accompany a disease. Commonly, the economist concentrates on measuring the first two elements. The third

* Department of Public Health Administration, School of Hygiene and Public Health, and Department of Political Economy, The Johns Hopkins University.

is likely to be disregarded; if mentioned at all, it is usually side-tracked.

Syphilis has certain characteristics that, for purposes of economic analysis, distinguish it from other diseases:

1. Both its symptoms and the economic losses associated with it are manifested in the early (primary and secondary) stages and the late (so-called tertiary) stage of the disease. There is a long intervening period—perhaps ten to thirty years—when the disease is latent.

2. Since syphilis is a communicable disease, a rise or decline in incidence in one period has an effect on the incidence in the next and subsequent periods. Large external benefits are present.

3. As a venereal disease, syphilis carries a continuing stigma that may impair a person's earnings even after symptoms have disappeared.

This paper will put some emphasis on the special characteristics of syphilis, while carrying out a calculation of economic benefits along lines that have become fairly well established in recent years. Specific objectives are to review the literature on the calculation of economic benefits (and costs) for a disease; estimate the incidence of syphilis for a current year and project the distribution of the total by the stage and year of discovery (and treatment); calculate the value of lost output and medical care expenditures associated with the new cases of syphilis by stage of discovery; and attempt to measure the loss of consumer benefit due to the syphilis infection and its late complications.

The paper consists of four sections, the first of which is a review and critique of the concepts and techniques employed by economists in measuring the costs of a disease and contains the decision rules adopted for this paper. The second measures the incidence of syphilis for a recent year, 1962, and the third calculates the several types of economic cost attributable to these new cases. In the final section the benefits of an eradication program are calculated, and some questions are raised concerning the calculation of economic benefits of control programs aimed at reducing incidence.

For policy purposes, that is, for evaluating alternative control programs, the costs per case of the several programs are highly relevant. These are lacking and no attempt has been made to develop them. It is understood, therefore, that this paper deals exclusively with the calculation of potential benefits.

Concepts and Techniques

A considerable body of work by economists has accumulated concerning the costs of specific diseases and the benefits of eliminating them.[1] This paper represents a further attempt to develop and refine the economist's concepts and techniques[2] and to apply them to calculating the benefits of syphilis control programs.

Elements of Economic Costs or Benefits

DIRECT AND INDIRECT COSTS. Expenditures for medical care to treat a disease (or injury) are not the total costs of that disease. The economic costs of a disease comprise at least two components: direct costs and indirect costs. Direct costs are the expenditures for health services attributable to the disease, reflecting the use of resources. Indirect costs are the loss of output attrib-

[1] A. G. Holtmann, *Alcoholism, Public Health, and Benefit-Cost Analysis* (paper presented before National Institute of Mental Health Seminar, Bethesda, Md., Dec. 6, 1963; processed); I. S. Blumenthal, *Research and the Ulcer Problem* (RAND Corporation, 1960); Earl E. Cheit, *Injury and Recovery in the Course of Employment* (Wiley, 1961); Louis L. Dublin and Alfred J. Lotka, *The Money Value of a Man,* rev. ed. (Ronald Press, 1946); Rashi Fein, *Economics of Mental Illness* (Basic Books, 1958); Howard Laitin, *The Economics of Cancer* (doctoral thesis, Harvard University, 1956); National Planning Association, *Good Health Is Good Business* (1948); D. J. Reynolds, "The Cost of Road Accidents," *Journal of the Royal Statistical Society,* Vol. 119 (1956, Part IV), pp. 393-408; Burton A. Weisbrod, *Economics of Public Health* (University of Pennsylvania Press, 1961).

[2] John Maurice Clark, *The Costs of the World War to the American People* (Yale University Press, 1931); Selma J. Mushkin and Francis d'A. Collings, "Economic Costs of Disease and Injury," *Public Health Reports,* Vol. 74 (September 1959), pp. 795-809; Selma J. Mushkin, "Health As an Investment," *Journal of Political Economy,* Vol. 70, Part 2; Supplement (October 1962), pp. 129-57; Jack Wiseman, "Cost-Benefit Analysis and Health Service Policy," *Scottish Journal of Political Economy,* Vol. 10 (February 1963), pp. 128-45; also literature cited below with reference to the discount rate.

utable to the disease, owing to premature death or disability. In the case of syphilis, there may be additional loss of output, owing to the "stigma" of having had the disease. Direct costs can be reduced by failing to provide services, but usually at the penalty of increasing the indirect costs.

The total (direct plus indirect) costs of a disease per case serve as the measure of benefits derived from preventing that case. In a cost-benefit calculation the comparison is between contemplated additional expenditures for health services, and the anticipated reduction in existing costs. This is the essential conceptual framework. Although Fein states that his study of mental illness deals with costs and Weisbrod states that his study of cancer, poliomyelitis, and tuberculosis is limited to benefits, both pursue the same problem in similar fashion.[3]

CONSUMER BENEFIT. A common difficulty in measurement is that few (if any) health services are pure investment goods or consumption goods that yield the same degree of health improvement. It is conventional to recognize the benefit in consumption derived from most health and medical care expenditures (such as reduction of pain, discomfort, etc.), to comment on the difficulty of measuring it, and then to dismiss it.[4] What receives weight (and space) is what is measurable; and that is not necessarily important. Since the measurable segments—output loss and medical care expenditures—are not equally important in all programs being evaluated, their sum is not likely to bear a consistent relationship to the loss of consumption benefit. Attaching a value to the latter, lest it be forgotten (or treated as zero), is both a sobering and challenging task.

For this purpose it is helpful to recall that consumers are frequently willing to incur expenditures, medical and other, that do not promise an increase in earnings or an offsetting saving in expenditures. Suppose a person's lifetime income were guaranteed to him and his heirs, and health and medical services were furnished free of charge. Notwithstanding, would not many per-

[3] Fein, *Economics of Mental Illness*, pp. 3, 5; Weisbrod, *Economics of Public Health*, p. 5.

[4] *Ibid.*, p. 29; Reynolds, "The Cost of Road Accidents," p. 393.

sons be willing to spend some money to avoid syphilis or to be cured of it in the early stages? It seems plausible to assume an affirmative answer.[5] The question is, how much would they be willing to spend?

To some persons the odds of contracting syphilis seem so low that, from an actuarial standpoint, only a small expenditure would be warranted. Others, more likely to contract the disease, are oblivious, or disdainful, of the consequences. In either case the insurance approach is not applicable, given the large degree of control over the outcome exercised by the prospective beneficiary—the factor of moral hazard. (I am indebted to Professor Mills for the ideas in this paragraph.)

Another approach is through the analogous disease, one which is deemed to produce nonpecuniary consequences no more dire than those of syphilis.[6] Such a judgment may be intuitive or represent a consensus among informed persons.

Consider a disease B (with symptoms that are somewhat similar to those of disease A), for which medical care expenditures are incurred both without any prospect of a return in increased output (either because the disease is not disabling or because the patient has retired from the labor force) and without prospect of an offsetting reduction in medical care expenditures in the future (because the disease is not curable and expenditures do not cease). These expenditures are incurred for consumption purposes only, and by analogy they may be held to indicate the value of the consumption benefit attached to avoiding the disease A under study. This is the approach adopted in this paper.

THE DISCOUNT RATE. In calculating benefits (or costs), the comparison of streams of income and expenditure over varying time intervals represents a technical problem which most economists recognize and many public health authorities overlook. A given amount of money has different values when it is realized (or spent) at different times. The process of discounting converts a stream of benefits into its present value: the higher the rate of discount,

---

[5] Jacques Thedie and Claude Abraham, "Economic Aspect of Road Accidents," *Traffic Engineering and Control*, Vol. 2 (February 1961), pp. 589-95.

[6] Wiseman, "Cost-Benefit Analysis . . . ," p. 139.

the lower the present value of a future income stream. Discounting is particularly important when a long time span is involved, as in a syphilis control program where some benefits accrue ten to thirty years after the outlay.

Nevertheless, some economists prefer not to discount. Blumenthal did not apply discounting in his study of ulcers, because he considered the procedure arbitrary; he also questioned its desirability on conceptual grounds.[7] Clark, in his study of the cost of World War I, developed some discounted figures, but wondered about their applicability to his problem; the point was that pure time preference alone is a weak reed to lean on when the capital involved cannot be said to be productive.[8] The National Planning Association did not employ discounting in estimating the cost of tuberculosis[9]—and was properly criticized for the omission.[10]

What rate of discount should be employed? Some economists employ the going market rate of interest.[11] This may not always be appropriate; the discount rate is meant to balance the productivity of an investment and the reluctance of society to sacrifice current for future consumption, and the individual's market calculations and collective calculations need not coincide.[12] In general, an individual's discount rate for the distant future tends to be higher than society's.[13] It may be necessary to employ a rate that synthesizes the social rate of discount and opportunity cost

[7] Blumenthal, *Ulcer Problem*, p. 20.

[8] Clark, *Costs of World War*, pp. 93-94, 222.

[9] National Planning Association, *Good Health Is Good Business*, p. 4.

[10] Weisbrod, *Economics of Public Health*, pp. 10-11.

[11] Fein, *Economics of Mental Illness*, p. 73; Reynolds, "The Cost of Road Accidents," p. 398; John V. Krutilla and Otto Eckstein, *Multiple Purpose River Development* (Johns Hopkins Press, 1958), Chap. 4; Roland N. McKean, *Efficiency in Government Through Systems Analysis* (Wiley, 1959), p. 79.

[12] Maynard M. Hufschmidt, *Standards and Criteria for Formulating and Evaluating Federal Water Resources Developments* (U.S. Bureau of the Budget, 1961), p. 11; and Stephen A. Marglin, "Economic Factors Affecting System Design," in Arthur Maass *et al.*, *Design of Water-Resource Systems* (Harvard University Press, 1962), p. 194.

[13] William J. Baumol, *Welfare Economics and the Theory of the State* (Harvard University Press, 1952), pp. 91-92; Harold M. Groves, *Financing Government*, 5th ed. (Holt, Rinehart, and Winston, 1960), p. 304.

(investment foregone) in the private sector.[14] Much work remains to be done in this area.

It has been noted that pure time preference is likely to vary inversely with the life expectancy of a population; in practice the discount rate may vary by socioeconomic class.[15] Readiness to sacrifice the present for some possible gain in the future may be less pervasive among lower-status people, who tend to accord priority to immediate rewards.[16]

Weisbrod sees the selection of a discount rate as tantamount to expressing a value judgment on the relative importance of successive generations. Acccordingly, he prefers to present two rates— 4 percent and 10 percent. He does not explicitly choose between them, but in the example illustrating the policy implications of his calculations he employs the higher rate.[17]

I believe that it is appropriate to present two or more rates when the criteria for selecting a particular rate are spelled out. Otherwise, a single rate seems preferable. In this paper I use a discount rate of 4 percent. It is intermediate along the range of available figures and is widely employed.[18]

EFFECTIVE NET DISCOUNT RATES. The present value of a future expenditure is the sum of money that would have to be set aside at present and cumulated at the social rate of discount to equal the monetary cost of the expenditure at the time it will be incurred. In making this calculation it is usually assumed that the general price level will remain constant. However, to the extent that changes take place in specific prices, relative to the general

[14] Hufschmidt, *Standards and Criteria*, p. 23; Martin S. Feldstein, "Review of Weisbrod's Economics of Public Health," *Economic Journal*, Vol. 73 (March 1963), pp. 129-30.

[15] Otto Eckstein, "A Survey of the Theory of Public Expenditure Criteria," in National Bureau of Economic Research, *Public Finances: Needs, Sources, and Utilization* (Princeton University Press, 1961), p. 457; and Eli Ginzberg, "Sex and Class Behavior," in Donald Porter Geddes and Enid Curie, *About the Kinsey Report* (New American Library, 1948), p. 136.

[16] Ozzie G. Simmons, *Social Status and Public Health* (Social Science Research Council, 1958), p. 26.

[17] Weisbrod, *Economics of Public Health*, pp. 57, 90.

[18] Cheit, p. 76; Clark, p. 220; Fein, p. 87; Hufschmidt, p. 23; Reynolds, p. 398; and Jack Hirshleifer, comments on Eckstein (see footnote 15, above), p. 499.

price level, they should be taken into account. Experience in the recent past indicates that prices charged by physicians are rising faster than the Consumer Price Index, at an extra rate of increase of 1.25 percent a year. Hospital cost has been rising at an extra rate of 4 percent a year. It is convenient, facilitating the calculations of present value, to combine the two rates of change that are simultaneously operative into a single rate. For example, by coupling a discount rate of 4 percent with a projected extra price rise of 1.25 percent a year we have the equivalent of an effective net discount rate of 2.72 percent ($1.04 \div 1.0125 = 1.0272$).

Similar reasoning applies to expected changes in productivity. Suppose a disease-control measure will reduce the absenteeism of a worker by one day one year hence. At present levels of wages, prices, and productivity, the worker earns $20 a day, which is also the value of the goods he produces. However, it is expected that productivity will rise 1.75 percent a year.[19] The goods produced by this worker a year from now will be worth $20.35 valued at current prices (20.00×1.0175). Discounted at 4 percent interest, $20.35 has a present value of $19.55. This is the current social value of averting a day of absenteeism one year later. The same result could be obtained by dividing $20 by $1.04/1.0175$ or 1.0221. Consequently, 2.21 is the effective net rate of discount for absenteeism prevented.

Calculating Medical Care Expenditures

The calculation of medical care expenditures presents fewer conceptual problems than the calculation of output loss associated with a disease. The statistical task may be just as formidable.

AVAILABILITY OF DATA. It is usually easier to calculate expenditures in the institutional setting than in the community at large. The hospital is likely to have data or maintain a mechanism for compiling them. Outside the hospital, one must interview physicians to ascertain prevailing practices regarding numbers of visits and charges. But physicians are usually reluctant to general-

[19] Based on Outdoor Recreation Resources Review Commission, *Projections to the Years 1976 and 2000: Economic Growth, Population, Labor Force and Leisure, and Transportation* (GPO, 1963), p. 121.

ize from their recollections or records to a representative cross section of patients; their thinking is geared to the diagnosis and treatment of individual patients, and the infinite variety of possible associated conditions and complications.

Study of a disease with a low incidence rate, or a declining long-run trend, poses the disadvantage of a dearth of data. Since most statistics on utilization and cost are compiled for operating or planning purposes, agencies lose interest when a disease-control program absorbs but a small proportion of their resources.

It is often easier to collect data on diseases that affect large numbers of persons, but greater accuracy is required of such data. To calculate long-run marginal cost, for example, it would be necessary to include the cost of services rendered by capital. It would also be in order to ascertain the differential use of various hospital services by patients with a given disease compared with other types of patient in the same institution.

Several diseases are sometimes associated in a person at the same time. In the case of a hospitalized patient, it may be that only the primary diagnosis (the final diagnosis for the condition for which the patient was admitted) is reported on the summary page of the medical record.

PRESENCE OF MULTIPLE DISEASES. The gain in production and saving in medical care expenditures due to the elimination of disease X could be added to the corresponding benefits from eliminating disease Y if every person with a disease had a single disease. If a person has two or more diseases simultaneously, eliminating one still leaves him with one or more diseases, which may also contribute to disability, premature death, or continued expenditures for medical care.

In some population groups certain diseases are so frequently associated that their simultaneous presence cannot be disregarded. Patients with syphilis may have gonorrhea. When the treatment of the two diseases is substantially the same, the calculated benefits for each program are not additive but neither does the calculated benefit for syphilis alone, the more important disease, represent an overstatement.

Syphilis may also be associated with alcoholism, tuberculosis, or certain other diseases that seem to be linked to adverse social conditions. When treatment and prognosis differ, the benefits calculated for one disease at a time constitute an overstatement, whatever may be the etiological relationships between diseases. The extent of overstatement is affected by the degree of interdependence in the origin of diseases, so that individuals and families with multiple problems present the greatest potential for overstatement. Methods for measuring the effect of the presence of an additional disease on medical care expenditures remain to be developed.

I have not pursued these matters with respect to syphilis. With a single exception, subsequent calculations in this paper make no allowance for the simultaneous presence of other diseases in persons with syphilis. The exception occurs in the latent stages of the disease when some of the costs of its discovery (diagnosis and initial treatment) may be assigned to the other diseases.

It is assumed that syphilis carries no extra mortality beyond that attributed to its own complications. Although a series of studies in Tuskegee, Alabama, concludes that persons with untreated syphilis do have a higher death rate from other causes than persons without syphilis, another study of major proportions (in Oslo, Norway) challenges and contradicts this finding.[20]

Since syphilis is a relatively minor cause of death today (accounting for less than 0.2 percent of all deaths), we need not deal with the problem posed by the presence of competing causes of death: that is, what would happen to a population's death rate if some single cause of death were removed? As Weisbrod has pointed out, there is no firm basis for recomputing the life table when a given disease is eliminated.[21]

In general, it is worth emphasizing that the presence of multiple diseases and of competing causes of death means that the total

[20] Sidney Olansky et al., "Untreated Syphilis in the Male Negro. X. Twenty Years of Clinical Observation of Untreated Syphilitic and Presumably Neurosyphilitic Groups," Journal of Chronic Diseases, Vol. 4 (August 1956), pp. 177-85; and Trygve Gjestland, The Oslo Study of Untreated Syphilis, Supplement 34 to Acta Dermato-Venereologica, Vol. 35 (1955), p. 355.

[21] Weisbrod, Economics of Public Health, pp. 34-35.

benefits of eliminating or controlling several diseases cannot be arrived at by summing the calculated economic benefits of individual programs.

Calculating Output Loss

Certain problems arise in calculating the indirect costs of a disease or injury. Among them are the treatment of transfer payments, taxes, consumption, the work of housewives, the appropriate measure of output loss, choice of assumptions regarding employment, as well as the discount rate. About some of these elements of cost, students of the economic costs of road accidents, mental illness, cancer, tuberculosis, poliomyelitis, ulcers, alcoholism, and job accidents are approaching a consensus. Differences of opinion and in approach persist regarding others.

TRANSFER PAYMENTS AND TAXES. The consensus revolves around the treatment of transfer payments, taxes, and the measure of output loss. It is possible to incur expenditures that entail no cost in resources; these are transfer payments. They reflect, not a change in resource cost to the community, but rather a redistribution of income within the community and a shift in command over resources. Once the resource cost (in this instance, loss of output) is taken into account, there is no reason to bring transfer payments into the calculation. If relief checks are replaced by earned income as the means of family support, it is the output, as measured by earnings, that constitutes society's economic gain. The desirability from a social viewpoint of having families live on their own earnings, in preference to relief grants, does not enter into this particular calculation; it may, however, warrant separate recognition.

The argument concerning the proper treatment of taxes is parallel. It is double counting to include tax receipts by government once earnings have been counted.[22] I believe that the amount of taxes and the distribution of the burden of taxes have no proper bearing on decisions concerning expenditures for

[22] Julius Margolis, *External Economies and the Justification of Public Investment* (Department of Economics, Stanford University, 1955), p. 16.

health programs, unless the beneficiaries of the programs are subject to a special tax. Yet it might be realistic to recognize that the proposed expenditures fall on one pocketbook, while the potential benefits accrue to another pocketbook or are diffused.

AVERAGE EARNINGS. Earnings of members of the labor force (wages and salaries and net income of the self-employed) are increasingly accepted as the appropriate measure of output loss.[23] To count total output per capita, as some do, seems to be tantamount to attributing to labor the entire output of the economy. (In commenting on this paper, Fein pointed out that under certain circumstances total output per capita is the appropriate measure of loss. This would be true during unexpected epidemics, when capital is idle.)

The proper measure of the expected earnings of a group of workers is the arithmetic mean. However, most sources of income data furnish the median; given the skewed distribution of income, the latter is the lower figure.

Earnings are defined by the Census Bureau exclusive of wage supplements. An upward adjustment may be indicated, which I did not make in this paper. I know of no data on income or other economic characteristics of persons who contract syphilis. It seems plausible, however, to suppose that a survey would reveal occupational and employment characteristics that connote a level of income below that of the entire population with similar demographic (sex, race, age, etc.) characteristics. Substituting a median earnings figure for the mean makes a partial allowance for this difference.

RATE OF EMPLOYMENT. The prevailing consensus among economists favors the assumption of full employment (or 4 percent unemployment). The argument is that if a health program is effective in preventing or curing a disease and prepares people for productive employment, then the program has achieved its objective, whether or not employment opportunities actually exist.[24]

---

[23] Blumenthal, *Ulcer Problem*, p. 18; Fein, *Economics of Mental Illness*, p. 69; Weisbrod, *Economics of Public Health*, p. 49.

[24] Mushkin and Collings, "Economic Costs of Disease and Injury," p. 801.

Notice should be taken, however, of the somewhat unfavorable employment prospects of rehabilitated persons in this country.[25] It is also likely that differences in employment potential (and in earnings potential) exist between a group in whom a disease has been prevented and one in whom the disease occurs and is cured. (That the difference may reflect irrational discrimination is of no consequence.) Although empirical data are lacking, developing illustrative measures of this difference for a disease like syphilis seems worthwhile. This is done under the heading of "stigma."

ALLOWANCE FOR CONSUMPTION. Diversity of opinion prevails regarding the treatment of consumption. Economists agree that insurance companies and families should deduct consumption in their calculations of the economic value of a man.[26] Unlike insurance companies or families, however, society as a whole is concerned with total output, of which consumption is the major component. Man is not a machine, and consumption is the ultimate goal of economic activity. Accordingly, net production after consumption is not relevant to the economist's central concern, and consumption should not be deducted.

Weisbrod's position is that the treatment of consumption depends on one's view of society—that is, whether the potential survivor is regarded as a member. If he is, consumption should not be deducted from earnings; and conversely. Although Weisbrod does not explicitly choose between the two views, he develops elaborate and carefully calculated estimates of consumption and employs them in his estimates of economic loss.[27]

Certain other economists treat consumption as a deduction from a person's contribution to output. (Clark deducts consumption because he is measuring the remaining interest of survivors of war; Laitin and Reynolds apply the deduction as a matter of course, and offer no explanation.[28]) Accordingly, the calculation

[25] Eli Ginzberg, "Health, Medicine, and Economic Welfare," *Journal of The Mount Sinai Hospital*, Vol. 19 (March-April 1953), pp. 734-43.
[26] Cheit, *Injury and Recovery in the Course of Employment*, pp. 76-82; Dublin and Lotka, *The Money Value of a Man*, p. 77.
[27] Weisbrod, *Economics of Public Health*, pp. 35-36, 52-55, 60-61, 64-69.
[28] Clark, *The Costs of the World War*, p. 214; Laitin, *The Economics of Cancer*, pp. 119, 130; Reynolds, "The Cost of Road Accidents," p. 396.

of benefits may result in a negative figure, such as Laitin's findings for cancer among the aged. The point is that not every health program pays off in economic terms. This finding has been erroneously interpreted to mean that killing may be desirable from an economic standpoint—a repugnant notion.

How to resolve the issue? As I see it, a major difference between a health program and most other types of expenditure is that the health program exerts an effect on the size of population, as well as on output. There is a distinction between a health program that saves people from death for useful labor and one that saves people from death to pursue an unproductive life.

In a poor nation it would seem necessary to pose a clear-cut choice among programs with respect to their effects on per capita output. In the United States, such a choice may not be necessary. Should it be necessary, separate weights might be attached to changes in per capita output and in aggregate output. The determination of relative weights falls within the purview of the political process. In dealing with a disease such as syphilis in this country, we may reasonably neglect its effect on per capita output.

The medical care expenditures of a person in an institution automatically replace his ordinary expenditures as a consumer. These should be deducted to the extent that his family reduces its own expenditures. The size of the reduction is likely to be greater for long-term than for short-term institutional care. In the short-term instance the numbers may be of a secondary order of magnitude and can perhaps be safely neglected.

SERVICES OF HOUSEWIVES. Economists differ on the treatment of housewives' services. Those who would exclude them from economic calculation acknowledge that the result is a serious understatement of the costs of disease, but the exclusion is justified on two grounds. One is the difficulty of measuring the contribution of housewives to national income. Because the contribution occurs outside the market, the imputation (attributing the equivalent of a market price where none exists) of economic value raises many statistical problems.[29] The other is that to include the con-

[29] Fein, *Economics of Mental Illness*, pp. 23-24, 143.

tribution of housewives is inconsistent with the accepted procedures of national income accounting.[30]

The second argument is not relevant; since the figures are not additive, the income loss associated with a particular disease is meant to be compared with a similar figure for another disease.

The first argument has merit. However, even when a factor is as difficult as this one is to measure accurately, trying to measure it does less harm than neglecting it. Specifically, to disregard the services of housewives in calculating output loss is to understate the benefits of any program designed to serve a preponderantly female population.

The question, then, is how to measure the value of a service not traded in the market. Two methods were considered:

1. *The opportunity cost of being a housewife.* Given her education, training, and other qualifications, how much would she earn in the market place? Arriving at this figure, then one would deduct income tax on the earnings and the extra expense of going to business. But elaborate calculations are entailed that go beyond the immediate requirements of this paper.

2. *The replacement cost of a housewife.* Weisbrod devised a refined measure of the cost of housekeeping help by taking into account the size of household to be cared for.[31] It saved a great deal of work to fall back on Kuznets' simple device of estimating the value of a housewife's services at the level of earnings of a domestic servant. This amount is admittedly on the low side since it makes no allowance for the housewife's longer work week; it offsets any tendency to overvalue a product not sold in the market by attributing to it the price of a counterpart.[32] (Since the earnings figure for domestic servants turned out to be close to the figure for all Negro women, the latter was substituted for convenience of calculation.)

[30] Mushkin and Collings, "Economic Costs of Disease and Injury," pp. 803-04.

[31] Weisbrod, *Economics of Public Health*, pp. 114-19.

[32] Simon Kuznets, *National Income and Its Composition, 1919-1938* (National Bureau of Economic Research, 1947), pp. 22-23, 432.

Incidence of Syphilis

The economic benefits of a control program are a function of two factors: the base line against which the results of the program are measured, and the anticipated results themselves. Let us consider the base line. What are the costs associated with syphilis today? To obtain these costs, the occurrence of syphilis in the population, as well as some of the relevant economic characteristics of the persons who incur syphilis, must be determined.

The public health literature states that 1.2 million Americans have syphilis, that 60,000 Americans acquire it every year, and that 124,000 are newly reported to the authorities in a single year, most of them years after the onset of the infection.[33] In the language of health statistics the first figure represents the prevalence of the disease; that is, the number of persons with a disease at a given time. The second figure is an estimate of the disease's incidence, that is, the number of new cases occurring in a specified interval. The third figure represents nothing clear-cut, but its several components testify that the efforts of the existing syphilis control programs are not completely effective. These components can, however, be employed in estimating the true number of new cases, which is the figure that concerns us.

When data on the incidence of syphilis are reported or cited, they usually pertain to infectious syphilis. The primary stage is the first symptomatic stage after the disease is acquired and lasts from two to thirteen weeks. The secondary stage follows, and lasts from six to twenty-five weeks. In both stages patients usually display definite clinical symptoms and are infectious to sex partners.

The figures for cases with infectious syphilis—reported by clinics and private physicians to local health authorities, consolidated at the state level, and compiled by the Communicable Disease Center (CDC) of the U. S. Public Health Service—rose from 68,200 in 1941 to a peak of 106,500 in 1947, declined rapidly to

[33] Communicable Disease Center, *V.D. Fact Sheet, 1962* (Atlanta, Georgia, 1962), pp. 2, 9.

12,000 in 1952 and 6,300 in 1957, rose slightly in 1958, and reached 12,500 in 1960 and 20,000 in 1962.[34] The number of persons who contract syphilis exceeds the number reported, because some persons receive treatment but are not reported to the authorities, while others fail to receive specific treatment for infectious syphilis and cannot be reported. Over the years the number reported is likely to represent a variable percentage of the true incidence, owing to fluctuations in the relative patient loads of clinics and private physicians and in the activity and effectiveness of case-finding programs.

Methods for Estimating Incidence

THE CDC PROCEDURE. The official estimate (as noted above) of the incidence of syphilis is 60,000. If we are to appraise the figure, it is helpful to know how it was obtained. A statistician (Oscar Jones) at the CDC kindly described the methodology in a letter to me, October 1963.

Cases reported in the primary and secondary stages of syphilis are clearly attributable to the year in which they are reported. (There is a lag in the figures at the end and beginning of the year, which is of no consequence in the absence of an epidemic or marked trend.) To be added to this number are the cases that escape detection in the primary or in the secondary stage but will be reported upon discovery in some future year in the early latent stage, late latent stage, late stage, stage not stated, or as a congenital case. If we knew how many cases in stages other than primary or secondary would be reported in future years, we could distribute them among past years according to some reasonable statistical procedure.

The curves plotting the numbers of cases reported in the past for each of the stages, which are apparently smooth, are projected by hand into the future for as many years as necessary for a given stage. The numbers are then read off and distributed back to the presumed years of occurrence.

[34] Leona Baumgartner, Chairman, Task Force to the U.S. Surgeon General, *The Eradication of Syphilis* (1961), p. 10; CDC, *V.D. Fact Sheet, 1962*, p. 9.

Estimating the incidence for year X illustrates the CDC procedure:

1.   The number of cases reported that year in the primary and secondary stages is employed intact.

2.   The contribution to incidence in year X of the cases discovered in the early latent stage is the sum of 25 percent each of the early latent cases estimated for the years X, $X + 1$, and $X + 2$, and of 12.5 percent each of the cases estimated for the subsequent two years.

3.   From the data for the late and late latent stages the formula attributes to year X 10 percent each of the cases estimated for years $X + 4$ and $X + 5$ and 5 percent each of the projected cases for the subsequent sixteen years. (The same formula is applied to the data for stage not stated.)

4.   The contribution of congenital cases is 5 percent of the cases reported in year X and 2 percent each of the cases projected for the subsequent forty-seven years and 1 percent of the cases projected for the forty-eighth year.

The incidence for year X is the sum of all these cases. Thus, for the year 1959 the calculated figure was 84,500 or 74,700 more than the number reported in the primary and secondary stages (9,800). The published figure of 60,000 represents an increase of 50,200 over the reported figure and reflects a downward adjustment of 24,500 from the calculated figure. (The last, it is noted, is one third of the calculated increment over the reported figure.)

The reasons for the downward adjustment are not given, but several explanations are plausible. Perhaps an allowance is being made for the upward bias in the formula, if incidence has been rising. Or it may be that the number of new cases in the future is expected to decline faster than the projected curves indicate. Included among the reported cases of syphilis in the historical series may be some so-called biological false positives. (One current estimate, according to unpublished data from CDC, is that 32 percent of persons with positive blood test do not have syphilis. Laboratory tests for the diagnosis of syphilis after the infectious stages are, however, steadily improving in reliability and specificity.) The reported figures may count some persons more than once, owing to two facts: patients travel between states, and some patients are treated by more than a single physician or clinic.

The CDC procedure is vulnerable to an increase in the reinfection rate. A person who incurs syphilis five times in five years and is not treated in an infectious stage can be discovered only once in a latent stage.

A COMPLETE, "HIGH" ESTIMATE. The CDC method does not purport to bring into the estimate of incidence one of the missing groups—the treated persons who are not reported to the authorities. If the treatment of syphilis with penicillin is almost 100 percent effective (as authorities agree), cases treated in the infectious stages but not then reported are lost to the authorities forever. They cannot be discovered and reported at a later date (unless they are again infected and become new cases).

Table 1 presents the elements for deriving a "high," as well as a "more reasonable," estimate of the total incidence of syphilis in the United States in the year 1962. The high figures are presented in column 1; it will be recalled that 74,700 additional new cases were projected by the CDC procedure (line 3a). The cases reported in the infectious stages—20,000—are taken as given (line 1).

TABLE 1. Number of New Cases of Syphilis,
"High" and "More Reasonable" Estimates, United States, 1962

Category of Cases	"High" (1)	"More Reasonable" (2)
1. Primary and Secondary Cases, Reported	20,000	20,000
a. By private physicians	7,300	7,300
b. By clinics	12,700	12,700
2. Treated in P. & S. Stages, Not Reported	54,600	33,400
a. By private physicians	51,000	31,200
b. By clinics	3,600	2,200
3. Not Treated in P. & S. Stages	128,400	66,200
a. Reported at Time of Discovery (CDC data)	74,700	50,200
b. Not Reported at Time of Discovery (ASHA data)	53,700	16,000
4. Perhaps Treated in P. & S. Stages, Not Reported (ASHA)	22,800	—
Total	225,800	119,600

Source: See text for bases of estimates.

Clinics report cases more completely than private physicians. On the basis of a recent nationwide survey the American Social Health Association (ASHA) estimates the number of cases in the primary or secondary stage who were treated by private physicians in fiscal year 1962 at 57,800; this compares with 6,800 cases reported by private physicians (the figure is four times the number reported in the ASHA survey, and is lower than the number reported to officials during the year).[35] At face value the adjustment for nonreporting of infectious cases by private physicians is 51,000 (line 2a). If allowance were made for the patients treated by the 28 percent of all physicians who did not answer the questionnaire (by assuming that the responding physicians were representative of the nonrespondents), an additional adjustment of 22,800 cases would be indicated (line 4).

Incompleteness of reporting is not confined to patients in the infectious stages. The ASHA survey estimates that private physicians treated 131,500 patients in the other stages of syphilis but reported only 50,000, leaving 80,500 unreported. When the last number is projected ahead and traced back (by analogy to the CDC method for reported cases), the estimate from this source of the number of persons who will have escaped treatment in the infectious stages is 53,700 (two thirds of 80,500, which is the ratio of nonreported cases calculated to have occurred in 1959 to the number reported in 1959 for all the noninfectious stages). These cases will presumably never be reported to the authorities.

There may be some under-reporting by clinics, too. If findings for the state of Minnesota are applied to nationwide statistics, an upward adjustment of 3,600 is indicated after the double counting of names is eliminated (line 2b). If only completely new names are added to the reported count, the required adjustment is 800.[36] The mid-value, which will be appropriate for the "more reasonable" estimate, is 2,200.

If we take each of the items enumerated above at face value, the

---

[35] American Social Health Association, *Findings of a National Study of VD Incidence* (1963; processed).

[36] D. S. Fleming *et al.*, "Syphilis and Gonorrhea in Minnesota," *Minnesota Medicine*, Vol. 46 (January 1963), pp. 34-38.

resulting estimate of the incidence of syphilis would be 225,800 ("total").

A "MORE REASONABLE" ESTIMATE. An alternative estimate of the incidence in 1962 can be developed, following these steps (Table 1, column 2):

1. The number of cases reported in the infectious stages remains at 20,000, as reported (line 1).

2. Adopt the CDC adjustment for the number infected in 1962 who were not then treated and will be reported in later years (substitute 50,200 for 74,700 in line 3a).

3. It may be that in responding to special questionnaires physicians overstate the total number of persons treated for syphilis and, therefore, the number not previously reported to the authorities. It seemed desirable to compare the findings of local health department surveys on the magnitude of nonre-porting with those of the ASHA survey for the same geographic areas. It was possible to do this for New York City.[37] (Data later became available for Cincinnati; they are not very different from New York City's.[38]) Physicians report 60 percent of the cases they treated in the primary or secondary stage, according to the New York Health Department survey, and 40 percent, according to the ASHA survey. If the Health Department's finding were taken as the more accurate and if the ratio between the two findings for New York City (65.7 percent) were applied as an adjustment factor to the ASHA nationwide data, then the number of cases with infectious syphilis treated by private physicians, but not reported, is reduced to 31,200 (57,800 × 65.7 percent = 38,000; less 6,800; line 2a). Since the NYC Health Department's survey incorporates an allowance for nonresponding physicians (21 percent), the 22,800 cases shown on line 4 are eliminated.

4. Similar adjustments can be applied to the ASHA data for the other stages of syphilis. First, a factor is applied to the total

[37] Anna C. Gelman, Jules E. Vandow, and Nathan Sobel, "Current Status of Venereal Disease in New York City: A Survey of 6,649 Physicians in Solo Practice," *American Journal of Public Health*, Vol. 53 (December 1963), pp. 1903-18.

[38] Aidan Cockburn, *The Evolution and Eradication of Infectious Diseases* (Johns Hopkins Press, 1963), p. 166.

number of cases treated, as indicated by the NYC Health Department's survey. The result is a reduced estimate (by one
third) of the number of cases treated by private physicians in
the noninfectious stages of syphilis. Next, the number of cases
reported is subtracted; it is presumably reflected in the CDC
method of estimating. To the figure of 36,000 not reported
(131,500 reported in the ASHA survey; 86,000 is the adjusted
figure from which the 50,000 cases reported are subtracted) are
applied two ratios derived from the CDC procedure. One ratio
is the quotient of the hypothetical number traced back to 1962
from future projections to the number currently reported in the
corresponding stage; the other is the number adopted by CDC
to the number obtained by applying its statistical procedure.
The resulting figure from this source is 16,000 (36,000 × 2/3 ×
2/3; line 3b).

   5.   The total is 119,600. Being the sum of a set of figures, each
of which is "more reasonable" than the corresponding "high"
figure, this is the estimate of the incidence of syphilis that will
be employed in this paper.

   It is recognized that the above procedure, which takes into account cases treated in the infectious stages but not reported and
cases to be discovered and treated in subsequent years (both reported and not reported), omits some persons who have acquired
syphilis. Among these are people who receive effective treatment
with penicillin for some other purpose while still in the presymp-
tom stage; there are others who escape all the complications of late
syphilis even though they remain untreated. An estimate of the
latter group could be made by applying the ratios developed in
Oslo, Norway, to the number who develop the complications of
late syphilis, but the estimate of the incidence of late syphilis is too
tenuous (see below) to warrant this step. I know of no basis for
estimating the number of persons who are cured of syphilis before
it can be diagnosed.

   INCIDENCE BY STAGE OF DISCOVERY. To calculate economic costs
(and benefits) it will be necessary to have a distribution of
total incidence by the stage in which the cases are discovered and
treated. Table 1 separates cases discovered in the infectious

stages (lines 1 and 2) from all other cases. It is now necessary to allocate the latter by stage of discovery.

To simplify this task three stages were recognized: early latent, late latent, and late. The early latent stage was assigned the same proportion of all other cases as it constituted in the reported data for the year 1962—25.7 percent. The late latent stage is a residual after the late stage is estimated.

There is no firm basis today for estimating the number of cases who are expected to develop the complications of late syphilis. There are the data on untreated syphilis from Oslo, Norway, where sixty and seventy years ago the physician in charge of the syphilis clinic at the university hospital (Dr. Caesar Boeck) believed that more harm came from treating patients with syphilis than from leaving them alone. According to the Oslo data, as reworked by Trygve Gjestland, 14 percent of all persons who are infected (and remain untreated) ultimately develop serious complications—6.5 percent in the central nervous system (leading to general paresis, tabes, and meningovascular syphilis) and 7.6 percent in the cardiovascular system (including aneurism of the thoracic aorta).[39] Based on 119,600 new cases, the number with late complications would be 16,900.

However, first admissions to mental hospitals in this country for syphilitic psychoses declined from 8,800 in 1939 to 3,850 in 1950—and to 780 in 1960.[40] The reason for the decline may be the known fact that in recent years many persons have received effective treatment prior to the late stage of the disease. However, it could be that the reinfection rate is much higher than 10 percent (thereby converting a given number of cases into a much smaller number of individuals); or the incidence of the disease could be vastly overestimated; or the virulence of the disease may have drastically abated.

It is known that syphilis declined in virulence in the Middle Ages after the great plague. Thus, it may also have altered a great deal more in the past century than we realize. Further al-

[39] Gjestland, *The Oslo Study of Untreated Syphilis*, pp. 209, 281.
[40] National Institute of Mental Health, Biometric Branch, *Age Specific First Admission Rates and Resident Patient Rates of Syphilitic Psychoses, United States, 1939, 1950, 1952, 1955, 1958, and 1960* (1963; typewritten).

teration in the recent past cannot be ruled out, but we have no evidence to support this thesis.

As for the cardiovascular complications of syphilis, obtaining any systematic data proved impossible. The large municipal hospital system in New York City could find scarcely any patients with these diagnoses in its general hospitals, and in Michigan a statewide representative sample of 10,700 discharges from general hospitals contained *none*.[41] Several physicians who were interviewed, including heart specialists, stated that they see few such patients today. Moreover, some of the diagnoses of cardiovascular syphilis are believed to be unreliable, having a low rate of confirmation in autopsy.[42]

In light of this, we adopted an arbitrary procedure. The number of cases with syphilitic psychoses was projected to continue at the level reported for the year 1960. On the one hand, a smaller number than is reported at present may be expected, since the estimated total incidence in a current year is probably lower than it was a generation ago; moreover, persons infected more than twenty years ago lacked access to penicillin in the early stages of the disease. On the other hand, as other infectious diseases decline in importance, life expectancy at the older ages is extended, thereby permitting the late complications of syphilis to develop; in addition, as physicians employ penicillin with greater selectivity, fewer patients will be cured of syphilis as a by-product of treatment for other diseases.

The total number of cases with late complications of syphilis was raised to 1,000 when allowance was made for persons who develop cardiovascular disease. This adjustment stems from the relationships found to exist between the numbers of persons with psychoses and with heart disease due to syphilis. A study of the Veterans Administration hospitals found that the latter comprised 17 percent of all patients with syphilis, and the Social Security Administration has reported that, of all applicants with syphilis who sought certification for disability in 1961, 21 percent had cardiovascular disease.[43]

[41] Unpublished data from Dr. Marta Fraenkel (New York); and see Walter J. McNerney and Study Staff, *Hospital and Medical Economics*, Vol. 1 (Hospital Research and Educational Trust, 1962), p. 534.

[42] Olansky *et al.*, "Untreated Syphilis in the Male Negro," p. 178.

[43] S. Ross Taggart *et al.*, "Report of Syphilis Follow-up Program Among Veterans

TABLE 2. Distribution of the Incidence of Syphilis by (Projected) Stage
of Discovery, United States, 1962

Stage of Discovery	Number	Percent
Primary and Secondary	53,400	44.7
Reported	20,000	16.7
Not Reported	33,400	28.0
Early Latent	17,000	14.2
Late Latent	48,200	40.3
Late	1,000	0.8
Total Incidence	119,600	100.0

Source: See text.

Given the estimate of 1,000 for the late complications of syphilis, the distribution of total incidence by the projected stage of
discovery is determined (Table 2).

Economic Characteristics of the New Cases

To calculate the economic benefits of a syphilis control program we must have data on the income and related characteristics of persons with the disease. In the absence of direct information, we rely on inferences from certain demographic characteristics, such as race, sex, and age of the reported cases. For this
procedure it was helpful to identify two populations—patients
who are discovered and treated by clinics; and patients who are
discovered and treated by private physicians.

Cases reported by clinics differ from the latter cases in sex,
race, and age. It seems plausible to assume that cases treated
by private physicians which are not reported to the authorities re-
semble most closely cases treated by private physicians which are
reported. It is, moreover, reasonable to attribute to persons who
escape treatment in the infectious stages the race and sex characteristics of cases reported in the latent and late stages of syphilis.

The estimated distribution of the 119,600 cases by sex, race,
stage of discovery, and source of treatment is shown in Table 3.

after World War II," *Journal of Chronic Diseases,* Vol. 4 (December 1956), p. 579;
and Social Security Administration, *Disability Applicants, 1961* (1963), pp. 15, 22.

TABLE 3. Distribution of Total Incidence of Syphilis by Sex, Race, Stage of Discovery, Source of Treatment, United States, 1962 (in thousands)

Stage of Discovery and Source of Treatment	Total			White			Nonwhite		
	Total	Male	Female	Total	Male	Female	Total	Male	Female
Primary and Secondary Stage, Reported									
Total	20.0	12.9	7.1	5.8	4.7	1.1	14.2	8.2	6.0
Private Physicians	7.3	5.0	2.3	3.5	2.9	0.6	3.8	2.1	1.7
Clinics	12.7	7.9	4.8	2.3	1.8	0.5	10.4	6.1	4.3
Primary and Secondary Stage, Not Reported									
Total	33.4	22.8	10.6	15.3	12.3	3.0	18.1	10.5	7.6
Private Physicians	31.2	21.4	9.8	14.9	12.0	2.9	16.3	9.4	6.9
Clinics	2.2	1.4	0.8	0.4	0.3	0.1	1.8	1.1	0.7
Early Latent Stage									
Total	17.0	8.9	8.1	8.1	4.5	3.6	8.9	4.4	4.5
Private Physicians	10.0	5.0	5.0	5.8	3.1	2.7	4.2	1.9	2.3
Clinics	7.0	3.9	3.1	2.3	1.4	0.9	4.7	2.5	2.2
Late Latent Stage									
Total	48.2	25.3	22.9	22.8	12.8	10.0	25.4	12.5	12.9
Private Physicians	28.6	14.3	14.3	16.5	8.8	7.7	12.1	5.5	6.6
Clinics	19.6	11.0	8.6	6.3	4.0	2.3	13.3	7.0	6.3
Late Stage	1.0	0.5	0.5	0.5	0.3	0.2	0.5	0.2	0.3

Source: See text.

All persons with syphilis were assigned the age distribution of cases reported in the primary and secondary stages. The median ages of the four sex and race groups are shown in Table 4.

TABLE 4. Median Age of Infection with Syphilis by Sex and Race, United States, 1962

Sex and Race	Age
White Male	29
White Female	24
Nonwhite Male	25
Nonwhite Female	22

Source: Calculated from U.S. Public Health Service, Communicable Disease Center, *Primary and Secondary Syphilis, United States, Morbidity Cases by Age Groups, Race, and Sex, Calendar Years 1956, 1959, 1960, 1961, 1962* (August 1, 1963, mimeographed).

Median ages were employed to simplify later calculations. Working with frequency distributions by age for each sex and race group would be very time consuming. The shortcut seemed permissible, since persons too young to join the labor force constitute small proportions of total cases—0.1 percent for white males, 0.8 percent for nonwhite males, 1.2 percent for white females, and 2.8 percent for nonwhite females. (It was not realized at the time that the procedure neglected differences by age in the rate of labor force participation.)

Selected Problems

RATE OF REINFECTION. The estimate of approximately 120,000 includes persons infected and reinfected during a single year, the reinfecting taking place after treatment and apparent cure. The percentage of all cases that represents reinfections is of the order of 10 percent in a twelve-month period.[44]

That some persons become reinfected is of no immediate consequence for an operating program, since infected persons must be treated when found. To postpone treatment on the ground that some persons will be promptly reinfected (or that a few will die before developing the late complications of syphilis) is to risk spreading the infection and losing track of the patient in the la-

[44] CDC, *V.D. Fact Sheet, 1962*, p. 19.

tent stages of the disease, which are of uncertain duration for the individual patient.

However, reinfection of a person within a short time interval serves to reduce cost. After initial treatment a person can undergo only one follow-up regime at a time. Moreover, reinfection reduces the reservoir from which the late complications of syphilis can develop.

TREND IN INCIDENCE. Data from the armed forces indicate a downward long-term trend in the incidence of syphilis.[45] For the civilian population, however, accurate figures are lacking.

For the postwar period, interpretation of the trend in reported figures falls within the realm of informed judgment. Information on the degree of reporting by private physicians is altogether lacking prior to 1962.

It has been suggested to me by Dr. Stanley Mayers that in the years before the almost instantaneously and completely effective treatment with penicillin was introduced, public health officials were largely preoccupied with the follow-up of patients involved in prolonged treatment, which was painful and inconvenient.[46] Few resources, if any, were left for following up contacts in the early, infectious stages of the disease. Today, follow-up of contacts is common, and it leads to the discovery of more cases. This chain of events would suggest that the reporting of existing cases has improved.

Nevertheless, authorities agree that the number of new cases of syphilis in this country has really increased in recent years. Some refer to a deterioration in moral standards, as evidenced by the rise in illegitimate births, and frequent reference has been made to the increased incidence of syphilis among homosexuals. Also noted is a reduction in the indiscriminate or "happenstance" administration of penicillin, which is almost uniquely effective in terminating syphilis at every stage of the disease.[47]

[45] William L. Fleming, "Syphilis through the Ages," The Medical Clinics of North America, Vol. 48 (May 1964), pp. 587-612.

[46] American Venereal Disease Association, "Penicillin and Syphilis," Journal of the American Medical Association, Vol. 144 (November 11, 1950), pp. 947-48.

[47] Ira Leo Schamberg, "Syphilis and Sisyphus," British Journal of Venereal Diseases, Vol. 39 (June 1963), pp. 87-96; and Jules E. Vandow, "Antimicrobial Therapy

In any attempt to control the spread of syphilis a real rise in the number of infectious cases is a source of serious concern. Such an increase need not result, however, in a proportionate increase in the number of late cases. It may be that some persons who are now infected would formerly have been immune to new infection because they were in the latent stage of the disease.

Economic Costs of Cases Infected in 1962

Let us continue with measuring the base line—that is, the economic costs attributable to the 119,600 cases of syphilis newly incurred in 1962. It will soon be apparent that some of the figures employed below lack firm support and are intended largely to serve illustrative purposes.

The present values of the following types of cost were calculated, where appropriate, for each group of cases, classified by stage of discovery and treatment:

1.   Medical care expenditures.

2.   Production loss due to time spent for treatment at the physician's office or outpatient clinic.

3.   Reduction in earnings due to the "stigma" of having had syphilis. Physicians interviewed in the course of this study stressed the economic and social cost—stigma—attached to having syphilis on one's record. Specifically they mentioned greater difficulty in getting a job and a tendency to remain fixed in a job lest the attempt to change it lead to disclosure of the disease. It seems intuitively plausible to attach a higher economic cost to a reported case than to a case treated but not reported—and the highest cost of all to a person discharged from a hospital with the diagnosis of late syphilis. For illustrative purposes the following percentages of all earnings after discovery of the disease were employed to measure "stigma": infectious or latent disease, reported, 1 percent; infectious or latent disease, not reported, 0.5 percent; late disease, 10 percent.

of Syphilis," *New York State Journal of Medicine*, Vol. 59 (November 15, 1959), pp. 4193-4202.

396 Syphilis Control Programs

4. Loss of consumer benefit. This is apart from medical care expenditures and output losses, as discussed earlier. The loss in consumption exists if the patient is aware of, and recognizes, his symptoms. The size of the loss depends on the severity of symptoms.

Cases Reported in Infectious Stages

In 1962, 20,000 cases were reported in the infectious stages. However, a reinfection rate of 10 percent reduces the number who require follow-up services.

MEDICAL CARE EXPENDITURES. Data are wanted on numbers of visits per case, cost per visit, and the cost of laboratory services and drugs. The net rate of discount is set at 2.72 percent (after adjustment for an extra rise in physicians' prices of 1.25 percent a year).

Measuring expenditures for medical care in the infectious stages proved difficult. For instance, data were lacking on the average cost of diagnosing and treating a patient with syphilis in the nation, and the patterns in the two cities—Baltimore and New York—that were selected for close observation in preparing this paper were very diverse. In Baltimore the treatment regime for early syphilis consisted of two visits; in New York, of five or more visits. These turned out to be true geographic differences, with the variation between cities exceeding that between private practice and outpatient clinic within a city. Moreover, the number of visits in New York City was just as large or even larger under conditions of prepaid medical group practice (HIP), in the absence of a direct fee for the physician, as under solo fee-for-service practice.

Accordingly, we calculated expenditures for medical care in the infectious stages on the following assumptions: that the average number of visits per case for diagnosis and treatment lay midway between prevailing practice in Baltimore and in New York City; that the average number of visits for follow-up was as stated in the standards recommended by the U.S. Public Health Service;[48] that medical charges to the patient (or costs to the clinic)

[48] U. S. Public Health Service, *Syphilis*, p. 25.

were moderate, say, at the level of private practice in Baltimore (fee schedules for the country as a whole do not furnish sufficient detail); and that all blood tests were performed by a health department laboratory free of charge to the patient and valued at cost to the laboratory. In accordance with the recommended standard, each person with infectious syphilis was assumed to have a dark-field examination.

The cost per physician visit was specifically assumed to be $7; per blood test performed by a laboratory, $1.50; per dark-field examination, $15; per spinal fluid test, $15; and drugs, $4. Units of service were specified separately by year of occurrence (first or second year):

	First Year (1962)	Second Year (1963)
Physicians' Visits	12	2
Laboratory Services	8	2
Dark-Field Examination	1	—

At current prices the cost per case amounted to $133 in Year 1 and to $17 in Year 2. When discounted at 2.72 percent, the former remains unchanged and the latter is reduced to $16.55. (The number of follow-up visits may in actuality be smaller than prescribed. If this is so, true cost is overstated.) The present value of medical care expenditures for 20,000 cases in Year 1 and 18,000 cases in Year 2 (reduced by the reinfection rate of 10 percent) is $3.0 million.

OUTPUT LOSS DUE TO TREATMENT TIME. Waiting time for the patient is longer in the clinic than in the private physician's office. With transportation time included, a visit to the clinic takes about half of a day and a visit to the physician's office a quarter of a day. (Numbers of visits are given in the preceding section.)

It was necessary to estimate for each sex and race group the fraction of the year spent at work, here called the work ratio (which is calculated as the product of the labor force participation rate and the full-employment rate). In calculating the work ratios for women, it was assumed that the proportion of housewives to all women is the difference between the labor-force participation rate for men and that for women in a given age class.

TABLE 5. Daily Earnings and Work Ratio, by Sex, Race, and
Labor Force Status

Sex and Race	Daily Earnings	Work Ratio
Male		
White	$24	0.93
Nonwhite	15	0.90
Female		
White		
Labor Force	14	0.40
Housewife	9	0.55
Nonwhite		
Labor Force	9	0.53
Housewife	9	0.39

Sources: Daily earnings: annual earnings in Table 6, divided by 243; work ratio: 0.96 (full-employment rate) times the labor force participation rate. The latter is from BLS; see Table 8.

Figures were also required for daily earnings. Both are shown in Table 5.

The net discount rate for calculating output loss is 2.21 percent (after dividing the gross rate of 4 percent by the projected rise in productivity of 1.75 percent a year—see footnote 19 above). Under these assumptions the present value of the output loss due to treatment time is calculated at $1.5 million.

REDUCED EARNINGS DUE TO "STIGMA." The money value of the "stigma" is not known; however, if estimated at 1 percent of the present value of future earnings by participants in the labor force, it amounts to $11.4 million. Among the assumptions underlying this estimate is a net discount rate of 2.21 percent (as above). Table 6 shows full-time earnings by members of the labor force; average work years remaining between the age of onset of infection and age 65, arbitrarily set as the date of retirement; and the present value of future earnings. The table illustrates the procedure for calculating the values of "stigma" for cases discovered in the other stages of the disease.

To the present value of all cases is applied an adjustment factor of 0.864, to allow for 4 percent unemployment and for 10 percent reduction in persons affected due to reinfection (.96 × .90 = .864). The result is $1,190 million, of which 1 percent amounts to $11.9 million.

TABLE 6. Calculation of Present Value of Future Earnings for Syphilis Cases Discovered and Reported in Infectious Stages, United States, 1962

Sex and Race	Average Full-Time Earnings (1)	Expected Labor Force Years (2)	Expected Income (before discount) (3)=(1)×(2)	One-Half Life Expectancy (4)	Discount Factor at 2.21 Percent (5)	Present Value per Case (6)=(3)÷(5)	No. of Cases (000) (7)	Present Value of All Cases (millions) (8)=(6)×(7)
Male								
White	$5,820	32	$186,200	21	1.585	$117,500	4.7	$ 552.2
Nonwhite	3,690	32	118,100	20.5	1.568	75,300	8.2	617.5
Female								
White	$3,430	16	$ 54,900	26	1.769	$ 31,100	1.1	$ 34.2
Nonwhite	2,260	21.5	48,600	24	1.694	28,700	6.0	172.2
Total							20.0	$1,376.1

Sources: Column 1, U. S. Bureau of the Census, Income of Families and Persons in the United States: 1961 (Series P-60, No. 39, 1963), p. 41; column 2, Table 8, line 11; column 4, U. S. Public Health Service, National Vital Statistics Division, Vital Statistics of the United States, 1960, Vol. 2, Section 2, Life Tables (1963), pp. 2-11; column 5, assumption that the entire loss occurs at the midpoint of remaining life expectancy: see Rashi Fein, Economics of Mental Illness (Basic Books, 1958), p. 74; column 7, Table 3.

CONSUMER BENEFIT LOSS. The loss of consumer benefit due to syphilis infection depends, not on the receipt of treatment, but only on the person's recognition of symptoms. To estimate this loss along the lines of an analogous disease, some physicians suggested psoriasis, a skin disease. Psoriasis usually entails no disability, but a permanent cure is not likely. Good data on the cost of treatment are not available without a special study. An estimate of four physician visits per episode appears to be conservative, as is a cost per episode of $50, including drugs. For 20,000 cases the calculated cost amounts to $1 million.

Cases Treated in Infectious Stages but Not Reported

The syphilis cases treated in the infectious stages but not reported are estimated at 33,400 (Table 1). The patients differ in sex, race, and source of treatment from those who are reported (Table 3). Procedures for calculating the present values of medical care expenditures, output loss due to treatment time, and consumer benefit loss are the same as above. The estimates of present value are (in millions): medical care expenditures, $5.0; output loss due to treatment time, $1.9; reduction in earnings due to "stigma," $11.1; consumer benefit loss, $1.7.

Even at ½ of 1 percent of the present value of future earnings, the "stigma" factor looms big, owing to the large number of white male cases (Table 3). Per case, the present value of future earnings is approximately 50 percent higher for white males than for white females or nonwhite males, and 130 percent higher than for nonwhite females (Table 6).

Cases Treated in Early Latent Stage

As noted earlier, it was estimated that 17,000 of the cases newly infected in 1962 would be discovered and treated in the early latent stage of syphilis. They were allocated year by year according to the CDC formula, and it was further estimated that three fourths of the cases would be reported to the authorities (Table 1, lines 3 and 3a).

The several assumptions concerning the infectious stages still govern, with three exceptions. With respect to medical care expenditures, the dark-field examination is no longer medically indicated. The number of physicians' visits in Year 1 of discovery is ten, rather than twelve, because a portion of the diagnosis and treatment occurs simultaneously with, and is attributed to, other diseases. It is further assumed that one half of the patients recognized that they had syphilis in the infectious stages but did nothing about it.

The estimates of present value are (in millions): medical care expenditures, $1.8; production loss due to treatment time, $0.8; reduction in earnings due to "stigma," $7.8; consumer benefit loss, $0.4. Loss due to "stigma" looms relatively larger here than it does for the cases treated and reported in the infectious stages, because the output losses are smaller.

Cases Treated in Late Latent Stage

The cases newly infected in 1962 that are likely to be discovered and treated in the late latent stage number 48,200. They were allocated to the year of treatment according to the CDC formula. Again, the estimate is that three fourths of the cases would be reported.

Almost all of the assumptions given for the early latent stage apply here. One exception is the number of physicians' visits, which is five in the first year rather than ten. The estimates of present value are (in millions): medical care expenditures, $2.5; production loss due to treatment time, $1.2; reduction in earnings due to "stigma," $19.3; consumer benefit loss, $1.2.

Late Complications of Syphilis

PSYCHOSES. The number of cases with syphilitic psychoses expected to result from the cases newly infected in 1962 was projected at 800. To calculate medical care expenditures (in this instance, hospital care), we must have the average duration of treatment, which is a function of two variables—the death rate in the hospital and the rate of discharge to the community. Although the former has declined, thereby prolonging average stay, the latter has increased. My best estimate of duration of stay, based on two sources, is four years.[49] For calculating output loss we must know the incubation interval of the complication, which is the same as the period for which output loss is postponed. According to the Oslo data this interval is fourteen years.[50]

For calculating the present value of hospital expenditures, the net discount rate is 0.9 percent (after combining the 4 percent discount with an extra increase in the cost of psychiatric hospital care of 3 percent a year). Current hospital expense per patient year is $2,100.[51] For 800 cases, the present value amounts to $5.8 million.

As for output loss, the present value is $35.3 million if all cases are forever lost to the labor force upon admission to the hospital. If some patients could be released and rehabilitated for productive work, the estimate of loss would be reduced. Since some may work for short intervals at reduced wages, Hahn's figure of one third is too high for this purpose.[52] Perhaps an adjustment factor of one fifth would be more appropriate, yielding an output loss of $28 million. In that event there would be some loss due to "stigma"—$0.7 million, calculated at 10 percent of the earnings of those who return to employment.

The earnings of persons with syphilitic psychoses who have been discharged from hospitals are not actually known. To be able to trace the work and earnings experience of such persons would be helpful; the research unit of the Social Security Administration has expressed interest in facilitating such statistical inquiry.

As for the loss of consumer benefit associated with the late

[49] Robert E. Patton, Director of Statistical Services, New York State Department of Mental Hygiene, personal communication to the author, September 17, 1963; Morton Kramer, Hyman Goldstein, Robert H. Israel, Nelson A. Johnson, *A Historical Study of the Disposition of First Admissions to a State Mental Hospital* (GPO, 1955), p. 18.

[50] Gjestland, *The Oslo Study of Untreated Syphilis*, pp. 211, 216, 220.

[51] American Hospital Association, *Hospitals*, Guide Issue, Vol. 37 (August 1, 1963), p. 451.

[52] Richard D. Hahn et al., "The Results of Treatment in 1,086 General Paralytics the Majority of Whom Were Followed for More than Five Years," *Journal of Chronic Diseases*, Vol. 7 (March 1958), p. 216.

complications of syphilis, an appropriate analogue is to be found in terminal cancer.[53] Here, too, medical care expenditures are incurred without prospect of averting loss of earnings or medical care expenditures in the future. The cost per case is estimated at $2,000.[54] If the net discount rate is zero (after allowance for an extra rise in the cost of general hospital care of 4 percent a year), the present value of a case remains at $2,000 and the present value of 800 cases is $1.6 million.

CARDIOVASCULAR DISEASES. The difficulty of obtaining data on the cardiovascular complications of syphilis has been noted. After several unsuccessful attempts, we decided to posit some plausible numbers for illustrative purposes. These were uniformly, if not arbitrarily, set at one fourth of the corresponding figures for syphilitic psychoses. However, a uniform ratio is difficult to justify, given differences between the two sets of complications in the incubation interval; course, source, and cost of treatment; and the ability to continue working.

Table 7 summarizes the results of the calculations so far presented on the present value of the several types of cost incurred by the 119,600 new cases of syphilis.

Appraisal of Recent Estimates

The Communicable Disease Center has estimated the output loss due to the late complications of syphilis at $8.6 billion. This estimate is far too high. An obvious reason is the employment of the Oslo ratios without any allowance for effective treatment. Other reasons, probably in declining order of importance, are that:

[53] Joseph Earle Moore, "The 'Rising Tide of Infection' of Syphilis," *ibid*. Vol. 6 (September 1957), pp. 280-83.

[54] Based on several sources. Duration of stay in hospital, 36 days, from Laitin, *The Economics of Cancer*, p. 139; patient day cost, $36.83, from American Hospital Association, *Hospitals*, Guide Issue, Vol. 37 (August 1, 1963), p. 448; surgeon's bill, $350, from House Committee on Interstate and Foreign Commerce, *Hearings*, Part 1 (October 1953), p. 212, and no rise indicated, Fee Guide, *Medical Economics*, Vol. 39 (February 12, 1962), p. 122; office and home calls, $50, from Fee Guide, pp. 85 and 88, and special study of medical care received by cancer patients (lethal sites), Health Insurance Plan of Greater New York (unpublished); stay in nursing home, one month (for lethal sites), Laitin, p. 232, at $275 a month (conservative estimate).

TABLE 7. Present Value of the Economic Costs of 119,600 New Cases of Syphilis by Type of Cost and Stage of Discovery and Treatment, United States, 1962

(dollar amounts in millions except for cost per case)

| Type of Cost | Total | P & S Stages | | Early Latent Stage | Late Latent Stage | Late Stage |
		Reported	Not Reported			
Medical Care Expenditures	$ 19.5	$ 3.0	$ 5.0	$ 1.8	$ 2.5	$ 7.2
Output Loss Due to Treatment Time or Disability	40.7	1.5	1.9	0.8	1.2	35.3
Reduction in Earnings Due to "Stigma"	51.0	11.9	11.1	7.8	19.3	0.9
Consumer Benefit Loss	6.3	1.0	1.7	0.4	1.2	2.0
Total Cost	$117.5	$ 17.4	$ 19.7	$ 10.8	$ 24.2	$ 45.4
Number of Cases (in thousands)	119.6	20.0	33.4	17.0	48.2	1.0
Cost per Case (dollars as shown)	$985	$870	$590	$635	$500	$45,400

Source: See text.

the number of work-years lost is based on the total prevalence reservoir, with an age distribution taken from the much younger population of newly infected and reported cases; the value of future output is not discounted to the present; the measure of output is taken as per capita adult income, rather than as earnings; and no allowance is made for reinfection in the first year, which serves to reduce the number of persons capable of developing the late complications of the disease.

Similarly, CDC estimates medical care expenditures for two psychoses due to syphilis (paresis and meningovascular syphilis) at $899 million. Again, the Oslo rates for untreated syphilis are applied to the estimated prevalence reservoir of syphilis in the United States. Furthermore, the average duration of stay in the hospital, ten years, is too long.

CDC routinely publishes a volume entitled *Venereal Disease Fact Sheet*, which contains a table (p. 4) "Estimated Annual Costs of Uncontrolled Syphilis." Three kinds of loss are presented: annual man years of disability due to the complications of syphilis; annual cost of maintaining certain categories of patients in institutions; and loss of life expectancy from deaths due to syphilis in one year, and loss of income to age 65. The first two are prevalence figures: disability in 1960 due to syphilis and the cost of hospital care in 1960 for patients with syphilitic psychoses. The third set of figures comes close to calculating the present value of future income losses, except that for persons dying in 1960 the infection occurred many years ago. Income loss stretching over many years is not discounted to a common base line.

Economic Benefits

Table 7 has presented the figures against which the benefits of a specified program can be measured.

Eradication

The present value of the benefits accruing from total eradication would be $117.5 million realized in perpetuity, or $2.95 billion (at a discount rate of 4 percent). Since the disease is communicable, it cannot recur in the absence of an external source of infection. An added benefit is the control and surveillance mechanism, which could presumably be abandoned and its averted cost realized in perpetuity. In 1962, $6 million were spent; discounted at 4 percent, this yields a present value of $150 million.

In sum, the present value of eradicating syphilis, on the above assumptions, would be $3.1 billion. More than 40 percent of this amount, or $1.3 billion, is due to "stigma" (evaluated mostly at 1 or 0.5 percent of earnings subsequent to the discovery of syphilis).

The distribution of total benefits by type is (in percent): medical care expenditures, 15.8; output loss due to treatment time or disability, 32.9; reduction in earnings due to "stigma," 41.3; consumer benefit loss, 5.1; abandonment of control mechanism, 4.9.

Direct costs are one fifth of the total (15.8 + 4.9 = 20.7 percent). Indirect costs, in the conventional sense, are one third but mount to three fourths when the reduction in earnings due to "stigma" is taken into account $(32.9 + 41.3 = 74.2$ percent).

Reduction and Control

In all fairness, it must be said that in the absence of radical technological innovations or social changes, eradication in the strict sense is not feasible. A surveillance and control mechanism would almost certainly be required to prevent infection by travelers from abroad and is also likely to be required in practice in a free civilian society as new generations grow up, to keep the disease at a tolerable level. It has been officially recommended that the annual expenditures for operating such a mechanism be raised to $10 million.[55] In perpetuity the present value of such an annual expenditure amounts to $250 million; this can be viewed as an increase in cost of $100 million over 1962, or as an equivalent reduction in prospective benefit.

ALTERNATIVE LEVELS OF CONTROL. Acceptable control of syphilis can be attained at alternative levels. The level of incidence achieved is in part a function of the amount spent on the surveillance and control mechanism. As the incidence of the disease declines, the reservoir from which infections are drawn is diminished, but tracing additional cases becomes increasingly difficult and costly.

The benefits of reducing incidence by a given amount or proportion could be uniquely determined if the distribution of cases were the same at each incidence level. Thus, if the number of new cases declined by 80,000 and the remaining 40,000 cases were distributed in the same way as the original 120,000 by sex, race, source of treatment, and stage of discovery, the annual benefit (apart from any changes in expenditures for the control mechanism) would be $78.5 million. It is unlikely, however, that the distribution of cases would be the same at each incidence level. Surely one factor would be the route by which a given level is attained.

The distribution of the resulting number of cases by stage of

[55] Baumgartner and Task Force, *The Eradication of Syphilis*, pp. 8, 19.

discovery is of obvious importance in calculating the benefits of a control program. On the face of it, a case discovered and treated in the latent stage of syphilis costs considerably less than a case treated in the primary or secondary stage (Table 7). Would it, then, not make sense to postpone treatment? The answer is "no," for the heaviest cost attaches to a late case. A person treated in the infectious stages is certain to avoid the late complications. The probability of developing neurosyphilis or syphilitic heart disease increases as the discovery and treatment of the patient is postponed to the latent stage. This is particularly true if the administration of penicillin is circumscribed.

Perhaps it would be appropriate to view the cost of a latent case as consisting of two components: the several types of cost shown for that stage in Table 7, and some fraction of the cost of a late case, corresponding to the probability of developing these complications. Unfortunately, in light of the discussion on the incidence of late cases, nothing more definitive on this score can be said.

COMMUNICABILITY. It was disappointing that we were unable to determine the ultimate effect on the incidence of syphilis of an initial reduction (or increase) of given amount, because the chain of communicability was broken. There is some doubt whether an adequate epidemiological theory exists that applies to the conditions of syphilis.[56] And if such a theory does exist, it certainly has not yet been applied. Letting the model operate over a large number of syphilis generations (the full interval of infectiousness) would be necessary before concluding whether a specified change in incidence would ultimately lead to stability at some other level, to fluctuations around that level, to a gradual decline (or rise) toward an asymptote, or to eradication (or explosion to an epidemic).

On the average, a case discovered and treated in the primary stage of syphilis is infectious to sex partners for four and a half

[56] Helen Abbey, "An Examination of the Reed-Frost Theory of Epidemics" and J. de Oliveira Coste Maia, "Some Mathematical Developments on the Epidemic Theory Formulated by Reed and Frost," *Human Biology*, Vol. 24 (September 1952), pp. 201-33 and 167-200, respectively; see pp. 202 and 168 for assumptions.

weeks; in the secondary stage, twice as long (nine weeks); and in a latent stage, almost three times as long (twelve weeks). To the extent that sexual activity is independent of the stage in which a case is discovered, there is a better chance of reducing the incidence of syphilis by finding infected persons early rather than later.

Conceivably the time element might also be important from another standpoint. If the infectious agent tends to acquire resistance to a drug, then speed in prosecuting a program becomes essential. There is, however, no evidence that the syphilis organism is becoming resistant to penicillin.

"STIGMA." However estimated, the "stigma" factor looms very large once it is recognized. Any change in procedures, whether technological or administrative, that would contribute to the elimination or reduction of "stigma" due to having had syphilis, would yield a large economic benefit. An obvious, partial remedy is to strengthen the safeguards surrounding the disclosure of official information concerning persons with a history of syphilis.[57] The value of "stigma" cannot really be measured without a survey of two populations under controlled conditions. Even then a difference in earnings could reflect other factors as well. To repeat, the estimates presented in this paper are illustrative and reflect only the stated assumptions.

Conclusions

1. Estimates are presented of the economic benefits (costs averted) of controlling syphilis. The largest amount would, of course, accrue from complete eradication. Short of that, a continuing surveillance and control mechanism would be required, which might well cost more in the future than it did in 1962. In addition, all the other costs would continue for the thousands of new cases of syphilis incurred each year. The magnitude of these costs would depend not only on the number of cases, but also on the stage in which they are discovered and treated.

[57] Cockburn, *The Evolution and Eradication of Infectious Diseases*, p. 171.

TABLE 8. Estimating Work-Life Expectancy of Participants in the Labor Force by Sex and by Race

Steps	White Male	White Female	Nonwhite Male	Nonwhite Female
1. Median age at infection (CDC)	29	24	25	22
2. Life expectancy at infection (VS)	41.7	52.3	41.1	48.0
3. Life expectancy at age 65 (VS)	12.9	15.9	12.7	15.2
4. Probability of surviving to age of infection (VS)	94.5	96.9	91.7	94.4
5. Number of years after infection (line 2×line 4×100)	3940	5070	3770	4600
6. Probability of surviving to age 65 (VS)	65.5	80.5	50.2	60.0
7. Number of years after age 65 (line 3×line 6×100)	845	1280	637	910
8. Number of years between infection and age 65 (line 5—line 7)	3095	3790	3133	3690
9. Average years between infection and age 65 (line 8÷line 4)	32.8	39.1	34.2	39.1
10. Labor force participation rate, percent (BLS)	97	42	94	55
11. Work-life expectancy at infection (line 9×line 10)	31.8	16.4	32.2	21.5

Sources: For CDC, see Table 4; for VS, U. S. Public Health Service, National Vita Statistics Division, *Vital Statistics of the United States, Vol. 2, Section 2, Life Tables* (1963), pp. 9, 10; for BLS, U. S. Department of Labor, Bureau of Labor Statistics, *Employment and Earnings*, Vol. 8 (July 1961), p. 65.

2. Among the costs calculated in this paper—in addition to medical care expenditures and output loss due to treatment time or disability—are reductions in earnings due to "stigma" and loss in consumer benefit. Since the latter two are seldom if ever calculated, they warrant close scrutiny. Cost due to "stigma" is particularly high, constituting almost 44 percent of the economic costs calculated for the 119,600 new cases of syphilis in 1962. The loss in consumer benefit does not reflect a cost in resources; rather, it represents an attempt to measure the value of avoiding a psychic cost.

3. It was unfortunately not possible to calculate realistically the costs associated with the communicability of syphilis. The probable outcome of an initial change in incidence of specified magnitude is a critical question and warrants intensive exploration.

4. The costs calculated in this paper are tantamount to potential benefits—that is, they represent benefits to the extent that they can be averted. No attempt was made to calculate the direct costs of alternative control programs capable of yielding specified benefits. It is virtually impossible to estimate such costs in the absence of an adequate epidemiological theory on the spread of syphilis.

Comments

SELMA MUSHKIN, *Council of State Governments*

Herbert Klarman's paper underscores the feasibility of measuring benefits from public health programs, by identifying and measuring classes of benefits for an illustrative disease category—syphilis. But the measurement is done without defining program content or even formulating the question being asked in the cost-benefit analysis. There are at least two possible questions: what are the benefits which would flow from the elimination of the disease, and what are the benefits from a defined control program which could achieve a reduction of the incidence of the disease to some specified level?

Each of these questions implies a different program cost and a different "benefit." Moreover, each has a different time-cost pattern. Elimination of a disease for which a preventive is not known calls for research with its implicit tradeoff of costs for time. Control of a disease using known therapeutics has another time-cost relationship. Low expenditures over long periods in the case of some diseases yield no gains, since reinfection occurs as rapidly as the disease control work proceeds. Thus, gains essentially cannot be measured except in terms of program content and program timing, and cannot be evaluated except when estimates are available for both costs and benefits and a time table is specified.

In the initial part of his paper, Klarman reviews the conceptual work which has been done on the "cost of sickness and the price of health" and improves the conceptual underpinning of the arithmetic. I agree with much that he writes (for he agrees with my earlier writing), but I have one main point of criticism and on this a rebuttal seems desirable because it concerns a general question of methodology.

Timing and Scope of Program

If the calculated payoff period for an investment is brief—for instance, if the investment, augmented by a profit, is returned within a year or two—no purpose is served by carrying out extensive computations of, for example, lifetime gains, discounting these gains to convert to a present-value basis, and adjusting for secular changes in earnings. Even in a single year, benefits would be many, many times the cumulative costs for some disease-eradication programs. By the same token, if preliminary analysis of costs and benefits suggests a negative finding, then extensive computations *must* be carried out. A good part of Klarman's discussion was concerned with what part of costs and of benefits is measurable, using tools readily at hand, what part can be measured only if new tools are forged, and what part will remain uncounted. But the more important the policy issue, the more vital it would seem to pursue the qualification—rather than label as "not measurable" that which can be measured only by additional hard work.

Especially is there an obligation to quantify when the gains or the costs in the "presently measurable" category account for a small share of the relevant totals and when the measured gains or costs patently are not significant indicators of overall program performance. Caveats and definitions of limitation are no substitute for such quantification. And the quantification must be directed to provide answers to precisely defined questions so that the estimates can furnish a guide to choice among methods of meeting a program objective, or among programs.

Public health happens to be a public expenditure with a long and impressive history of applications of cost-benefit comparisons, a use which can be traced back to the beginnings of social statistics and modern medicine. At the turn of this century, cost-benefit computations became part of the effective campaigns for city water supplies and sewage disposal systems, both here and abroad. Many years later, the technique fell into disrepute among health officials and their biostatistics experts, because costs and gains which were essentially not relevant to the program issues were injected into the public program debate. At this stage, when a new interest has been aroused in both Congress and the Administration concerning cost-benefit analysis in health and related fields, it seems urgent that even preliminary findings be carefully assessed to make sure that what is being measured relates to the public program issues and that the measurement is sufficiently complete to provide reasonable results.

Within the scope of his paper, Klarman has done a careful job of measurement. Perhaps even more important, the illustrative disease category he selected strongly emphasizes the links among the several public programs discussed at this conference. One link is a common problem of measurement, another the essential interrelations of public services.

A number of public programs require the valuation of a human life. Benefits from a health program include the value of saving a life, preventing a disability, and reducing a debility. Assessments of the gains and costs of alternative highway or airport projects similarly require a valuing of a life saved or disability prevented. As the Eleventh International Road Congress concluded after a review of the costs of accidents in cost-benefit analysis of highways: "Decisions [on highway construction] attribute unconsciously in each case a value to human life and suffering. It seems preferable to make this more conscious and systematic." Nor is the valuation of human life limited to safety aspects of construction projects; it is also a part of the valuation of gains from such endeavors as housing projects and education.

Measurement and Program Interrelation

Program objectives frequently necessitate a packaging of meaningful end-product activities. And the question arises of how one allocates costs and gains among public programs with closely interrelated objectives. What part of the costs and gains of a group of programs directed to enhance opportunities for equality by elevating human capacity should be charged, for example, to a syphilis control program, a program reducing the number of school dropouts, a public housing program, a program of outdoor recreation, and so on? More specifically, if through a school dropout program an individual's sense of failure is lessened and he gains a greater respect for the dignity of his own being that can serve to reduce the number of "repeaters" under a syphilis control program, where are the costs and benefits to be assigned and how much to each program?

The simple fact is that, over a broad range of public programs, services are provided for people; furthermore, people live, not in isolation, but as related members of families. The output for the beneficiaries of the interdependent public programs for which we seek measurement is of two kinds: (1) their increased efficiency as producers in the economy, and (2) their increased effectiveness as consumers both of public and private goods. It is the total influence of the public service "package" on the efficiency of the family and the individuals who make it up with which we necessarily become concerned in economic measurements. Gains are measured by the enlarged earning capacity, or increased productivity, and by consumer "benefits" assessed either by consumer market expenditures for closely similar products, or by the potential release of funds for alternative uses. While distributional considerations are important in many items of this package of public services, the allocation of resources to its production must be assessed in terms of the multiple products and of the scale of interdependent products that are required for carrying out the efficiency and effectiveness objectives.

Over the broad spectrum of public services there are complementarities and substitutabilities. A complementary production of health and education services is required, for example, to gain a given productivity of a work force member, and a mix of scientific education and research outlays is needed to gain new scientific and technological advances. However, health can be improved by increasing educational outlays—outlays on training in nutrition and in personal hygiene, for example. By higher research outlays the competitive position of the universities in retaining faculty can be improved. Not only are there complementarities and substitutabilities in the program mix for a direct beneficiary group, but also a single project or program may serve a number of different beneficiary groups. The pulls and hauls of the political process create this multibeneficiary, multipurpose program determination: for example, we have education for national defense, and national defense also provides training for civilian life pursuits. How best to fashion a yardstick for expenditure choices, within this complex scheme of meshing relationships, is an important measurement issue.

The general class of problem is not unfamiliar. It is similar to quantification for multiple-purpose projects calling for physical capital outlays. There is often complementary production of a number of goods and services, and there are frequently joint costs. But a major difference must be emphasized. Alternative physical investment projects—the possible ways to connect two places by building roads, to provide airport facilities, to develop the waterways—have been formulated, and costs and benefits of the various possibilities have been measured. But for investments in people, the alternatives have not been described. As a backdrop for cost-benefit analysis, research is needed on combinations of education, health, and welfare services which could enlarge the productive capacity of people, and the effectiveness of their consumption outlays.

Paths to human investment are many. The efficiency per one dollar of allocated resource undoubtedly differs depending on the combination of services to the people selected, and the physical project supports for these services—housing, roads, hospitals, schools. And perhaps most important, because of the neglect in the past, is the time-cost tradeoff that is appropriate for each combination of services. Certainly we need cost-benefit comparisons for syphilis control, and for other specific disease-control work, but we also need a mechanics for judging syphilis control (or alternative disease-control work) as part of a "bundle" of investments in people.

List of Conference Participants

Howard Ball
   Department of Interior

Francis Bator
   National Security Council

William G. Bowen
   Princeton University

Philip H. Burch, Jr.
   Rutgers University (New Brunswick)

Robert D. Calkins
   Brookings Institution

William Capron
   Bureau of the Budget

Marion Clawson
   Resources for the Future

Samuel M. Cohn
   Bureau of the Budget

Lyle E. Craine
   University of Michigan

Robert Dorfman
   Harvard University

Anthony Downs
   Real Estate Research Corporation

Otto Eckstein
   Council of Economic Advisers

Rashi Fein
   Brookings Institution

Allen R. Ferguson
   Department of State

Lyle Fitch
   Institute of Public Administration

Gary Fromm
   Brookings Institution

Kermit Gordon
   Bureau of the Budget

Frederick O'R. Hayes
   Office of Economic Opportunity

Walter W. Heller
   University of Minnesota

Maynard Hufschmidt
   Harvard University

Herbert E. Klarman
   Johns Hopkins University

William Kolberg
   Bureau of the Budget

Roland N. McKean
   University of California (Los Angeles)

Fritz Machlup
   Princeton University

415

Ruth P. Mack
  Institute of Public Administration

M. Cecil Mackey
  Federal Aviation Agency

Edwin Mansfield
  University of Pennsylvania

Michael S. March
  Bureau of the Budget

Julius Margolis
  Stanford University

John Meyer
  Harvard University

Herman Miller
  Bureau of the Census

Herbert Mohring
  University of Minnesota

Gordon M. Murray
  Bureau of the Budget

Selma Mushkin
  Council of State Governments

Sumner Myers
  National Planning Association

Joseph A. Pechman
  Brookings Institution

Maurice Peston
  London School of Economics

Joseph Reeve
  Bureau of the Budget

Alice M. Rivlin
  Brookings Institution

William Ross
  Bureau of the Budget

Jerome Rothenberg
  Northwestern University

Frederick M. Scherer
  Princeton University

Charles Schultze
  University of Maryland

Willis Shapley
  Bureau of the Budget

Paul L. Sitton
  Department of State

William Vickrey
  Columbia University

Burton Weisbrod
  University of Wisconsin

Index